Green Energy and Technology

Climate change, environmental impact and the limited natural resources urge scientific research and novel technical solutions. The monograph series Green Energy and Technology serves as a publishing platform for scientific and technological approaches to "green"—i.e. environmentally friendly and sustainable—technologies. While a focus lies on energy and power supply, it also covers "green" solutions in industrial engineering and engineering design. Green Energy and Technology addresses researchers, advanced students, technical consultants as well as decision makers in industries and politics. Hence, the level of presentation spans from instructional to highly technical.

Indexed in Scopus.

Indexed in Ei Compendex.

Muhammad Adly Rahandi Lubis · Seng Hua Lee ·
Efri Mardawati · Souvia Rahimah · Petar Antov ·
Robi Andoyo · Ľuboš Krišťák · Bambang Nurhadi
Editors

Biomass Conversion and Sustainable Biorefinery

Towards Circular Bioeconomy

Springer

Editors
Muhammad Adly Rahandi Lubis
Research Center for Biomass and Bioproducts
National Research and Innovation Agency
Cibinong, Indonesia

Efri Mardawati
Department of Agro-Industrial Technology
Padjadjaran University
Bandung, Indonesia

Petar Antov
Department of Mechanical Wood Technology
Faculty of Forest Industry
University of Forestry
Sofia, Bulgaria

Ľuboš Krišťák
Department of Physics, Electrical Engineering and Applied Mechanics
Technical University of Zvolen
Zvolen, Slovakia

Seng Hua Lee
Faculty of Applied Sciences, Department of Wood Industry
Universiti Teknologi MARA (UiTM)
Jengka, Pahang, Malaysia

Souvia Rahimah
Department of Food Industrial Technology
Padjadjaran University
Bandung, Indonesia

Robi Andoyo
Faculty of Agro-industrial Technology
Padjadjaran University
Bandung, Indonesia

Bambang Nurhadi
Department of Food Technology
Padjadjaran University
Bandung, Indonesia

ISSN 1865-3529 ISSN 1865-3537 (electronic)
Green Energy and Technology
ISBN 978-981-99-7768-0 ISBN 978-981-99-7769-7 (eBook)
https://doi.org/10.1007/978-981-99-7769-7

This Springer imprint is published by the registered company Springer Nature Singapore Pte Ltd.
The registered company address is: 152 Beach Road, #21-01/04 Gateway East, Singapore 189721, Singapore

Preface

The energy crisis, caused by limited natural resources, compels industry and research society to seek alternative energy sources. Biomass, as a renewable and abundant material on the planet, offers numerous appealing benefits. In recent decades, there has been a growing interest in converting biomass into energy. Biomass stores solar energy, which can then be released as heat through burning or combustion. Typically, biomass is classified according to the type of biomass found in nature, such as wood and woody biomass, herbaceous biomass, aquatic biomass, animal and human biomass wastes, and mixtures of those biomass. The wood and woody biomass could include all of the woody trees on the land, as well as their trunks, branches, and sawdust after processing. Meanwhile, herbaceous biomass is also known as non-woody biomass, which includes bamboo (a type of grass), bagasse, straws, and so on. Algae and lake weed, on the other hand, are examples of aquatic biomass, whereas manures and bones are examples of animal and human biomass. These biomasses are high in organic matter and thus have the potential to be used as energy sources. Different conversion methods have been used due to the differences in the nature of these biomasses. The biorefinery process is what is needed to convert these biomasses into energy as well as other useful products in an effective manner. In a nutshell, it can be described as an integrated system of conversion processes and equipment that transforms biomass into useful products such as biofuels, energy, and other chemical compounds. The International Energy Agency's Bioenergy Task 42 describes biorefining as "the sustainable processing of biomass into a spectrum of bio-based products and bioenergy." Biorefinery, which is an environmentally friendly method, has the potential to make significant contributions to the circular bioeconomy through its promotion of the recyclable use of resources as well as value-adding characteristics. In fact, it is one of the most important strategies that enables the circular economy, which closes loops of raw biomass materials. The creation of a cleaner and more environmentally friendly environment is one of the potential contributions that biorefineries could make to the development of a sustainable bio-based circular economy. It is possible to form a circular economy as opposed to a linear economy by adhering to the principles of the philosophy of reuse and recycling. The realization of these ideologies, however, is not without limitations. Many constraints lie ahead, and strategies to make it work are critical. The success of biorefinery technologies is highly dependent on several factors, including the process's technical feasibility and cost-effectiveness. It must be cost effective and, most importantly, environmentally friendly to start with, but social acceptance is also a deciding factor to determine the effectiveness of the technologies. This book is organized to present the Biomass Conversion and Sustainable Biorefinery: Towards Circular Bioeconomy, consisting of 14 chapters discussing the current progress on biomass resources, biomass pretreatment, biomass conversion, and their applications.

Cibinong, Indonesia	Muhammad Adly Rahandi Lubis
Jengka, Malaysia	Seng Hua Lee
Bandung, Indonesia	Efri Mardawati
Bandung, Indonesia	Souvia Rahimah
Sofia, Bulgaria	Petar Antov
Bandung, Indonesia	Robi Andoyo
Zvolen, Slovakia	Ľuboš Krišťák
Bandung, Indonesia	Bambang Nurhadi

Contents

Current Progress on Biomass Pretreatment: The Key for Its Valorization

Roni Maryana, Eka Triwahyuni, Muryanto Muryanto, Joni Prasetyo, Oktaviani Oktaviani, Sri Sugiwati, Teuku Beuna Bardant, Atanu Kumar Das, and Yanni Sudiyani

Abstract It has become a global concern for reducing the utilization of fossil sources for energy and chemical purposes. Not only environmental issues such as greenhouses gases increment but also depleting fossil reserves, which became the reason to quest for new and renewable raw materials. Lignocellulosic biomass that is abundantly available is being studied as a potential material to provide both fuels and biochemicals. However, to fractionate and disintegrate its main components: cellulose; hemicellulose and lignin is a very challenging step. Different pretreatment approaches that considering energy consumption, cost effectiveness, percentage of lignin removal, further utilization of the cellulose, hemicellulose, lignin, and environmentally friendly process have been applied. The review covers the recent pretreatment technology that applied physicochemical approach; chemicals pretreatment and application of Deep Eutectic Solvent (DES); biological pretreatment; and high energy of radiation. This study explained a deeper understanding of the pretreatment technologies for developing biorefinery concepts to support the sustainable development of utilization of lignocellulosic biomass.

Keywords Lignocellulosic · Pretreatment · Physicochemical treatment · Biological pretreatment

R. Maryana (✉) · E. Triwahyuni · M. Muryanto · J. Prasetyo · S. Sugiwati · T. B. Bardant · Y. Sudiyani
Research Center for Chemistry, National Research and Innovation Agency, Building 452 Kawasan PUSPIPTEK, Serpong Tangerang Selatan, Banten, Indonesia
e-mail: roni002@brin.go.id

O. Oktaviani
Research Center for Radiation Process Technology, National Research and Innovation Agency, Pasar Jumat, Jakarta Selatan, Indonesia

A. K. Das
Department of Forest Biomaterials and Technology, Swedish University of Agricultural Sciences, 901 83 Umeå, Sweden

M. A. R. Lubis et al. (eds.), *Biomass Conversion and Sustainable Biorefinery*, Green Energy and Technology, https://doi.org/10.1007/978-981-99-7769-7_1

1 Introduction

Pretreatment of lignocellulosic biomass is the key for the separation of cellulose, hemicellulose, and lignin. Several factors such as the energy required and the economical value of the biomass pretreatment will be the key for industrial application. Another consideration is the end product of the fractionation. For instance, lower molecular weight or lower density cellulose will not be suitable for dissolving pulp application. Moreover, the enzymatic accessibility will be the key factor for the saccharification of cellulose into the sugar on the bioethanol production process.

The main objective of the pretreatment process on lignocellulosic biomass is lignin removal. Determination of the pretreatment methods depends on the type of lignocellulosic biomass. Therefore, no universal pretreatment can be applied to all lignocellulosic biomass (Costa Sousa et al. 2009). Therefore, it can be summarized that pretreatment has to be: (a) considering energy consumption, (b) cost-effectiveness, (c) percentage of lignin removal, (d) further utilization of the cellulose, hemicellulose, lignin, and (e) environmentally friendly process.

Generally, the pretreatment of biomass can be divided into chemical, physical, biological, physicochemical, and high-energy radiation treatment. However, combination of these methods is mostly applied to get desirable results. This study aims to provide recent information and discuss the above-mentioned biomass pretreatment methods as the information for choosing suitable pretreatment conditions to get optimum fractionation for wood and non-wood biomass components.

2 Pretreatment Methods

2.1 *Physicochemical Pretreatment*

Physicochemical pretreatment is believed to cause more well lignin removal and cellulose crystallinity decrement efficiency as compared with physical, chemical, and biological pretreatment (Zhao, et al. 2021). The most widely used physicochemical pretreatment involved liquid-hot water (LHW), steam explosion, alkali explosion or alkali-heat pretreatment, and ammonia fiber expansion (AFEX) (Wang et al. 2015; Elliston et al. 2015; Jacquet et al. 2015; Yu et al. 2022; Sudiyani et al. 2020; Nurani et al. 2020; Emam 2013; Chundawat 2020; Bensah and Mensah 2018).

2.2 *Liquid-Hot Water (LHW) Pretreatment*

Liquid-hot water (LHW) is one of the pretreatment technologies utilizing liquid water without other chemicals at high temperature and pressure (Yan et al. 2016; Zhuang et al. 2016). Water serves as a solvent and also a catalyst accompanied by removed organic acids from biomass to aid in disrupting the cell wall matrix (Mosier 2013). Moreover, LHW brings increased cellulose accessibility and produces minimal inhibitory products (Kim et al. 2009). After LHW pretreatment, the major changes in lignocellulose involve the limited deconstruction of cellulose, dissolution of hemicellulose, partial removal of lignin, and carbohydrate degradation. Nitsos et al. has been demonstrated that hemicellulose is almost completely solubilized and deconstructed from biomass in hot water pretreatment at ~200 °C for 50 min (Nitsos et al. 2016). Different from hemicellulose, cellulose was less affected by LHW pretreatment. Less than 22% by weight of cellulose is degraded in wood and herbaceous biomass treated with LHW at 200–230 °C (Mok and Antal 1992).

2.2.1 Steam Explosion

Steam explosion is an effective, environmentally safe, and industrially scalable pretreatment method. This pretreatment results in the combination of structural disruption of biomass by high-temperature steaming and explosive decompression as well as autohydrolysis (Singh et al. 2015; Kumari and Singh 2018). Generally, steam explosion was carried out under high-pressure steam (0.5–5 MPa) and temperature intervals between 160 and 250 °C (Paudel et al. 2017). This method uses less chemicals and does not result in unnecessary dilution of hydrolyzates, but the steam explosion has some disadvantages, such as energy intense, inhibitory compounds generation, and causing the partial ruin of carbohydrate-lignin matrix (Agbor et al. 2011). To improve the effectiveness of the steam explosion method, the addition of a catalyst or impregnating agent such as H_2SO_4, SO_2, or CO_2 can decrease the production of inhibitors and improve the enzymatic hydrolysis on the biomass (Mosier et al. 2005; Sun and Cheng 2002).

The most attractive catalyst is CO_2 due to its cheap cost of CO_2, low corrosion, low toxicity, and the possibility of biomass having high solids content (Kucharska et al. 2018). In most cases, CO_2 explosion applies supercritical CO_2 or high-pressure CO_2 to help the biomass digestibility (Zheng et al. 1998; Morais et al. 2015; Hendriks and Zeeman 2009). Moreover, adding CO_2 in mild condition before steam explosion pretreatment also has been studied can increase the surface area of the material which can make the enzymes more easily to enter the material during the hydrolysis process. Pretreatment of oil palm empty fruit bunch (EFB) using steam explosion with CO_2 as an impregnating agent gives more slightly lower crystallinity index, more disrupt biomass, and increasing enzymatic hydrolysis of EFB as compared to pretreatment EFB using conventional steam explosion (Triwahyuni et al. 2021). The changes in the surface morphology of EFB using SEM results in the untreated EFB fiber had a rigid surface, with a layer of matrix material, meanwhile, the surface of EFB after CO_2-added steam explosion, appeared uniform pores in the surface of EFB. It indicated that some silica was removed from EFB (Triwahyuni et al. 2021).

2.2.2 Alkali Explosion

Alkali explosion is one of chemical pretreatment which can provide a high yield of delignification. Alkali-heat pretreatment uses heat and the alkali such as sodium hydroxide, sodium carbonate, and alkaline peroxide to dissolve lignin and reduce the crystallinity of lignocellulose by swelling, and also enlarge the specific surface area of cellulose (Rodrigues et al. 2016). For example, the delignification of EFB using NaOH explosion reached 58.36% and the percentage of cellulose could increase from 30.16% (untreated EFB) to 63.82% (Muryanto et al. 2015). Pretreatment of corn straw with NaOH-heat pretreatment obtained lignin and hemicellulose removal efficiency of 54.09% and 67.67%, meanwhile, the relative content of cellulose enhanced to 51.65% (Zhao, et al. 2021; Lopez et al. 2019; Shahabazuddin et al. 2018). The ability of NaOH to dissolve lignin is due to the opening of aromatic rings of lignin caused by the use of high temperatures and pressure, then, the resulting explosive effects can dissolve these components (Rezania et al. 2020). However, this method requires high temperature and pressure and takes a long time as well as high treatment cost (Muryanto et al. 2015; Kumar et al. 2020).

2.2.3 Ammonia Fiber Explosion

Ammonia fiber explosion is one of physicochemical pretreatment under high pressure with ammonia and oxyhydrogen ions released from liquid ammonia, resulting in a prompt rise in temperature and breaking the ester bond and ether bond between hemicellulose lignin in lignocellulose (Zhao et al. 2020) as well as reducing the cellulose crystal. AFEX has been demonstrated to be effective in low lignin-contained lignocellulose pretreatment. The application of AFEX shows lignin degradability of barley straw (Beauchemin et al. 2019) and corn stover (Mankar et al. 2021) achieved 24–1.3%.

2.2.4 Chemi-Extrusion Method

Application of the combination of alkaline and thermochemical can reduce energy consumption and remove lignin (Chen et al. 2019). However, these combinations still cannot meet the requirements for industrial scale that need efficient operational cost. The extrusion method then be considered as one of the solutions because it applies both chemical and physical methods. Moreover, in the biomass processing temperature can be applied together with compression forces.

Extruder screws are used in the extrusion method and the single or twin screws are usually used. The screws rotate to generate shear forces, an unaligned force acting on one part of a reactor in a particular direction, among the biomass, screw, and barrel. These forces generally increase the temperature and the pressure inside the reactor which will change the biomass's physical properties. It has been studied that shear force will strengthen the rotation of biomass, the screw, and the barrel. Pressure

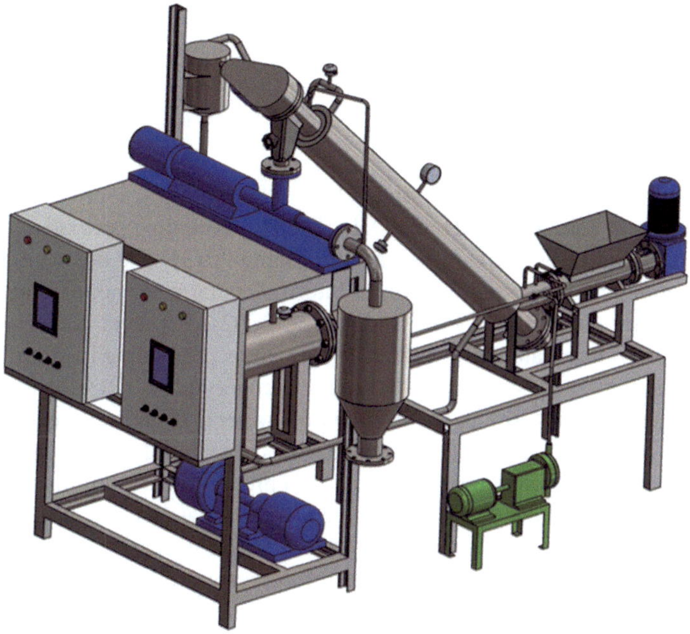

Fig. 1 Screw continuous reactor model that has been installed at research center for chemistry—BRIN

and temperature increased during the process will lead to the breakage of biomass lignocellulosic fibers (Khan et al. 2021).

It has been reported that screw rotation speed, residence time in the screw reactor, and temperature influenced the lignin removal percentage by using 10% NaOH (Maryana et al. 2022) (Fig. 1).

Moreover, screw speed is responsible for the reduction of the length of fiber and increasing of the surface area of biomass that improve sugar yield after the saccharification process (Karunanithy and Muthukumarappan 2010). The screw extrusion method is usually combined with chemical and thermo-chemical methods. The second generation of bioethanol production has used the extrusion method. For instance, the application of a single screw extruder that used a corn stover by varying different screw speeds and temperatures (Karunanithy and Muthukumarappan 2010; Jorge et al. 2006).

2.3 Chemicals Pretreatment and Application of Deep Eutectic Solvent (DES)

2.3.1 Chemicals Pretreatment

Chemical pretreatment is generally divided into alkaline and acid pretreatment. Alkaline pretreatment can use NaOH, $Ca(OH)_2$, KOH, or alkaline peroxide, a combination of NaOH and H_2O_2 (Alvarez-vasco and Zhang 2013). The mechanism of alkaline pretreatment is the saponification of intermolecular ester bonds that cross-link silane and other components (Sun and Cheng 2002). However, alkaline pretreatment has some disadvantages, such as longer time pretreatment, more water for biomass washing, and more expensive waste recovery costs (Rezania et al. 2017). Acid pretreatment uses low concentrations of mineral acids such as sulfuric acid (H_2SO_4) or hydrochloric acid (HCl). Sometimes concentrated acid can be used to conduct pretreatment at low temperatures. Acid pretreatment commonly hydrolyzes cellulose and hemicellulose to get monomeric sugar that can be converted to some biobased chemical. Produce inhibitor products, hard to separate, and corrosive are the disadvantages of acid pretreatment. Therefore, many studies have been conducted to develop green solvents in the delignification process. The ionic liquid is widely developed because it is environmentally friendly compared to conventional solvents (Nguyen et al. Oct. 2010; Pena-Pereira and Namieśnik 2014). However, the ionic liquid is still as expensive as solvent pretreatment.

2.3.2 Deep Eutectic Solvent (DES)

Deep eutectic solvent (DES) has been introduced as the new green solvent (Abbott et al. 2001). DES is a solvent consisting of a mixture of a high melting point salt (i.e., hydrogen bond acceptor or HBA) with a molecular hydrogen bond donor (HBD) in a specific ratio to form a liquid. DES has properties like an ionic liquid, although the DES ion content is lower. DES began to be used in the pretreatment process because of several advantages such as cheaper material costs, non-toxicity, ease of manufacturing process, and more biodegradable and biocompatible (Haldar and Purkait 2021). Besides that, most DES have high solubility for lignin and poor solubility for cellulose and hemicellulose (Francisco et al. 2012). DES can be explicitly made according to the required properties that are needed in the process. Table 1 shows the DES from a variation of HBD and HBA.

DES in the pretreatment process is divided into several types depending on the HBD used. Several types of DES include carbohydrate-based DES, such as CHCl-glucose and CHCl fructose. Polyalcohol-based DES consisted of CHCl-glycerol, CHCl-ethylene glycol, and CHCl-Propanediol. Acidic DES is commonly used in biomass processing, containing HBD such as acetic acid, lactic acid, glycolic acid, oxalic acid, and citric acid. Basic-based DES contains amide, amine, and imidazole

Table 1 Deep Eutectic solvent in the variation of HBA and HBD

HBA	HBD	Ratio HBA:HBD	Molar masses (kg/mol)	T melting point (°C)
Choline chloride	Urea	1:2	0.2597	12
	Malonic acid	1:1	0.24368	10
	Lactic acid	1:2	0.31978	−78
	Oxalic acid dehydrate	1:1	0.26569	−40
	Ethylene glycol	1:2	0.26376	−66
	Glucose	2:1	0.43237	15
Choline acetate	Urea	1:2	0.27168	<−20
Choline fluoride	Urea	1:2	0.24329	1
$Acetylcholine^{+}Cl^{-}$	Urea	1:2	0.26627	−14
	Glucose	1:1	0.36382	−7
Proline	Oxalic acid	1:1	–	−14.5
	Lactic acid	2:1	–	−36.7
Betaine	Urea	1:2	–	−42.5
	Oxalic acid	1:1	–	−17.2
	Lactic acid	2:1	–	−46.9
Glucose	Citric acid	1:1	–	9.8
	Tartaric acid	1:1	–	−18.3
$AlCl_3$	Urea	1:1	–	<25
K_2CO_3	Ethylene glycol	1:10	–	−122

Source Marcus and Solvents (2019)

compounds. CHCl-Urea is a basic DES commonly used in biomass processing (Tan et al. 2019).

The pretreatment performance may be impacted by pretreatment parameters, including temperature, duration, or the solid-to-liquid ratio, and DES type also affects DES pretreatment. For example, DES Pretreatment conducted by Majová et al. (2017) used hardwood kraft pulps as raw materials and DES solution from a mixture of ChCl-Oxalic Acid with a 1:1 molar ratio (Majová et al. 2017). The pretreatment process was carried out for 1 h at a temperature of 60 °C. The results obtained from DES pretreatment were able to degrade lignin in hardwood kappa pulp by 38.7%. Another study conducted by Okur and Koyuncu (2020) regarding the delignification of rice husks using a DES solution of ethylene glycol-citric acid obtained optimal results at a temperature of 120 °C and a processing time of 4 h showed that 57.33% of lignin was degraded (Okur and Koyuncu 2020). DES pretreatment consisting of K_2CO_3 and glycerol was conducted on wheat straw and corn stalks at a temperature of 80–100 C for 8–24 h. This pretreatment can reduce the lignin content in wheat straw and corn

Table 2 DES pretreatment

DES	Molar ratio	Biomass	Process condition	Delignification (%)	Sources
CHCl:ethylene glycol	1:2	Switchgrass	130 °C, 30 min, S/L 10	24	Chen and Wan (2018)
CHCl:glycerol	1:2	Oil palm trunk	100 °C, S/L 5	49	Zulkefli et al. (2017)
CHCh:lactic acid		Poplar wood	145 °C, 9 h, S/L 10	79	Alvarez-Vasco et al. (2016)
CHCl:lactic acid	1:5	OPEFB	120 °C, 8 h, S/L 10	88	Tan et al. (2018)
CHCl:formic acid	1:2	Akebia herbal residue	120 °C, 8 h, S/L 10	41	Yu et al. (2018)
CHCl:urea	1:2	OPEFB	120 °C, 8 h, S/L 10	34	Tan et al. (2018)
CHCl: monoethanolamine	1:6	Wheat straw	90 °C, 5 h, S/L 5	81	Zhao et al. (2018)
CHCl:4-catechol	1:1	Switchgrass	160 °C, 3 h, S/L 5	49	Kim et al. (2018)

stalks from 22% to 12.1% and 24% to 16.3%, respectively (Suopajärvi et al. 2020). DES pretreatment in variation DES type and operation condition is shown in Table 2.

DES is used in the pretreatment process in various delignifications. Some types of DES can reduce the lignin content above 60%. However, some are still low. Like the chemical pretreatment process, DES pretreatment can also be combined with other technologies to improve its performance. Some integrations can include combining microwave, ultrasonic, and sequential pretreatment with hydrothermal, biological, or another chemical (Tan et al. 2019).

2.3.3 Organosolv Pretreatment

Organic solvents such as methanol, ethanol, acetone, ethylene glycol, and tetrahydrofurfuryl alcohol are commonly used as pretreatment chemicals in the organosolv method. Moreover, catalysts like salicylic acid and organic acid as well as sodium hydroxide are generally used in this method (Cheah et al. 2020). It was reported by Mirmohamadsadeghi et al., that pretreated Elmwood, pinewood, and rice straw resulted in lignin removal of 27, 21, and 37.7% respectively, after pretreated by using 1% (w/w) sulfuric acid at 150–180 °C for 30–60 min in 75% aqueous ethanol (Mirmohamadsadeghi et al. 2014).

Ethanol is the most common solvent due to its economical price, not harmful properties, ease of mixing with water, and good soluble solvent for lignin. In the case of softwood delignification, ethanol can enhance the impregnation and support lignin fragments in the biomass to the liquor solution (Rinaldi et al. 2016). It was studied that lignin removal of 16 for 50% water–ethanol system at 175 °C for southern yellow pine that was pretreated for 80 min (McGEE and APRIL 1982).

2.4 *Biological Pretreatment*

The effectiveness of a biological pretreatment is determined by several factors including composition of biomass, moisture content, incubation time, temperature, pH, and type of microorganism (Sindhu et al. 2016). Based on the composition of biomass, a suitable microbial consortium must be used for the effective removal of lignin and hemicellulose from the biomass.

2.4.1 Microorganism Used in Biological Pretreatment

Many researcher-published articles revealed that common microorganisms used in biological pretreatment are bacteria including *Bacillus*, Actinomycetes and some known fungi are able to degrade organic matter (Poszytek et al. 2016). Microbial consortia consisting of cellulolytic bacteria of the genus *Bacillus, Streptomyces, Candida*, and *Aspergillus* exhibit broad-spectrum biodegradation. It was also that many potent brown rot fungus, white rot fungus, and softening fungi. Among the three fungi, the most effective for upstream treatment of lignocellulosic materials is white rot fungi. The white rot fungus helps in delignification which in turn improves the enzymatic saccharification rate, especially, *Phanerochaete chrysosporium, Ceriporiopsis subvermispora, Pleurotus ostreaus, Ceriporia lanceolata* and *Cyanthus stercoreus*, which selectively degrading lignin by producing a wide variety of enzyme, i.e., laccase, lignin peroxidase (LiP), and manganese peroxidase (MnP) (Maurya et al. 2015).

Currently, there is a need for unique consortia for biological pretreatment. Effective biodegradation of lignocellulosic biomass takes place by biodegradation by synergistic action of microbial consortiums including various bacteria and fungi. There are several advantages of using microbial consortium which include increased adaptability, improved productivity, improved enzymatic saccharification efficiency, control of pH during sugar utilization, and increased substrate utilization (Kalyani et al. 2013).

The development of an eco-friendly simultaneous pretreatment and saccharification (SPS) methodology using a cocktail of hydrolytic and oxidizing enzymes from a fungal consortium was reported by Dhiman et al., 2015 (Dhiman et al. 2015). The novel laccase effectively functioned as a detoxifying agent. This process completely eliminates the use of hazardous chemicals. Pretreatment and saccharification conducted in the same vessel make the process economically viable, reduce energy consumption, and generate a simple process for the removal of residual biomass.

Table 3 Microorganism used in the biological pretreatment

Microorganism	Biomass	Major effects
Punctualaria sp. TUFC20056	Bamboo culms for bioethanol production	50% of lignin removal
Irpex lacteus	Corn stalks	82% of hydrolysis yield, 28 days pretreatment
Fungal consortium	Straw	Seven-fold increase in hydrolysis
Pleurotus ostreatus/P. pulmonarius	Eucalyptus grandis saw dust	Twenty-fold increase in hydrolysis
Phanerochete chrysosporium	Rice husk	Cellulase, xylanase, lignin peroxidase, glyoxidase, and aryl alcohol oxidase were produced during 18 days pretreament
Fungal consortium	Corn stover	43.8% lignin removal/ seven-fold increase in hydrolysis
Ceriporiopsis subvermispora	Wheat straw	Minimal cellulose loss and highest sugar yield up to 44% after 10 weeks
Ceriporiopsis subvermispora	Corn stover	2–threefold increase in reducing sugar yield
Fungal consortium	Plant biomass	Complete elimination of the use of hazardous chemicals

Different biological pretreatment strategies are involved in the pretreatment of lignocellulosic biomass and its advantages (Sindhu et al. 2016) (Table 3).

Biological pretreatment is considered as an inexpensive process when compared to other pretreatment processes such as AFEX and organo solvent. Large-scale operation leads to high operational costs since pretreatment is to be carried out in sterile conditions and this increases the cost of the process. One approach to reducing the pretreatment time is by applying a combination of biological processes with physical and chemical methods. Employing more potent microbial consortia as well as the development of specific reactors for biological pretreatment can reduce the pretreatment time as well as the loss of carbohydrate significantly.

2.5 *Pretreatment of Biomass with High Energy of Radiation*

In order to produce a digestible raw material and prevent the target material from degrading and the production of harmful by-products, an efficient pretreatment should break down the lignocellulose structure. To enhance the available surface area and the size of pores in lignocellulosic materials, cellulose is physically separated by removing lignin and hemicellulose as well as breaking down the crystallinity of the substance. Any pretreatment must be economically advantageous (Maryana et al. 2014, 2017, 2020).

Among other physical methods, high-energy radiation is regarded as an attractive method for the pretreatment process of biomass (Duarte et al. 2012; Khan et al. 2006; Mohd Asyraf Kassim et al. 2016). In general, radiation processing technology onto lignocellulosic material is defined as a radiolysis reaction that induces lignocellulosic material breakdown using radioisotopes like cobalt-60 (Co-60) or cesium-137 (Cs-137), or an electron beam generated by an electron accelerator. Generally, biomass materials degrade similarly under gamma and electron beam irradiation. As electromagnetic radiation, an electron beam has a significantly lower penetration but with a higher dose rate compared to gamma rays. The wide application of electron beams to the degradation of biomass is due to a number of its inherent benefits. Since electron particle energy is immediately transferred to molecules of biomass, thus no need for heating and catalyst during the irradiation. Because the electron beam can be precisely controlled by adjusting the accelerator's operating condition, the irradiation process by using EBM is highly reproducible, controllable, and precise.

High energy radiation from electron beams, X-rays, or gamma rays can create ions and/or radicals that initiate chemical processes that primarily cause chemical bond cleavages and decrease the molecular weight of the biomass by increasing the irradiation dose. The high-energy radiation produced during this process has the potential to change the properties of cellulosic biomass, due to 1,4-glycosidic linkages being ruptured by gamma radiation and electron beams. Subsequently increasing the lignocellulosic biomass' specific surface area, decreasing the polymerization and crystallinity of cellulose, hydrolyzing hemicellulose, and partially depolymerizing lignin (Al-Masri and Zarkawi 1994). The irradiation of biomass with an electron beam causes the complex carbohydrates to become soluble, which facilitates processing, such as enzymatic hydrolysis for the production of bioethanol (Danu et al. 2012).

The effectiveness of high-energy radiation as a pretreatment technique on lignocellulosic material has been studied intensively (Hyun Hong, et al. 2014). Irradiation of Jute fibers by gamma irradiation with the doses up to 100 kGy and by an electron beam with a dose up to 400 kGy for oil palm empty fruit bunch (OPEFB) resulted in a decrease in the weight fraction of large particles and an increase in the weight fraction of small and medium particles both of Jute fibers and OPEFB (Khan et al. 2006; Danu et al. 2012). It was revealed that the chemical stability of irradiated fibers gradually decreased with an increase in irradiation dose. Studies using thermogravimetry and differential scanning calorimetry revealed a considerable loss in thermal stability. The wide-angle X-ray diffraction analysis revealed that lignocellulose's chemical response to radiation caused structural alterations in cellulose. The solubility of irradiated OPEFB in water and 1% sodium hydroxide solution also increases. Based on the bands shown in the fingerprint area and the degradation found using FTIR analysis, cellulose, hemicelluloses, and lignin were identified as the primary constituents of OPEFB and were subject to C–O stretching and C–H deformation (Danu et al. 2012).

Combining irradiation with additional pretreatment techniques, both physical and chemical, can improve the effectiveness of gamma pretreatment. Pretreatment is more effective in relatively milder conditions when multiple treatments are used because they increase each other's effects (Duarte et al. 2012; Wang et al. 2012; Abo-State et al. 2014; Chung et al. 2012; Saini et al. 2015). The yield of total reducing sugar following saccharification and pretreatment of rice straw during the production of ethanol was enhanced by combining milling, autoclaving, and gamma irradiation (700 kGy) (Abo-State et al. 2014). When sugarcane bagasse was pretreated by using electron beam processing (EBP), after 24 and 48 h of incubation, the conversion yield of cellulose to glucose improved from 8 to 12% becoming 15%. The yield of the enzymatic hydrolysis of cellulose improved with EBP and thermal treatment (60 min at 180 °C), reaching 71.55%. After EBP, hemicelluloses were completely hydrolyzed (Kucharska et al. 2018).

The effect of gamma irradiation on the saccharification of Undaria sp. for bioethanol production was evaluated by Yoon et al. (2012). The irradiation has been employed as a pretreatment in conjunction with dilute acid hydrolysis. Undaria sp. wet samples were exposed to radiation at 22 °C with irradiation dose between 10 and 500 kGy and a dose rate of 10 kGy/h. By increasing the gamma irradiation dose, the concentration of reducing sugars increased. The reducing sugar content was 0.017 g/L for unirradiated samples, but it rose to 0.048 g/L for a sample irradiated by the dose of 500 kGy. This reveals that the combined method of gamma irradiation and diluted acid hydrolysis considerably enhances the saccharification process for the generation of bioethanol from materials derived from marine algae.

To improve enzymatic hydrolysis for the manufacture of bioethanol, poplar bark was pretreated with 3% w/w sulfuric acid and gamma radiation (0–1000 kGy). With an increasing irradiation dose, the yields of reducing sugar were slightly increased, going from 35.4 to 51.5%. After pretreatment by the addition of dilute acid, the yield of reducing sugar reached 56.1%. These results clearly demonstrated that soluble sugars were released more quickly and to a higher amount in poplar bark that had been prepared with diluted acid as opposed to bark that had been exposed to gamma radiation. A significant increase in reduced sugar yield (83.1%) was found after combination pretreatment as compared to individual pretreatment, suggesting that the convertibility of poplar bark may be improved after combined pretreatment (Chung et al. 2012).

On the other hand, Henniges et al. (2013) found that tuning the irradiation process, independent of the dosage, is possible to increase oxidation by introducing iron(II) ions. They observed the same amount of oxidized cellulose functionalities at better preservation of molar mass (less cellulose chain degradation). They found that the addition of hydrogen peroxide speeds up the irradiation-induced degradation. Another finding is that wet irradiation samples are more impacted than dry irradiated samples. When wet straw was utilized instead of dry straw, the required dose of gamma irradiation was decreased from 500 to 10 kGy.

According to some studies and reports, irradiating biomass for pretreatment uses less energy than other methods (Mosier et al. 2005; Mohd Asyraf Kassim et al. 2016; Henniges et al. 2013). This method is more effective and efficient for producing the intended product since it is selective and easy to control. On the other hand, there are certain points that require to be considered due to the studies conducted in recent years only focusing on biomass irradiation at a lab scale. Meanwhile, validating the result requires pilot-scale investigations, and the cost of the commercial deployment must be investigated throughout technology development (Saini et al. 2015).

3 Conclusion

The Biorefinery concept is a potential approach to respond to the depletion of fossil resources by utilizing lignocellulosic biomass. Eco-friendly and efficient biomass pretreatment methods in order to fractionate cellulose, hemicellulose, and lignin are continuously developed. By studying several recent pretreatment methods, it can be concluded that the process has to overcome the recalcitrant of lignin in the chemical bonding complex with cellulose and hemicellulose. The cellulose crystallinity and sugars recovery are also the main concern of the pretreatment methods. Different pretreatment approaches that considering energy consumption, cost-effectiveness, percentage of lignin removal, further utilization of the cellulose, hemicellulose, lignin, and environmentally friendly process can be chosen for addressing the challenges in the biomass pretreatment method.

References

Abbott AP, Capper G, Davies DL, Munro HL, Rasheed RK, Tambyrajah V (2001) Preparation of novel, moisture-stable, Lewis-acidic ionic liquids containing quaternary ammonium salts with functional side chains electronic supplementary information (ESI) available: plot of conductivity versus temperature for the ionic liquid formed from z. Chem Commun 19:2010–2011

Abo-StateMA, Ragab AME, EL-Gendy NS, Farahat LA, Madian HR (2014) Bioethanol Production from Rice Straw Enzymatically Saccharified by Fungal Isolates, Trichoderma viride F94 and Aspergillus terreus F98. SciRes 03(02):19–29. https://doi.org/10.4236/soft.2014.32003

Agbor VB, Cicek N, Sparling R, Berlin A, Levin DB (2011) Biomass pretreatment: fundamentals toward application. Biotechnol Adv 29(6):675–685. https://doi.org/10.1016/j.biotechadv.2011.05.005

Al-Masri MR, Zarkawi M (1994) Effects of gamma irradiation on cell-wall constituents of some agricultural residues. Radiat Phys Chem 44(6):661–663. https://doi.org/10.1016/0969-806X(94)90227-5

Alvarez-Vasco C et al (2016) Unique low-molecular-weight lignin with high purity extracted from wood by deep eutectic solvents (DES): a source of lignin for valorization. Green Chem 18(19):5133–5141

Alvarez-vasco C, Zhang X (2013) Bioresource technology alkaline hydrogen peroxide pretreatment of softwood: hemicellulose degradation pathways. Bioresour Technol 150:321–327. https://doi.org/10.1016/j.biortech.2013.10.020

Beauchemin KA et al (2019) Recombinant fibrolytic feed enzymes and ammonia fibre expansion (AFEX) pretreatment of crop residues to improve fibre degradability in cattle. Anim Feed Sci Technol 256:114260

Bensah EC, MensahMY (2018) Emerging physico-chemical methods for biomass pretreatment. Intech 11:13

Cheah WY et al (2020) Pretreatment methods for lignocellulosic biofuels production: current advances, challenges and future prospects. Biofuel Res. J. 7(1):1115

Chen Z, Wan C (2018) Ultrafast fractionation of lignocellulosic biomass by microwave-assisted deep eutectic solvent pretreatment. Bioresour Technol 250:532–537

Chen J, Adjallé K, Barnabé S, Perrier M, Paris J (2019) Mechanical and thermal pretreatment processes for increasing sugar production from woody biomass via enzymatic hydrolysis. Waste and Biomass Valoriz 10(7):2057–2065. https://doi.org/10.1007/s12649-018-0217-x

ChundawatSPS et al (2020) Ammonia fiber expansion (AFEX) pretreatment of lignocellulosic biomass. JoVE 158:e57488. https://doi.org/10.3791/57488

Chung BY et al (2012) Enhanced enzymatic hydrolysis of poplar bark by combined use of gamma ray and dilute acid for bioethanol production. Radiat Phys Chem 81(8):1003–1007. https://doi.org/10.1016/j.radphyschem.2012.01.001

da Costa Sousa L, Chundawat SPS, Balan V, Dale BE (2009) 'Cradle-to-grave' assessment of existing lignocellulose pretreatment technologies.Curr Opin Biotechnol 20(3):339–347

Danu OOS, Harsojo H, Darsono D, Kardha MS, Marsongko M (2012) Electron beam degradation of oil palm empty fruit bunch. Int. J. Environ. Bioenergy 3(3):168–179

Dhiman SS, Haw J-R, Kalyani D, Kalia VC, Kang YC, Lee J-K (2015) Simultaneous pretreatment and saccharification: green technology for enhanced sugar yields from biomass using a fungal consortium. Bioresour Technol 179:50–57

Duarte CL, Ribeiro MA, Oikawa H, Mori MN, Napolitano CM, Galvão CA (2012) Electron beam combined with hydrothermal treatment for enhancing the enzymatic convertibility of sugarcane bagasse. Radiat Phys Chem 81(8):1008–1011. https://doi.org/10.1016/j.radphyschem.2011.11.008

Elliston A, Wilson DR, Wellner N, Collins SRA, Roberts IN, Waldron KW (2015) Bioresource technology effect of steam explosion on waste copier paper alone and in a mixed lignocellulosic substrate on saccharification and fermentation. Bioresour Technol 187:136–143. https://doi.org/10.1016/j.biortech.2015.03.089

Emam SF (2013) Ammonia fiber expansion (Afex) treatment of wheat straw for production of bioethanol

Francisco M, Van Den Bruinhorst A, Kroon MC (2012) New natural and renewable low transition temperature mixtures (LTTMs): screening as solvents for lignocellulosic biomass processing. Green Chem 14(8):2153–2157. https://doi.org/10.1039/c2gc35660k

Haldar D, Purkait MK (2021) A review on the environment-friendly emerging techniques for pretreatment of lignocellulosic biomass: mechanistic insight and advancements. Chemosphere 264:128523. https://doi.org/10.1016/j.chemosphere.2020.128523

Hendriks ATWM, Zeeman G (2009) Pretreatments to enhance the digestibility of lignocellulosic biomass. Bioresour Technol 100:10–18. https://doi.org/10.1016/j.biortech.2008.05.027

Henniges U, Hasani M, Potthast A, Westman G, Rosenau T (2013) Electron beam irradiation of cellulosic materials-opportunities and limitations. Materials (Basel) 6(5):1584–1598. https://doi.org/10.3390/ma6051584

Hyun HongS et al (2014) Improved enzymatic hydrolysis of wheat straw by combined use of gamma ray and dilute acid for bioethanol production. Radiat Phys Chem 94(1):231–235. https://doi.org/10.1016/j.radphyschem.2013.05.056

Jacquet N, Maniet G, Vanderghem C, Delvigne F, Richel A (2015) Application of steam explosion as pretreatment on lignocellulosic material: a review. Ind Eng Chem Res. https://doi.org/10.1021/ie503151g

JorgeM, Roberto G, Karina P, Oliva C, Benjamin R, Cuauhtemocc R (2006) The optimization of the extrusion process when using maize flour with a modified amino acid profile for making tortillas. Int J Food Sci Technol (41)

Kalyani D et al (2013) Microbial consortia for saccharification of woody biomass and ethanol fermentation. Fuel 107:815–822

Karunanithy C, Muthukumarappan K (2010) Influence of extruder temperature and screw speed on pretreatment of corn stover while varying enzymes and their ratios. Appl Biochem Biotechnol 162(1):264–279. https://doi.org/10.1007/s12010-009-8757-y

Khan F, Ahmad SR, Kronfli E (2006) γ-Radiation Induced Changes in the Physical and Chemical Properties of Lignocellulose. Biomacromol 7(8):2303–2309. https://doi.org/10.1021/bm060168y

KhanMU, Usman M, Ashraf MA, Dutta N, Luo G, Zhang S (2020) A review of recent advancements in pretreatment techniques of lignocellulosic materials for biogas production: opportunities and limitations. Chem Eng J Adv 10(November 2021):100263. https://doi.org/10.1016/j.ceja.2022.100263

Kassim MA, Khalil HP, Serri NA, Kassim MH, Syakir MI (2016) Irradiation pretreatment of tropical biomass and biofiber for biofuel production

Kim KH, Dutta T, Sun J, Simmons B, Singh S (2018) Biomass pretreatment using deep eutectic solvents from lignin derived phenols. Green Chem 20(4):809–815

Kim Y, Hendrickson R, Mosier NS, Ladisch MR (2009) Liquid hot water pretreatment of cellulosic biomass. In: Biofuels, pp 93–102

Kucharska K, Rybarczyk P, Hołowacz I, Łukajtis R (2018) Pretreatment of lignocellulosic materials as substrates for fermentation processes. Molecules 23(2937):1–32. https://doi.org/10.3390/molecules23112937

Kumar B, Bhardwaj N, Agrawal K, Chaturvedi V, Verma P (2020) Current perspective on pretreatment technologies using lignocellulosic biomass: an emerging biorefinery concept. Fuel Process Technol 199:106244

Kumari D, Singh R (2018) Pretreatment of lignocellulosic wastes for biofuel production: a critical review. Renew Sustain Energy Rev 90:877–891

Lopez MDCF, Rigal M, Rigal L, Vilarem G, Vandenbossche V (2019) Influence of temperature and soda concentration in a thermo-mechano-chemical pretreatment for bioethanol production from sweet corn co-products. Ind Crops Prod 133:317–324

Majová V, Horanová S, Škulcová A, Šima J, Jablonský M (2017) Deep eutectic solvent delignification: Impact of initial lignin. BioResources 12(4):7301–7310. https://doi.org/10.15376/biores.12.4.7301-7310

Mankar AR, Pandey A, Modak A, Pant KK (2021) Pretreatment of lignocellulosic biomass: a review on recent advances. Bioresour Technol 334:125235

MarcusY (2019) Deep eutectic solvents

MaryanaR, Ma'rifatun D, Wheni IA, K.w S, Rizal WA (2014) Alkaline pretreatment on sugarcane bagasse for bioethanol production. In: Energy Procedia, vol 47, pp 250–254

MauryaDP, Singla A, Negi S (2015) An overview of key pretreatment processes for biological conversion of lignocellulosic biomass to bioethanol. 3 Biotech 5(5):597–609

Maryana R, Nakagawa-izumi A, Kajiyama M, Ohi H (2017) Environment-friendly non-sulfur cooking and totally chlorinefree bleaching for preparation of sugarcane bagasse cellulose. J Fiber Sci Technol 73(8):182–191. https://doi.org/10.2115/fiberst.2017-0025

MaryanaR, Anwar M, Hasanah SU, Fitriana E, Suwanto A, Rizal WA (2020) Cellulose acetate production from paddy rice straw and oil palm empty fruit bunch : trichloroacetate catalyst cellulose acetate production from paddy rice straw and oil palm empty fruit bunch: trichloroacetate catalyst. IOP Conf Ser Earth Environ Sci 462. https://doi.org/10.1088/1755-1315/462/1/012054

MaryanaR et al (2022) Reducing sugars and bioethanol production from oil palm empty fruit bunch by applying a batch and continuous pretreatment process with low temperature and pressure. Biomass Convers Biorefinery (0123456789). https://doi.org/10.1007/s13399-022-03138-0

McGee JK, AprilGC (1982) Chemicals from renewable resources: hemicellulose behavior during organosolv delignification of southern yellow pine. Chem Eng Commun 19(1–3):49–56

MirmohamadsadeghiS, Karimi K, Zamani A, Amiri H, Horváth IS (2014) Enhanced solid-state biogas production from lignocellulosic biomass by organosolv pretreatment. Biomed Res Int (2014)

Mok WSL, Antal MJ Jr (1992) Uncatalyzed solvolysis of whole biomass hemicellulose by hot compressed liquid water. Ind Eng Chem Res 31(4):1157–1161

Morais ARC, Da Costa Lopes AM, Bogel-ŁukasikR (2015) Carbon dioxide in biomass processing: contributions to the green biorefinery concept. Chem. Rev. 115(1):3–27. https://doi.org/10.1021/cr500330z

Mosier N et al (2005) Features of promising technologies for pretreatment of lignocellulosic biomass. Bioresour Technol 96(6):673–686. https://doi.org/10.1016/j.biortech.2004.06.025

MosierNS (2013) Fundamentals of aqueous pretreatment of biomass. In: Aqueous pretreatment of plant biomass for biological and chemical conversion to fuels and chemicals, pp 129–143

Muryanto ET, Wahyuni, Abimanyu H, Cahyono A, Cahyono ET, Sudiyani Y (2015) Alkaline delignification of oil palm empty fruit bunch using black liquor from pretreatmnet.Procedia Chem 16:99–105. https://doi.org/10.1016/j.proche.2015.12.032

Nguyen T-AD et al (2010) Pretreatment of rice straw with ammonia and ionic liquid for lignocellulose conversion to fermentable sugars. Bioresour Technol 101(19):7432–7438. https://doi.org/10.1016/j.biortech.2010.04.053

Nitsos CK, Choli-Papadopoulou T, Matis KA, Triantafyllidis KS (2016) Optimization of hydrothermal pretreatment of hardwood and softwood lignocellulosic residues for selective hemicellulose recovery and improved cellulose enzymatic hydrolysis. ACS Sustain Chem Eng 4(9):4529–4544

NuraniDA, Krisnandi YK, Simanjuntak MB, Rahayu DUC, Saepudin E (2020) Conversion of cellulose from sweet sorghum (Sorghum bicolor) bagasse to levulinic acid over hierarchical Mn/ZSM-5 heterogeneous catalyst. IOP Conf Ser Mater Sci Eng 763(1). https://doi.org/10.1088/1757-899X/763/1/012043

Okur M, Koyuncu DDE (2020) Investigation of pretreatment parameters in the delignification of paddy husks with deep eutectic solvents. Biomass Bioenerg 142:105811

Paudel SR, Banjara SP, Choi OK, Park KY, Kim YM, Lee JW (2017) Pretreatment of agricultural biomass for anaerobic digestion: current state and challenges. Bioresour Technol 245:1194–1205

Pena-Pereira F, Namieśnik J (2014) Ionic liquids and deep eutectic mixtures: sustainable solvents for extraction processes. Chemsuschem 7(7):1784–1800. https://doi.org/10.1002/cssc.201301192

Poszytek K, Ciezkowska M, Sklodowska A, Drewniak L (2016) Microbial consortium with high cellulolytic activity (MCHCA) for enhanced biogas production. Front Microbiol 7:324

Rezania S et al (2017) Review on pretreatment methods and ethanol production from cellulosic water hyacinth. BioResources 12(1):2108–2124. https://doi.org/10.15376/biores.12.1.rezania

Rezania S et al (2020) Different pretreatment technologies of lignocellulosic biomass for bioethanol production: an overview. Energy 199:117457. https://doi.org/10.1016/j.energy.2020.117457

Rinaldi R et al (2016) Paving the way for lignin valorisation: recent advances in bioengineering, biorefining and catalysis. Angew. Chemie Int. Ed. 55(29):8164–8215

Rodrigues CIS, Jackson JJ, Montross MD (2016) A molar basis comparison of calcium hydroxide, sodium hydroxide, and potassium hydroxide on the pretreatment of switchgrass and miscanthus under high solids conditions. Ind Crops Prod 92:165–173

Saini A, Aggarwal NK, Sharma A, Yadav A (2015) Prospects for irradiation in cellulosic ethanol production. Biotechnol Res Int 2015:1–13. https://doi.org/10.1155/2015/157139

Shahabazuddin M, Chandra TS, Meena S, Sukumaran RK, Shetty NP, Mudliar SN (2018) Thermal assisted alkaline pretreatment of rice husk for enhanced biomass deconstruction and enzymatic saccharification: physico-chemical and structural characterization. Bioresour Technol 263:199–206

Sindhu R, Binod P, Pandey A (2016) Biological pretreatment of lignocellulosic biomass—An overview. Bioresour Technol 199:76–82

Singh J, Suhag M, Dhaka A (2015) Augmented digestion of lignocellulose by steam explosion, acid and alkaline pretreatment methods: a review. Carbohydr Polym 117:624–631

SudiyaniY, Wahyuni ET, Muryanto M, Marno S, Putri N (2020) Bioethanol production from alkali steam explosion of oil palm of empty fruit bunch fiber. IOP Conf Ser Mater Sci Eng 854(1). https://doi.org/10.1088/1757-899X/854/1/012030

Sun Y, Cheng J (2002) Hydrolysis of lignocellulosic materials for ethanol production: a review q. Bioresour Technol 83(1):1–11. https://doi.org/10.1016/S0960-8524(01)00212-7

Suopajärvi T, Ricci P, Karvonen V, Ottolina G, Liimatainen H (2020) Acidic and alkaline deep eutectic solvents in delignification and nanofibrillation of corn stalk, wheat straw, and rapeseed stem residues. Ind Crops Prod 145:111956

Tan YT, Ngoh GC, Chua ASM (2018) Evaluation of fractionation and delignification efficiencies of deep eutectic solvents on oil palm empty fruit bunch. Ind Crops Prod 123(July):271–277. https://doi.org/10.1016/j.indcrop.2018.06.091

Tan YT, Chua ASM, Ngoh GC (2020) Deep eutectic solvent for lignocellulosic biomass fractionation and the subsequent conversion to bio-based products—A review. Bioresour Technol 297(October 2019):122522. https://doi.org/10.1016/j.biortech.2019.122522

TriwahyuniE, Miftah AK, Muryanto M, Maryana R, Sudiyani Y (2021) Effect of CO2-added steam explosion on oil palm empty fruit bunch for bioethanol production. Cellul Chem Technol 55(7–8):839–847

Wang KQ, Xiong XY, Chen JP, Chen L, Su X, Liu Y (2012) Comparison of gamma irradiation and steam explosion pretreatment for ethanol production from agricultural residues.Biomass Bioenergy 46:301–308. https://doi.org/10.1016/j.biombioe.2012.08.013

Wang K, Chen J, Sun S, Sun R (2015)Steam explosion. Elsevier B.V.

Yan L, Ma R, Li L, Fu J (2016) Hot water pretreatment of lignocellulosic biomass: an effective and environmentally friendly approach to enhance biofuel production. Chem Eng Technol 39(10):1759–1770

Yoon M, Choi JI, Lee JW, Park DH (2012) Improvement of saccharification process for bioethanol production from Undaria sp. by gamma irradiation.Radiat Phys Chem. https://doi.org/10.1016/j.radphyschem.2011.11.035

Yu Q et al (2018) Deep eutectic solvents from hemicellulose-derived acids for the cellulosic ethanol refining of Akebia' herbal residues. Bioresour Technol 247:705–710. https://doi.org/10.1016/j.biortech.2017.09.159

Yu Y et al (2022) Steam explosion of lignocellulosic biomass for multiple advanced bioenergy processes: a review.Renew Sustain Energy Rev 154(October 2021):111871. https://doi.org/10.1016/j.rser.2021.111871

Zhao Z, Chen X, Ali MF, Abdeltawab AA, Yakout SM, Yu G (2018) Pretreatment of wheat straw using basic ethanolamine-based deep eutectic solvents for improving enzymatic hydrolysis. Bioresour Technol 263:325–333

Zhao C, Shao Q, Chundawat SPS (2020) Recent advances on ammonia-based pretreatments of lignocellulosic biomass. Bioresour Technol 298:122446

ZhaoL et al (2022) Advances in pretreatment of lignocellulosic biomass for bioenergy production: challenges and perspectives. Bioresour Technol 343(October 2021):126123. https://doi.org/10.1016/j.biortech.2021.126123

Zheng Y, Lin HM, Tsao GT (1998) Pretreatment for cellulose hydrolysis by carbon dioxide explosion. Biotechnol Prog 14(6):890–896. https://doi.org/10.1021/bp980087g

Zhuang X et al (2016) Liquid hot water pretreatment of lignocellulosic biomass for bioethanol production accompanying with high valuable products. Bioresour Technol 199:68–75

Zulkefli S, Abdulmalek E, Rahman MBA (2017) Pretreatment of oil palm trunk in deep eutectic solvent and optimization of enzymatic hydrolysis of pretreated oil palm trunk. Renew Energy 107:36–41

Recent Updates on Biopolymers: Precursors, Process, Properties, Challenge, and Future Perspectives

Aisyah Hanifah, Arfiathi, Melbi Mahardika, Riska Sumirat, Rossy Choerun Nissa, and Yeyen Nurhamiyah

Abstract Fossil sources are common raw materials in the polymer industry because they are cost-effective and ensure a straightforward manufacturing process. However, the insufficient supply of fossil sources failed to afford adequate feedstock for polymer production in the future. Fossil sources are projected to reach a saturation point where supply would be less than demand due to the increasing human population. Another important concern is the fact that fossil-based polymer creates several environmental problems, such as non-degradable products, air pollution, and wastewater contamination (Okkerse and Bekkum 1999). These two main reasons are the main factors why the switch of the raw materials of polymers from fossil to renewable materials is necessary, and the research on biobased polymeric materials becomes an interesting yet urgent topic. In this chapter, we review the current updates on the development of biopolymers. The precursors, technological processes, and updates on the currently available biopolymers are being reviewed. Challenges and future perspectives are also being discussed.

Keywords Biopolymers · Starch · Seaweed · Plant oil · Polymerization

A. Hanifah · M. Mahardika · R. C. Nissa · Y. Nurhamiyah (✉)
Research Center for Biomass and Bioproducts, National Research and Innovation Agency (BRIN), Jl. Raya Bogor KM. 46 Cibinong, Bogor, Indonesia
e-mail: yeye001@brin.go.id

Arfiathi · R. Sumirat
Department of Agricultural Industrial Technology, Universitas Padjajaran, Jl. Raya Bandung Sumedang KM.21, Bandung, Indonesia

M. Mahardika · Y. Nurhamiyah
Research Collaboration Center for Biomass and Biorefinery between BRIN and Universitas Padjadjaran, Bandung, Indonesia

M. A. R. Lubis et al. (eds.), *Biomass Conversion and Sustainable Biorefinery*, Green Energy and Technology, https://doi.org/10.1007/978-981-99-7769-7_2

1 Introduction

Polymer materials are the most widely used applications in everyday life. The production of synthetic polymers continues to grow, such as the use of polycarbonate, polyvinyl chloride, polyethylene (PE), polyethylene terephthalate (PET), polypropylene (PP), poly(methyl methacrylate), and polystyrene, with an estimated 1.1 billion tonnes in 2050 (Geyer 2020). Increasing awareness of the environment and limited fossil resources have led to the importance of replacing biobased materials (biopolymer) to reduce greenhouse gas emissions. Biopolymers are polymers derived from natural and renewable sources obtained from chemical polymerization reactions (Das et al. 2022; Mohanty et al. 2022). There are classifications of biopolymers: polysaccharides, nucleic acids, polyamides (PAs), polythioesters (PTEs), polyanhydrides, polyphenolic biopolymers, polyisoprenoids, and polyoxoesters (Sharma and Dhingra 2021). The most widely available biopolymers on earth are cellulose, lignin, rubber, and protein (Bayón et al. 2018; Mikus and Galus 2022). The advantages of biopolymer materials are low cost, easy to form, superior properties such as resistance to water, biodegradable, and mechanical properties that can be improved for various applications. However, the increased use of biopolymers is only about 1% of the total industrial materials produced each year (Ranganathan et al. 2020).

This chapter book provides comprehensive information on biopolymers, current research being developed, process, properties, production on a lab or commercial scale, future, and perspective. The first is a summary of the sources of raw materials derived from biomass for the production of biopolymers and bioprocesses. Then, the precursors of biopolymers will be discussed in detail with a focus on current production. The mass production of biopolymers raises concerns about economic and environmental sustainability (Chang et al. 2016; Maraveas 2020). Consequently, the production of biopolymers from renewable biological sources has received great attention during the last decade. Currently, most of the precursors of biopolymers are obtained from biomass sources such as starch, seaweed, and plant oil, which are in direct competition with food production (de Jong and Jungmeier 2015; Popa 2018; Gajula and Reddy 2021). To build a biobased economy, non-edible raw materials must be used for the production of biopolymers. All petrochemical-derived chemicals can be replaced with biopolymer materials produced from biobased materials, which will significantly contribute to the economic progress (Isikgor and Becer 2015; Nanni et al. 2021).

2 Precursor of Biopolymers

2.1 Starch

Starch is a polysaccharide that plants produce to store energy. It is stored intracellularly as spherical granules ranging in size from 2 to 100 μm. The majority of market-available starches are derived from grains such as wheat, rice, and corn or from tubers such as potato and cassava (tapioca) (Jiang et al. 2020). Starch is an amorphous, soft white powder without any sweetness and insoluble in water, alcohol, and ether. The starch granules are densely packed in a semicrystalline structure with a density of about 1.5 g cm^{-3}. Specific shapes, sizes, and surfaces (smoothness or roughness) of starch granules from various botanical sources play a crucial role in functionality and digestibility (Magallanes-Cruz et al. 2017).

Amylose and amylopectin are glucans that combine to form starch; most starches consist of 10–20% water-soluble amylose and 80–90% water insoluble amylopectin (Pokhrel 2015). Amylose is a linear polymer polysaccharide with 1–4-D-glucose units. Amylase's structure contributes to the gelling properties of heated and cooled starches. Amylopectin has a backbone of α-(1 $\rightarrow$ 4) linkages as well, but it is also branched through α-(1 $\rightarrow$ 6) linkages to the extent of 4–5%. Amylopectin is responsible for the thickening during starch preparations, but it does not make a contribution to the formation of a gel (Pokhrel 2015).

Starch has traditionally played a significant role as a food ingredient, but it is finally starting to be used in other applications, such as textiles, paper, pharmaceuticals, and pharmaceuticals. Starch can be applied in various ways, including pure starch as thermoplastics, starch as blends formed by combining other polymers, and copolymers formed by combining synthetic polymers (Pokhrel 2015; Encalada et al. 2018). There have been many studies conducted (Sriroth and Sangseethong 2006; Ibrahim et al. 2017; Area et al. 2019; Mesias and Murillo 2020) related to the development of starch-based products, one of which is thermoplastic starch that can be applied as a raw material for bioplastics. As a result, the use of starch in biopolymers is very promising, especially given that natural starch comes from agricultural sources, which has the advantages of abundant supply, ease of filling, low cost, and a wide range of applications and modifications in non-food products.

2.2 Seaweed

Seaweeds are macroscopic, multicellular, and benthic algae. Seaweed has a rapid growth rate, resulting in a fast accumulation of biomass (Carina et al. 2021; Perera et al. 2021). Edible marine macroalgae, or seaweed, is well-known for being extremely nutrient-rich as both food and culinary additives. Coastal populations all across the world, including Asians (Chinese, Japanese, Indonesian), Europeans (Irish, Icelandic), and South Americans, among others, have harvested wild seaweed for food and other uses. Although wild seaweeds are still gathered from the coastal areas of several nations, farmed seaweeds make up the majority of the seaweed and seaweed products produced today (Padam and Chye 2020). More than US$6 billion is thought to be generated by the international trade of seaweed each year. China continues to be the world's largest seaweed producer, producing 35.7 million tons, or roughly 56.75% of the total, followed by Indonesia (27.86%) and Korea (5%) (FAO 2021).

Around the world, there are more than 10,000 different varieties of seaweed. 10 seaweed species are intensively grown globally, out of the 221 species of seaweed that have been commercially exploited (Padam and Chye 2020). Seaweeds remain an unexplored resource with huge potentials such as food ingredients, cosmetics, agri-chemicals, edible foods, fishmeal, biomaterials, and bioenergy molecules, yet also having an important aspect in the ecosystem and economic profits (Padam and Chye 2020; Carina et al. 2021; Chudasama et al. 2021). Based on their color, seaweeds can be divided into three groups: red seaweeds (*Rhodophyceae*), green seaweeds (*Chlorophyceae*), and brown seaweeds (*Phaeophyceae*) (Perera et al. 2021).

Three major polysaccharides that are present in seaweed are alginate, agar, and carrageenan. The majority of polysaccharides can swell in water under ambient conditions or are water soluble (at high temperatures), producing colloidal, highly viscous solutions or dispersions with pseudoplastic flow properties. Due to their natural functional properties, such as thickening, stabilizing suspensions and emulsions, water retention and binding, and gelling, polysaccharides are advantageous in a range of applications. Terrestrial plants do not possess polysaccharides with specific and distinct characteristics that are received from a significant oceanic source (Chudasama et al. 2021).

2.3 *Plant Oil*

Plant oil is a potential source for biopolymers since it offers availability, low cost, easy processing route, and chemical functionality. Plant oil is a promising source for polymer building blocks as it resembles the hydrocarbon composition of petrochemicals. Globally, plant oil production capacity was 209.42 million metric tons in 2019/2020, and only around 20 million tonnes were used in the chemicals industry (Stempfle et al. 2016; USDA Foreign Agricultural Services 2021).

The total carbon atoms in the chain and the way carbon atoms are joined control the properties of fatty acids. Single carbon atoms may join each other and connect to the hydrogen atom by a single atom. Carbon atoms are also linked by a double bond and joined to only one hydrogen each. The double bond increases the ratio of carbon atoms to hydrogen because of the absence of two hydrogen atoms. The disappearance of hydrogen generates fatty acids that are less "saturated" with hydrogen. This kind of bond is called saturated, while the double bond fatty acid is called unsaturated (Pond 1998). Unsaturated fatty acids are unstable compared with saturated fatty acids. The presence of a double bond makes the carbon atoms in this fatty acid more reactive.

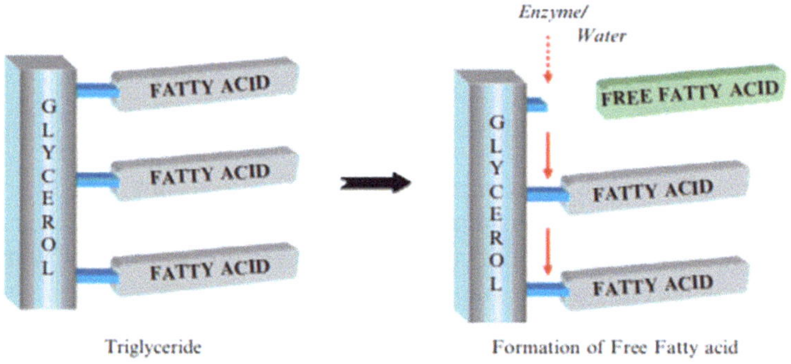

Fig. 1 Ester linkage hydrolysis (Mallakpour and Rafiee 2012)

Double bonds adjust the crystal packs; they limit the rotation of a carbon atom and bend the chain (Gurr et al. 2002) (Fig. 1).

Polymer made from plant oil offers material with beneficial properties such as hydrophobicity, flexibility, low melting temperature, and biodegradability. Polyamide (Fan et al. 1997; Hablot et al. 2010), polyester (Quinzler and Mecking 2010; Testud et al. 2017), polyurethane (Hojabri et al. 2010; More et al. 2013), and self-healing materials (Cordier et al. 2008) are examples of polymers that have been synthesized from fatty acids.

2.4 Others

Classification biopolymers are (1) the agro-polymers that comprise polysaccharides, proteins, and lipids; and (2) bio-polyesters (biodegradable polyesters), such as polyhydroxyalkanoate (PHA), polylactic acid (PLA), and aromatic and aliphatic co-polyesters. Biopolymers classified as agro-polymers are biomass products obtained from agricultural materials, such as polysaccharides, proteins, and lipids. Bio-polyester is subdivided based on its source. The polyhydroxyalkanoate (PHA) group is obtained from activity-derived microorganisms obtained by extraction. Examples of PHAs include Poly(hydroxybutyrate) (PHB) and Poly(hydroxybutyrate co-hydroxy valerate) (PHBV). Another group is the polyesters obtained from biotechnology applications, namely by the conventional synthesis of monomers obtained biologically, called the polylactide group. An example of a polylactide is polylactic acid. The last group is obtained from conventionally synthesized petrochemicals made from synthetic monomers. This group consists of polycaprolactone (PCL), polyester amides, aliphatic co-polyesters, and aromatic co-polyesters.

3 Production Process

3.1 Plasticization

Plasticizers are substances used to improve the flexibility, extensibility, and processability of polymeric materials. Plasticizers have the property of lowering the melting temperature, melt viscosity, and glass transition temperature, as well as having high boiling points, molecular weights ranging from 300 to 600 and elastic modulus of polymers without changing their chemical nature (Drobny 2015; Subramanian and Varade 2017). Plasticizers are classified into three types: polyol (glycerol, polyethylene glycols, and propylene glycol), oils/glycerides (acetylated monoglycerides, castor oil, and fragmented coconut oil), and organic ester (phthalate ester, citrate ester, triacetin, and dibutyl sebacate) (Foroughi-dahr et al. 2017). The amount of plasticizer added to a polymer is determined by the intended result. To improve the workability of the polymer melt, a small amount of plasticizer can be added, contrary to the impression that large additions can completely transform the product's properties (Harrison 2002).

When starch is plasticized, the plasticizers break the inter- and intra-molecular hydrogen bonds existing in the starch, causing it to become thermoplastic. The basic crystalline structure of the starch granules is broken by the plasticizer, which permeates them. Thermoplastic starch (TPS) becomes extremely brittle at low moisture contents while becoming more flexible and softer at high moisture amounts. When wet, it also rapidly loses strength. The effectiveness of TPS in situations with extremely low or high humidity is obviously greatly constrained by its sensitivity to moisture (Janssen and Moscicki 2009).

Plasticizers work by decreasing interactions between molecules and spreading them out. Low viscosities and temperature coefficients of viscosity are characteristics of efficient plasticizers. A polymer with a low molar mass and a higher free volume is frequently used as a plasticizer. A crucial need is that the plasticizer be perfectly blended at the molecular level, either in a homogenous mixture with the polymer or with the plasticizer itself (Janssen and Moscicki 2009). A plasticizer's principal function is to improve process efficiency and flexibility. The glass transition temperature is lowered by the plasticizer (T_g). Due to different environmental issues and rising petroleum product prices, plasticizers are becoming more and more important. So that the process can be made both affordable and biodegradable, plasticizer should be natural, affordable, and renewable (Khan et al. 2017).

The research about TPS and the effect of plasticizer processes obtained by melt processing was carried out by (Carvalho et al. 2006) using 1,4-butanediol (BUT), 1,6-hexanediol, 2,5-hexanediol, glycerol, ethylene glycol (EG), diethylene oxide glycol (DEG), trimethylene oxide glycol (TEG), ethylene glycol monomethyl ether, D-sorbitol (SOR), propylene glycol (PG), polyethylene oxide glycol (PEG) 300 and 600, and polypropylene-oxide glycol (PPG). The results showed that compounds containing OH had a good impact on the characteristics of TPS prepared by melt mixing.

3.2 Polymerization

Similar to fossil-based polymers, polymerization is a common technique to produce biobased polymers. Polycondensation and ring-opening polymerization (ROP) are examples of polymerization to get aliphatic biopolymers. Condensation polymerization is a chemical reaction with the elimination of a small molecule, which results in the repeat units in polymers having less atoms than when they are in the presence of monomers (Young and Lovell 2011). This technique was found by Wallace Carothers who discovered the condensation polymerization technique and linear polymers such as polyamides and polyesters which are still useful today (Carothers 1936, 1937). An example of polycondensation in biopolymers is the polycondensation of poly glycerol sebacate (PGS) which starts from glycerol and sebacic acid. PGS is a biodegradable and bioresorbable polymer which suitable for biomedical applications such as tissue engineering (Rai et al. 2012). Glycerol is a by-product of the saponification process from oleochemical plants meanwhile sebacic acid is originated from fatty acid (Tan et al. 2013). According to Wang et al. (2002), PGS was prepared in two stages: pre-polycondensation and crosslinking. The equimolar amount of sebacic acid and glycerol was mixed under argon at 120 °C for 24 h with the pressure reduced over 5 h. For the crosslinking stage, the pre-polymer is kept reacted at 40 mTorr and 120 °C for 48 h (Fig. 2).

Meanwhile, ring-opening polymerization (ROP) is a polymerization technique of cyclic monomers that produces a polymer with identical molecular formulae to those of the monomers (Young and Lovell 2011). For example, biobased polyamides can be prepared from monomers with cyclic structure (lactams). An example of lactams is ε-caprolactone as the precursor of caprolactam for PA6 and ω-laurolactam for PA12. Evonik produces PA12 on an industrial scale with the tradename Vestamid E Even though the monomer is mostly produced from fossil-based laurin lactam, currently, the alternative laurin lactam is being developed by Evonik from palm kernel through biotechnological process shows the ring-opening polymerization of 12-aminodecanoic acid (laurin lactam) to obtain PA12.

Even though ring-opening polymerization can obtain high-molecular-weight polymers, however, due to few available cyclic monomers (Zhao 2018; Santoro et al. 2020) only limited polymers can be obtained. On the other hand, the polycondensation process brings possibilities and flexibility that are incomparable since could use a variety of monomers.

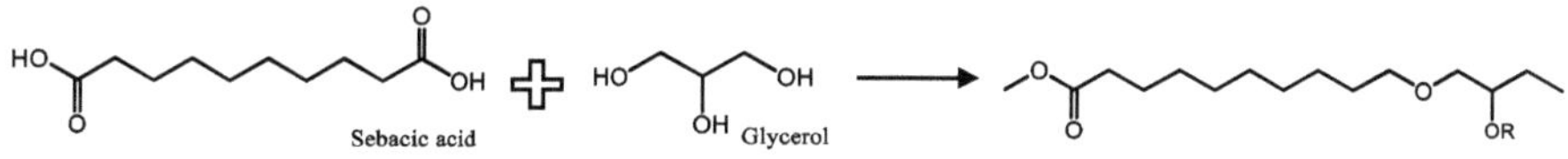

Fig. 2 The polymerization scheme of poly glycerol sebacate (PGS) (Wang et al. 2002)

3.3 Blending

To modify the properties of biopolymers to the intended use, various ways are employed for their modifications, including absorption of fillers and reinforcements, blending, plasticization, and impact modification (Rajeswari et al. 2021). Blending is the simple mix of polymeric components without chemical reaction to generate novel materials with appropriate qualities. From an industrial perspective, this procedure can be conducted by employing regular machines, so a costly investment is unneeded. Polymer blends are applied in a wide variety of industrial applications. In comparison to alternative polymerization processes and the production of novel monomers, this technique can obtain a wide range of characteristics, which meet the standard of the targeted application at an efficient cost and time.

The aim of blending might be the optimization of the material performance, reduction of the sensitivity to water, cutting down the cost, and improvement of the application properties. The most significant motivation for mixing could differ from other methods.

3.4 Fermentation

Biopolymers which are constructed from microorganisms need certain nutrients and restricted environmental conditions. They are created directly by fermentation or indirectly by chemical polymerization of the monomers produced by fermentation. Most biopolymers are biocompatible and do not negatively affect biological systems. Biopolymers derived from bacteria are considered to be produced due to their defense mechanisms or as storage materials (Mohan et al. 2016). Biopolymers are synthesized by biological organisms and produced by processive enzymes that connect building ingredients such as sugars, amino acids, and hydroxy fatty acids to achieve a high-molecular-weight molecule. Bacteria can produce various ranges of biopolymers, including polysaccharides (composed of sugars and sugar acids linked by glycosidic linkages), polyesters (composed of hydroxy fatty acids linked by ester bonds), polyamides (consisting of amino acids linked by peptide bonds), and polyphosphates (polyPs; composed of inorganic phosphates linked by anhydride bonds) (Fig. 3).

PLA and PHAs are examples of biopolymers that belong to the polyester group and are formed when microorganisms are used in the processing step. PLA is produced from a low molecular weight organic acid synthesized through microbial fermentation called lactic acid (LA). This acid is utilizing renewable sources such as cane sugar, corn, and sugar beets. PHAs are an extensive family of bio-polyesters manufactured by diverse bacteria to store carbon and energy. Fermentation is influenced by several parameters, including the substrate, temperature, pH, oxygen, and bacteria used. The term "substrate" refers to the fermentable material that includes the essential nutrients for the growth and production of fermented products by bacteria. The

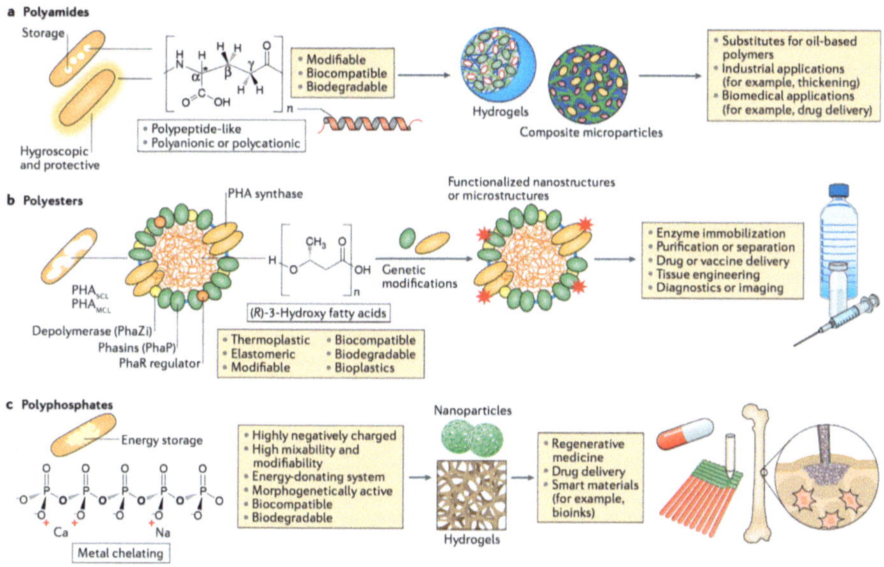

Fig. 3 Bacterial polymer granules as biomaterials (Moradali and Rehm 2020)

bacterial strains used in the fermentation process can be divided into heterofermentative and homofermentative methods. This homofermentative process is constantly utilized in industry.

Lemoigne discovered PHAs in 1943 (Lemoigne and Girard 1943); they are a family of naturally occurring polyesters generated intracellularly by many prokaryotes as carbon and energy storage polymers. In general, PHAs can be synthesized via three distinct carbon source-based steps: step I is mainly applied by poly(hydroxybutyrate) (PHB) PHB-producing organisms such as *C. necator* and *Bacillus sp.*, whereas steps II and III exist in mcl-PHA-producing *Pseudomonas sp.* Different monomers can be produced from different bacterial strains with varying carbon substrates of the microorganisms; consequently, co-polymer proportions can be modified based on the targeted applications of the final product (Fig. 4).

For the synthesis of PLA, raw material starch/sugar is extracted and fermented using lactic acid bacteria (LAB), which results in the generation of lactic acid. The fermentation of starch and other polysaccharides (sugar-containing materials), which are readily available from corn, sugar beet, sugar cane, potatoes, and other biomasses, accounts for 90–95% of the global lactic acid generation. Normal fermentation provides 85–90% of L(+) lactic acid and 70–80% of D(−) lactic acid, depending on the carbon source. During fermentation, anaerobic lactic acid bacteria (LAB) consume pyruvic acid, the last product of the Embden–Meyerhof–Parnas (EMP) route, via glycolysis. Prior to this, the carbon source sugars are converted to pyruvic acid. L-lactate dehydrogenase or D-lactate dehydrogenase can convert pyruvic acid to lactate while influencing the stereospecificity of the lactic acid generated. Lactic acid

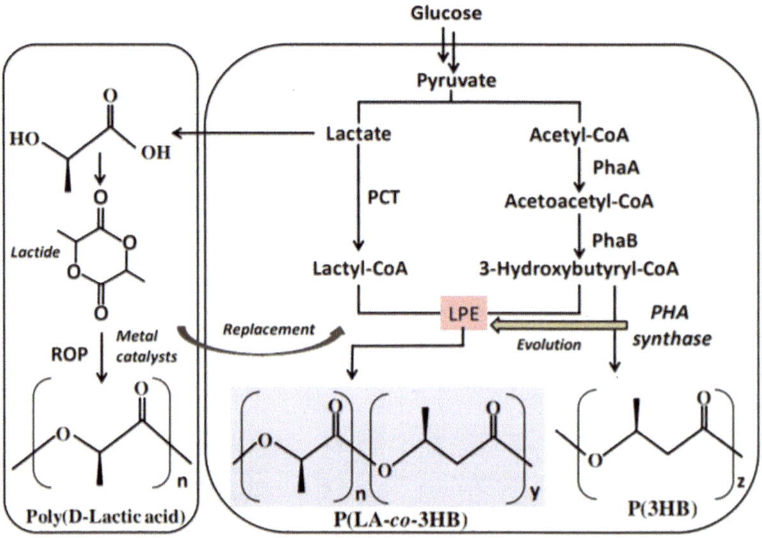

Fig. 4 Schematic illustration of the biosynthetic steps for producing lactate-based polymers (Nduko and Taguchi 2021)

is chemically processed and polymerized to create the final product by polycondensation reaction, ring-opening polymerization, and azeotropic dehydrative condensation. Ring-opening polymerization produces a product with a high molecular weight, making it the most practical method for producing PLA (Fig. 5).

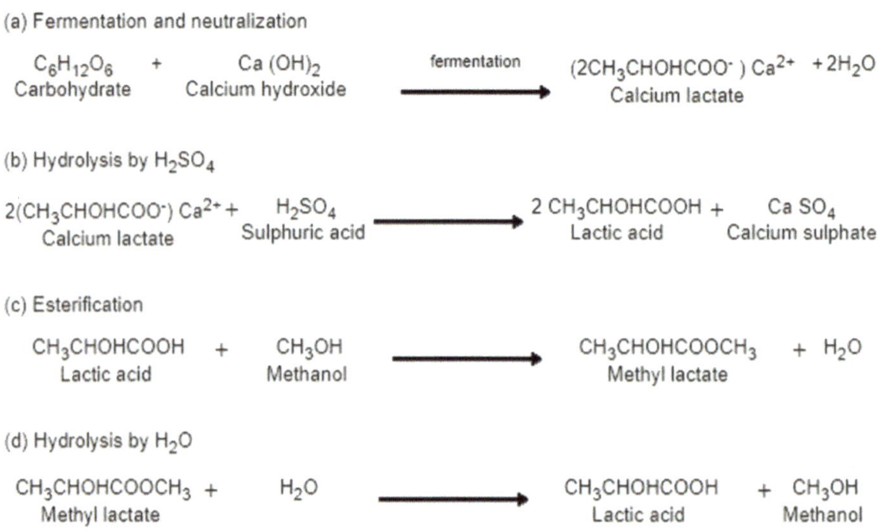

Fig. 5 Lactic acid production by fermentation procedure (Moradali and Rehm 2020)

4 Biobased Polymers

In contrast to fossil fuel plastics, which are produced from petroleum, biopolymers or organic plastics are generated from renewable biomass resources such as starch, seaweed, and proteins. Biopolymers offer the additional advantages of conserving fossil resources and reducing CO_2 emissions, making them an essential invention for sustainable development.

4.1 Thermoplastic Starch

Native starch exists in granular form. Pure starch is hydrophilic, brittle, has a lower thermal decomposition temperature, poor thermal processability, a high melting point (Khan et al. 2017), and rapidly degrades when exposed to water (Janssen and Moscicki 2009). Starch can be processed into thermoplastics by processing under shear of melt extrusion and high temperatures with the addition of plasticizers (Ma et al. 2009). Low-molecular-weight plasticizers are typically used during the gelatinization process to help break up the starch granules (Martin and Gonz 2017). Thermoplastic is a material that may be molded into desired shapes because of its weak physical crosslinks, which can be melted and reshaped in a heating and cooling cycle (Sjoo and Nillsson 2018).

Nowadays, research on the development of TPS continues to increase using various materials due to advances in TPS production technology. The raw materials include wheat starch (Moghaddam et al. 2018), cassava bagasse (Edhirej et al. 2017), cornstarch (Baran et al. 2022), potato starch (Niazi et al. 2015), mango starch (Agwamba 2021), cassava starch (Chotiprayon et al. 2020), rice starch (Prachayawarakorn et al. 2010), and pea starch (Cao et al. 2009). TPS can be processed by extrusion, blowing, thermocompression, or injection molding (Castillo et al. 2013). The advantages of TPS are its low cost, renewable properties, and wide availability. However, TPS has various drawbacks, such as being hygroscopic, hydrophilic, having a high sensitivity to moisture, poor mechanical properties compared with conventional polymers, low gas permeability, and inadequate water barrier characteristics (Bangar et al. 2021). Blending TPS with a different biodegradable polymer and renewable filler are examples of methods of overcoming these drawbacks. TPS can be used for various applications like food packaging, plastic bags, disposable cutlery, etc. (Khan et al. 2017).

4.2 Seaweed Based Polymers

Polysaccharides from seaweed can be used as a potential active agent, a raw material that is rich in polysaccharides, or extracts. The usefulness, sensory qualities, and sustainability of materials can all be enhanced by seaweed-based products. Seaweed has drawn attention because of its biodegradability, non-toxicity, antioxidant properties, and great film-forming ability. Seaweeds are applied in active packaging (Yildirim et al. 2018), edible films (Tran et al. 2020), and edible coatings (Aayush et al. 2022). Different seaweed polysaccharides' capacities for biocompatibility, gel formation, emulsification, gelation, and foaming are based on their distinctive structural features. In addition to the seaweed's natural qualities, its nutritional value, which includes vitamins, antioxidants, minerals, and calories, is helpful in creating edible coatings and films. When seaweeds are mixed with other polysaccharides, nanoparticles, essential oils, or plant extracts, their barrier, thermal, mechanical, antioxidant, and antibacterial characteristics are improved (Perera et al. 2021).

Active packaging combining sodium alginate and lemongrass oil can inhibit *E. coli* and *L. monocytogenes* growth (de Oliveira et al. 2019). κ-Carrageenan and mulberry polyphenol extracts can improve antioxidant and pH-sensitive properties (Liu et al. 2019). Two seaweed polysaccharides, alginate and carrageenan, are extensively used in the production of edible films and coatings. Film homogeneity and transparency were favored by alginate and glycerol (Paula et al. 2015). Glycerol and k-carrageenan enhanced the moisture barrier and tensile characteristics (Paula et al. 2015).

Food coatings serve a number of purposes, including changing the functional characteristics of foods, serving as a barrier between the environment and food products, and managing the moisture on the food's surface (Perera et al. 2021; Aayush et al. 2022). The use of agar-based coatings has been beneficial in increasing the storage shelf life of banana fruits by reducing fruit hardness and weight loss (Hussein Ziedan et al. 2018). Carrageenan and chitosan, which are added to dragon fruit that is kept at a temperature of 10 °C and a relative humidity of 90–95%, have been shown to reduce weight loss. Bract chlorophyll concentration and freshness retention for 30 days (Thi et al. 2021).

4.3 Polylactic Acid

Poly(lactic acid) (PLA) is a type of biodegradable thermoplastic polyester that is projected to replace conventional petrochemical-based polymers. PLA is a desirable biopolymer due to its processability, sustainability, and eco-friendliness; thus, it has gained popularity in packaging, textile, automotive composites, and biomedical applications. In terms of future industrial uses, the physical characteristics of PLA are crucial, including glossy and translucent, stable at low temperatures, moderately permeable to oxygen and water, and resistant to grease and oil. These characteristics make it an excellent material for the production of film, bottles, cups, and trays (Ranakoti et al. 2022).

PLA can be synthesized through direct lactic acid polycondensation, ring-opening polymerization of lactide, and a lactic acid cyclic dimer. The ring-opening polymerization method is a technique that combines a metal catalyst with lactide to produce larger PLA molecules (Ebnesajjad 2012). Direct polycondensation requires extreme

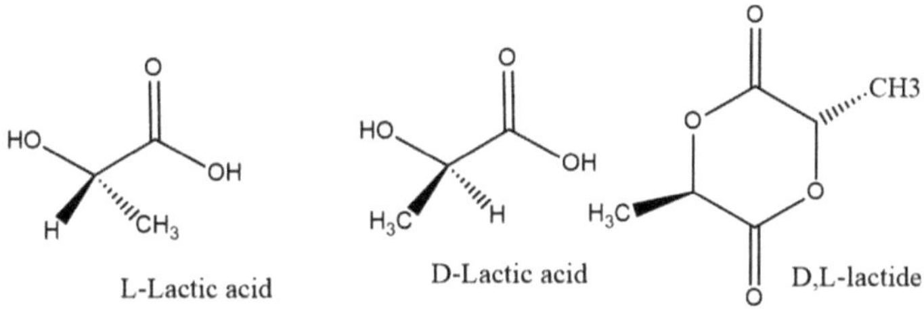

Fig. 6 Chirality structures of PLA

conditions to generate polymer with a high molecular weight at a short reaction time, low pressure of 5 mm Hg, and temperatures as high as 180–200 °C, meanwhile ROP can generate a high-molecular-weight PLA with a controlled molecular weight distribution under mild reaction conditions (temperature of 130 °C at a short reaction time) (Ashothaman et al. 2021). PLA is a kind of polyester that is often made from hydroxy acid, one of the few polymers whose stereochemical form can be easily modified. This is obtained by polymerizing a restricted mixture of L and D isomers, which produces polymers with a high molecular weight that is either crystalline or semicrystalline. Lactic Acid (LA) or 2-hydroxy propionic acid is the main monomer component for PLA which includes an asymmetric carbon that generates two enantiomeric forms, L-LA and D-LA (Fig. 6), which make PLLA and PDLA, respectively. The combination of the two optical isomers results in PDLLA. The proportion of D- and L-enantiomers results in various PDLLA stereo blocks. The ratio of these isomers, which is the specific chemical composition of PLA, regardless of whether it is PDLA or PLLA, and the structure of the PLA stereo block, may affect essential aspects, such as crystallization extent and thermal properties. However, PLA can be made with many qualities by modifying the polymer chains' molecular weight, composition, and stereoisomeric distribution (Jiménez et al. 2019).

PLA is one of the most potential biodegradable polymers due to its thermoplastic processibility, mechanical properties, and biological properties, such as biocompatibility and biodegradability. Glass transition temperature (T_g) is the most important parameter for amorphous PLA because significant changes in polymer chain mobility occur at the glass transition temperature. The behavior of semicrystalline PLA can be anticipated to influence critical physical characteristics such as T_g and crystalline melting temperature (T_m) (Revati et al. 2016) (Table 1).

4.4 *Polyhydroxyalkanoates*

Polyhydroxyalkanoates (PHAs) are polyesters of hydroxylic acid (HA) monomers linked by an ester group. PHAs are bacteria-based biodegradable polymers that encourage carbon neutrality and sustainable industry. The monomer composition determines the elastomeric/thermoplastic characteristics of these polymers. Despite

Table 1 Mechanical properties of PLA (Jiménez et al. 2019)

Property	PLA
Density (g/cc)	1.25
Haze (%)	2.1
Tensile strength (MPa)	109.97
Tensile modulus (MPa)	3299.26
Ultimate elongation (%)	160
Tear resistance (g/mm)	0.3810

the significant potential and efforts to build cost-effective fermentative methods, the PHAs production costs are still relatively high (5–10 $/kg), which hinders the commercialization of these biopolymers as commodities materials (Ortelli et al. 2019). According to ASTM standard, PHAs such as poly(3-hydroxybutyrate) (PHB) and poly(3-hydroxybutyrate-co-3-hydroxy valerate) (PHBV) are defined as biodegradable in all aerobic and anaerobic environments. They can be used to make soil, marine biodegradable, and compostable products, which is an advantage compared to synthetic non-degradable plastics (Mohapatra et al. 2021).

The main PHAs subcategories are short (3–5 carbons), medium (6–14 carbons), and long (15+ carbons) (Meereboer et al. 2020). Unsaturated fatty acids form short and medium-chain PHAs with double bonds. The common short-chain PHAs are poly(3-hydroxybutyrate) (PHB) and poly(3-hydroxybutyrate-co-3-hydroxy valerate) (PHBV). Other biodegradable PHAs include poly(3-hydroxybutyrate-co-4-hydroxybutyrate) (PH4B), poly(3-hydroxybutyrate-co-3-hydroxyhexanoate) (PHBHx), and poly(3-hydroxybutyrate-co-3-hydroxyoctanoate) (PHBO) (Fig. 7).

PHAs, such as poly(3-hydroxybutyrate-co-3-hydroxy valerate (PHBV) and poly(3-hydroxybutyrate-co-3-hydroxy valerate (PHB), are brittle due to their high crystalline degree. The chemical structure of PHAs is responsible for giving it these characteristics. Therefore, given that these numerous varieties of PHAs each have specific structural, physical, and chemical qualities, it is necessary to categorize them according to those properties and then modify them so that they are simple to use for the purposes that have been described. The general properties of the common types of PHAs are shown in Table 2.

PHAs have various excellent properties, including a high volume-to-surface ratio, a tiny pore size with a high chance of being recycled, biodegradability, and biocompatibility. Recent attention has been drawn to PHAs due to their multiple favorable qualities, such as ease of processing, resistance to UV rays, and insolubility in water. The biodegradable nature and other advantages of polyhydroxyalkanoates, such as high-temperature stability, low degree of surface porosity, improved toughness, and elasticity, result in their use in a variety of industries, including medicine, agriculture, etc. (Chai et al. 2021).

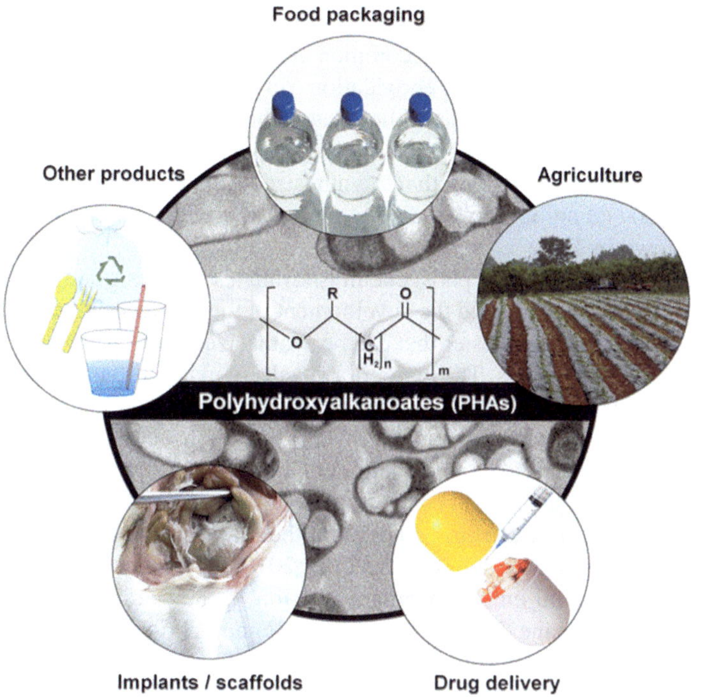

Fig. 7 General chemical structure and various applications of PHAs (Sharma et al. 2021)

Table 2 The general properties of the common types of PHAs (Chai et al. 2021)

Property	P(3HB)	P(3HB-co-4HB)
Glass transition temperature, T_g (°C)	2–4	− 48–4
Melting temperature, T_m (°C)	160–175	50–175
Tensile strength, σ (MPa)	15–40	17–104
Young's modulus (GPa)	1–2	0.07–1.5
Elongation at break (%)	1–15	14–1320
Crystallinity (%)	50–80	34–60

4.5 *Plant Oil-Based Polymers*

Plant oil has several functional groups coming from fatty acids that are useful for polymer synthesis. Carboxylic acid and alkyl ester are the standard functional groups in the fatty acid. There are also additional groups such as epoxy group (vernolic acid) and hydroxyl (ricinoleic acid). Plant oil can be directly used for polymerization due to these functional groups. Direct polymerization usually converts plant oil to polyol through epoxidation.

Junming and his co-workers (Junming et al. 2012) reported the preparation of polyester polyols from oleic acid, a common unsaturated fatty acid in plant oils. The synthesis consists of three steps: epoxidation, ring-opening reaction, and esterification. The polyols appear as a viscous liquid at ambient temperature with OH number ranging from 307 to 425 mg KOH g^{-1}. Polyester polyol was then used to synthesize polyurethane foams. The polyurethane synthesized from the oleic-based polyols and isocyanate showed an acceptable result and reached the standard of rigid foam in China as shown in Table 3.

The hazardous reaction of phosgene in isocyanates synthesis and the sustainable demand in chemistry generated the development of the phosgene-free method. Plant oils have double bonds and ester that can be synthesized into various new structures including "bio" isocyanates. The major drawbacks of plant oil-based isocyanates are that they are still not able to substitute current isocyanates on the market. Despite that, Henkel Corporation Company and General Mills have commercialized dimer fatty acid diisocyanates from fatty acid containing 36 carbon atoms in the chain. Other raw materials, such as soybean oil, azelaic acid, and oleic acid, also have been attempted to make diisocyanates and used for the synthesis of "green" thermoplastic PU (Caylı and Kusefoglu 2008; Hojabri et al. 2010; More et al. 2013).

Rix et al. used mini-emulsion polymerization in their attempt to prepare non-isocyanates polyurethane (NIPU). A fatty acid diamine, Priamine 1075, was reacted with fatty acid-based bis-cyclic carbonates at 60 °C for several hours. The mini-emulsion process was then prepared to obtain waterborne NIPU latexes with the addition of surfactants and hydrophobic additives. The result obtained was NIPU with a solid content reaching of 30wt%. The molar masses were low compared to bulk NIPUs, and this is because of the existence of partial hydrolysis of the carbonates. The NMR result of PHUs from the mini-emulsion process showed signals at 3.5–4 ppm that are considered equal to protons of hydrolysis compounds (Rix et al. 2016).

Despite direct polymerization being an easy route to obtain plant-oil-based polymer, only few polymers can be obtained through this route, such as bio-polyurethane and epoxy. Other routes such as the dimerization of fatty acid, and self-metathesis of methyl ester are needed to obtain various types of monomers with double functional groups.

Table 3 Physical properties of polyurethane foams (Junming et al. 2012)

Properties	Foam 1[a]	Foam 2[a]	Foam 3[a]	China standard[b]
Thermal conductivity W/(m K)	0.028	0.037	0.076	0.024
Density (kg/m^3)	31.3	40.6	58.1	30–50
Compressive strength (MPa)	150	210	230	150–300
Modulus of compression (MPa)	3.44	4.58	5.66	–
Bending strength (MPa)	0.28	0.34	0.41	–

[a] *Foam 1 used polyols with hydroxy number 425; foam 2 used polyols with hydroxy number 361; foam 3 used polyols with hydroxy number 307*
[b] *China standard (JC/T 998-2006): Spray polyurethane foam or thermal insulation*

Most of the fatty acids have only one functional group and act as a chain terminator in polymer synthesis. Polymerization of fatty acid only occurs in dimer or trimer fatty acids which have two or three functional groups. So, the fatty acid has to be dimerized or trimerized before use (Mallakpour and Rafiee 2012). CRODA has developed various functional dimer and trimer fatty acids for polymer synthesis such as fatty acid, fatty diol, and fatty diamine.

Mecking and his groups (Quinzler and Mecking 2010; Stempfle et al. 2011; Trzaskowski et al. 2011) have investigated the conversion of fatty acid into saturated α,ω-diacid and diol as a preparation of long-chain aliphatic semicrystalline polyester. Oleic acid, erucic acid, or 10-undecenoic acid are modified into diacid through carbonylation and olefin metathesis. Carbonylation converts the internal double bond in the hydrocarbon to terminal ester groups while olefin metathesis coupling of two $=CH(CH_2)nCOOR$ part and then the saturated product being hydrogenated. The reduction mechanism of dicarboxylic acid is used to obtain diols. One of their results has similar properties to low-density polyethylene (LDPE) and polyethylene. The material has a melting temperature (T_m) of 103 °C and crystallize temperature (T_c) of 87 °C while LDPE has T_m 110 °C and T_c 94.09 °C. This material was prepared by the linear incorporation of oleic acid and erucic acid. Equal amounts of dimethyl-1,19-nonadecanoate and nonadecane-1,19-diol were prepared for polycondensation of novel polyester catalyzed by titanium alkoxides (Quinzler and Mecking 2010; Gaska et al. 2017).

Vilela and his groups used erucic acid to generate long-chain aliphatic polyester via self-metathesis for monomer synthesis. Hydrogenation of erucic acid to obtain dicarboxylic acid and reduction of the acid to get diols (Fig. 8). Polycondensation of hexacosane-1,26-diol with α,ω,-dicarboxylic acid generates long-chain aliphatic polyesters 26,26 with excellent properties. The polyester has the highest T_m at 104 °C, and the highest degradation temperature (T_d) at 386 °C. The DSC graph of polyester 26,26 exhibits a sharp endotherm and crystallization peak indicating a high degree of the polyesters (Vilela et al. 2012).

A long carbon chain of fatty acids is beneficial for the synthesis of hyperbranched polyester. Testud et al. reported the preparation of hyperbranched polyester (HPBE) from fatty acid methyl ester (FAME) with tunable properties. They use various plant

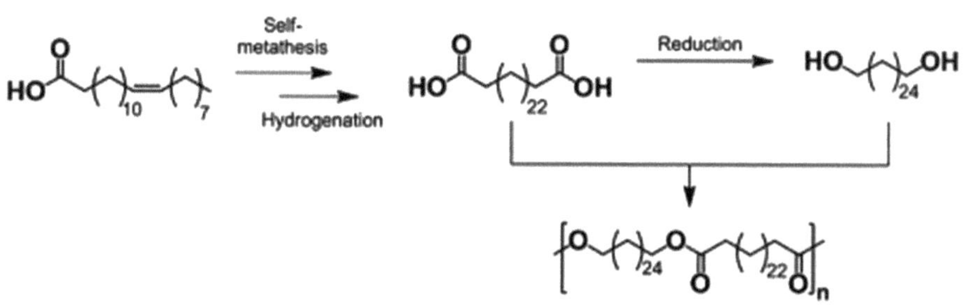

Fig. 8 Conversion of erucic acid to 1,26 diacid and 1,26 diol and polymerization of both (Vilela et al. 2012)

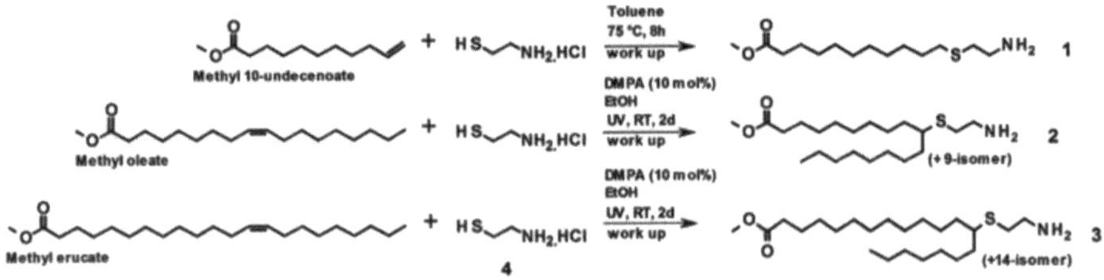

Fig. 9 Thiol-ene scheme of polyamide monomer (Türünç et al. 2012)

oils as starting material for the synthesis of monomers of AB_n type ($n = 2$ or 3) where A is acid and B is diol moieties. The most efficient catalyst system that achieves high molar masses (3000–10000 g mol^{-1}) is shown by zinc acetate, sodium methoxide, and 1,5,7-triazabicyclo [4.4.0] dec-5-ene (TBD). The glass transition of the samples varies from −33° to 9 °C, with a degree of branching ranging from 0.07 to 0.45 and thermal stability above 300 °C. A significant amount of hydroxyl group in HPBE can have further modifications to reach desired properties (Testud et al. 2017).

Falkenburg et al. reported the preparation of polyamide-based fatty acid for the first time in 1945. They synthesized polyamide from difunctional and polyfunctional amines and polymeric fat acid (dimer and trimer). The polyamides result has unique properties including strong adhesion to various surfaces, good water resistance, soluble in alcohol, excellent mechanical properties, and flexible as described in Table 2. It is suitable for resin, coating, elastomer, and hot melting adhesives application (Falkenburg et al. 1945).

Thiol-ene addition was introduced in the synthesis of fatty acid monomers. Thiol-ene additiona is an effective method for various transformations because it shows a click reaction behavior. Türünç and his co-workers prepared a fatty acid-based amine monomer through thiol-ene addition to obtain polyamide. Cysteamine hydrochloride addition to double bonds of Methyl-10 undecenoate, methyl erucate, and methyl oleate produced a good multifunctional monomer (Fig. 9). The products were used for polyamides with adipic acid and 1,6-hexamethylene. 1,5,7-triazabicyclo [4.4.0] dec-5-ene (TBD) was used as a catalysts in copolymerization. The result yields polyamide with excellent thermal and solubility properties. The resulted polyamide gave the highest T_m at 138 °C (Türünç et al. 2012).

Nurhamiyah et al. have been synthesized a series of fully biobased polyamides from a fatty acid biobased dicarboxylic acid and biobased diamine, PA36,36, and PA36,9 (Nurhamiyah et al. 2021a, 2021b). PA36,36 was prepared from a facile condensation of Pripol 1009 and Priamine 1075 at 220 °C at various times. It was found that the optimum time to synthesize this biobased polyamide is at 24 h. PA36,36 has excellent properties, for example, zero water absorption, high toughness (14.21 ± 4.58 MJ m^{-3}), and large elongation at break (up to 2286%), and shows autonomous self-healing behavior at room temperature (Nurhamiyah et al. 2021a). Meanwhile, PA36,9 is a semicrystalline fully biobased polyamide elastomer that shows similar properties to the synthetic medium hardness, prospecting as an alternative substitution. The example is a melting temperature (T_m) of 83.6 °C, a glass transition temperature (T_g) of 17.6 °C, a large elongation at break (1220%), a high tensile strength (31.8 MPa), a medium hardness (Shore A/Shore $D = 90/35$), and excellent hydrophobicity.

5 Challenges and Future Perspectives

Biopolymer production is growing rapidly, and the production process of precursors of biopolymers (starch, seaweed, and plant oil) from renewable resources continues to increase. This is due to the high demand for biopolymer materials from various industries ranging from packaging, the automotive industry, the agricultural sector, bioplastics, and electronics, as well as biopolymer research and development to increase its capacity in industrialization. In the early stages of the biopolymer industry, biodegradable materials were developed for short-term applications to solve the problem of synthetic polymers. Currently, biopolymer production focuses on the application of durable materials with sustainable bioprocesses using biomass-derived materials in the hope of reducing greenhouse gas emissions and limiting fossil resources. Development of new applications with the advantage of even better properties. In general, the expected development and sustainability of biopolymer materials in the future depend on a variety of applications and the quantity of biobased materials. Particularly, advances in bioprocessing and the utilization of lignocellulosic biomass as a waste product for the production of biopolymers with low production costs, sustainability, and material properties that can be modified according to application based on renewable energy sources are becoming more attractive.

References

Aayush K, McClements DJ, Sharma S, Sharma R, Singh GP, Sharma K, Oberoi K (2022) Innovations in the development and application of edible coatings for fresh and minimally processed apple. Food Control 141:109188. https://doi.org/10.1016/j.foodcont.2022.109188

Agwamba EC (2021) Taguchi optimization of plasticisation process for glycerol and TEA based thermoplastic mango starch biofilms. Res Sq 1–13. https://doi.org/10.21203/rs.3.rs-932373/v1

Area MR, Rico M, Montero B, Barral L, Bouza R, López J, Ramírez C (2019) Corn starch plasticized with isosorbide and filled with microcrystalline cellulose: processing and characterization. Carbohydr Polym 206:726–733. https://doi.org/10.1016/j.carbpol.2018.11.055

Ashothaman A, Sudha J, Senthilkumar N (2021) A comprehensive review on biodegradable polylactic acid polymer matrix composite material reinforced with synthetic and natural fibers. Mater Today Proc. https://doi.org/10.1016/j.matpr.2021.07.047

Bangar SP, Whiteside WS, Ashogbon AO, Kumar M (2021) Recent advances in thermoplastic starches for food packaging: a review. Food Packag Shelf Life 30:100743. https://doi.org/10.1016/j.fpsl.2021.100743

Baran A, Fričová O, Vrábel P, Popovič Ľ, Peidayesh H, Chodák I, Hutníková M, Kovaľaková M (2022) Effects of urea and glycerol mixture on morphology and molecular mobility in thermoplastic starch/montmorillonite-type nanofiller composites studied using XRD and NMR. J Polym Res 29:1–12. https://doi.org/10.1007/s10965-022-03110-6
Bayón B, Berti IR, Gagneten AM, Castro GR (2018) Biopolymers from wastes to high-value products in biomedicine. In: Waste to wealth. Springer, pp 1–44
Cao N, Yang X, Fu Y (2009) Effects of various plasticizers on mechanical and water vapor barrier properties of gelatin films. Food Hydrocoll 23:729–735. https://doi.org/10.1016/j.foodhyd.2008.07.017
Carina D, Sharma S, Jaiswal AK, Jaiswal S (2021) Trends in food science & technology seaweeds polysaccharides in active food packaging: a review of recent progress. Trends Food Sci Technol 110:559–572. https://doi.org/10.1016/j.tifs.2021.02.022
Carothers WH (1936) Linear polyamides and their production
Carothers WH (1937) Linear condensation polymers
Carvalho AJF, Gandini A, Curvelo AAS, Ro AL Da (2006) The effect of plasticizers on thermoplastic starch compositions obtained by melt processing 63:417–424. https://doi.org/10.1016/j.carbpol.2005.09.017
Castillo L, López O, López C, Zaritzky N, García MA, Barbosa S, Villar M (2013) Thermoplastic starch films reinforced with talc nanoparticles. Carbohydr Polym 95:664–674. https://doi.org/10.1016/j.carbpol.2013.03.026
Caylı G, Kusefoglu S (2008) Biobased polyisocyanates from plant oil triglycerides: synthesis, polymerization, and characterization. J Appl Polym Sci 109:2948–2955. https://doi.org/10.1002/app.28401
Chai JM, Amelia TS, Mouriya GK, Bhubalan K, Amirul A-AA, Vigneswari S, Ramakrishna S (2021) Surface-modified highly biocompatible bacterial-poly(3-hydroxybutyrate-co-4-hydroxybutyrate): a review on the promising next-generation biomaterial. Polymers (Basel) 13
Chang I, Im J, Cho G-C (2016) Introduction of microbial biopolymers in soil treatment for future environmentally-friendly and sustainable geotechnical engineering. Sustainability 8:251
Chotiprayon P, Chaisawad B, Yoksan R (2020) Thermoplastic cassava starch/poly(lactic acid) blend reinforced with coir fibres. Int J Biol Macromol 156:960–968. https://doi.org/10.1016/j.ijbiomac.2020.04.121
Chudasama NA, Sequeira RA, Moradiya K, Prasad K (2021) Seaweed polysaccharide based products and materials: an assessment on their production from a sustainability point of view. Molecules 26. https://doi.org/10.3390/molecules26092608
Cordier P, Tournilhac F, Soulié-Ziakovic C, Leibler L (2008) Self-healing and thermoreversible rubber from supramolecular assembly. Nature 451:977–980. https://doi.org/10.1038/nature06669
Das A, Ringu T, Ghosh S, Pramanik N (2022) A comprehensive review on recent advances in preparation, physicochemical characterization, and bioengineering applications of biopolymers. Polym Bull 1–66
de Jong E, Jungmeier G (2015) Biorefinery concepts in comparison to petrochemical refineries. In: Industrial biorefineries & white biotechnology. Elsevier, pp 3–33
de Oliveira JP, Bruni GP, Fabra MJ, da Rosa ZE, López-Rubio A, Martínez-Sanz M (2019) Development of food packaging bioactive aerogels through the valorization of Gelidium sesquipedale seaweed. Food Hydrocoll 89:337–350. https://doi.org/10.1016/j.foodhyd.2018.10.047
Drobny JG (2015) Handbook of thermoplastic elastomers, 2nd edn. Elsevier
Ebnesajjad S (2012) Handbook of biopolymers and biodegradable plastics: properties, processing and applications. William Andrew
Edhirej A, Sapuan SM, Jawaid M, Zahari NI (2017) Preparation and characterization of cassava bagasse reinforced thermoplastic cassava starch. Fibers Polym 18:162–171. https://doi.org/10.1007/s12221-017-6251-7

Encalada K, Aldás MB, Proaño E, Valle V (2018) An overview of starch-based biopolymers and their biodegradability Una revisión sobre biopolímeros con base en almidón y su biodegradabilidad. Cienc e Ing 39:245–258
Falkenburg LB, Teeter HM, Skell PS, Cowan JC (1945) Polyamides from polymeric fat acids. Oil Soap 22:143–148
Fan X, Deng Y, Waterhouse J, Pfromm P (1997) Synthesis and characterization of polyamide resins from soy-based dimer acids and different amides 305–314
FAO (2021) Global status of seaweed prduction, trade and utilization. Food Agric rganization United Nations
Foroughi-dahr M, Mostoufi N, Sotudeh-gharebagh R, De Montre P, Chaouki J (2017) Particle coating in fluidized beds. Elsevier Inc.
Gajula S, Reddy CRK (2021) More sustainable biomass production and biorefining to boost the bioeconomy. Biofuels Bioprod Biorefining 15:1221–1232
Gaska K, Xu X, Gubanski S, Kádár R (2017) Electrical, mechanical, and thermal properties of LDPE graphene nanoplatelets composites produced by means of melt extrusion process. Polymers (Basel) 9:30–40. https://doi.org/10.3390/polym9010011
Geyer R (2020) Production, use, and fate of synthetic polymers. In: Plastic waste and recycling. Elsevier, pp 13–32
Gurr M., Harwood J, Frayn K. (2002) Lipid biochemistry: an introduction
Hablot E, Matadi R, Ahzi S, Avérous L (2010) Renewable biocomposites of dimer fatty acid-based polyamides with cellulose fibres: thermal, physical and mechanical properties. Compos Sci Technol 70:504–509. https://doi.org/10.1016/j.compscitech.2009.12.001
Hojabri L, Kong X, Narine SS (2010) Novel long chain unsaturated diisocyanate from fatty acid: synthesis, characterization, and application in bio-based polyurethane. J Polym Sci Part A Polym Chem 48:3302–3310. https://doi.org/10.1002/pola.24114
Hussein Ziedan ES, El Zahaby HM, Maswada HF, El Rafh Zoeir EHA (2018) Agar-agar a promising edible coating agent for management of postharvest diseases and improving banana fruit quality. J Plant Prot Res 58. https://doi.org/10.24425/122938
Ibrahim N, Wahab MKA, Uylan DN, Ismail H (2017) Physical and degradation properties of polylactic acid and thermoplastic starch blends—Effect of citric acid treatment on starch structures. BioResources 12:3076–3087. https://doi.org/10.15376/biores.12.2.3076-3087
Isikgor FH, Becer CR (2015) Lignocellulosic biomass: a sustainable platform for the production of bio-based chemicals and polymers. Polym Chem 6:4497–4559. https://doi.org/10.1039/c5py00263j
Janssen L, Moscicki L (2009) Termoplastic starch (A green material for various industries). Wiley-VCH, Germany
Jiang T, Duan Q, Zhu J, Liu H, Yu L (2020) Starch-based biodegradable materials: challenges and opportunities. Adv Ind Eng Polym Res 3:8–18. https://doi.org/10.1016/j.aiepr.2019.11.003
Jiménez L, Mena MJ, Prendiz J, Salas L, Vega-Baudrit J (2019) Polylactic acid (PLA) as a bioplastic and its possible applications in the food industry. J Food Sci Nutr 5:2–6
Junming X, Jianchun J, Jing L (2012) Preparation of polyester polyols from unsaturated fatty acid. J Appl Polym Sci 126:1377–1384. https://doi.org/10.1002/app.36740
JS, Harrison ZO (2002) Introduction Types of Plasticizers. Encycl Polym Sci Technol 3:518
Khan B, Bilal Khan Niazi M, Samin G, Jahan Z (2017) Thermoplastic starch: a possible biodegradable food packaging material—A review. J Food Process Eng 40. https://doi.org/10.1111/jfpe.12447
Lemoigne M, Girard H (1943) Reserves lipidiques-hydroxybutyriques chez Azotobacter chroococcum. C. Rmcad Sci Paris 217:537–558
Liu Y, Qin Y, Bai R, Zhang X, Yuan L, Liu J (2019) pH-sensitive and antioxidant packaging fi lms based on κ -carrageenan and mulberry polyphenolic extract. Int J Biol Macromol 134:993–1001. https://doi.org/10.1016/j.ijbiomac.2019.05.175

Ma X, Chang PR, Yu J, Stumborg M (2009) Properties of biodegradable citric acid-modified granular starch/thermoplastic pea starch composites. Carbohydr Polym 75:1–8. https://doi.org/10.1016/j.carbpol.2008.05.020
Magallanes-Cruz PA, Flores-Silva PC, Bello-Perez LA (2017) Starch structure influences its digestibility: a review. J Food Sci 82:2016–2023. https://doi.org/10.1111/1750-3841.13809
Mallakpour S, Rafiee Z (2012) Green solvents fundamental and industrial applications BT—Green solvents I: properties and applications in chemistry. Springer, Netherlands
Maraveas C (2020) Production of sustainable and biodegradable polymers from agricultural waste. Polymers (Basel) 12:1127
Martin L, Gonz A (2017) D-isosorbide and 1, 3-propanediol as plasticizers for starch-based films: Characterization and aging study. J Appl Polym Sci 44793:1–10. https://doi.org/10.1002/app.44793
Meereboer KW, Misra M, Mohanty AK (2020) Review of recent advances in the biodegradability of polyhydroxyalkanoate (PHA) bioplastics and their composites. Green Chem 22:5519–5558. https://doi.org/10.1039/D0GC01647K
Mesias R, Murillo EA (2020) Properties of the thermoplastic starch/polylactic acid blends compatibilized by hyperbranched polyester. J Phys Conf Ser 1587. https://doi.org/10.1088/1742-6596/1587/1/012001
Mikus M, Galus S (2022) Biopolymers from agriculture waste and by-products. In: Biopolymers. Springer, pp 111–128
Moghaddam RA, Mohammad, Mohammad Ali Razavi S, Jahani Y (2018) Optimization of the effects of thermoplastic starch and glycerol concentration on physicomechanical properties of polylactic acid/thermoplastic starch blend by response surface methodology. J Res Innov Food Sci Technol 7:309–322. https://doi.org/10.22101/JRIFST.2018.10.20.736
Mohan S, Oluwafemi OS, Kalarikkal N, Thomas S, Songca SP (2016) Biopolymers—Application in nanoscience and nanotechnology. In: Perveen FK (ed) IntechOpen, Rijeka, p Ch. 3
Mohanty AK, Wu F, Mincheva R, Hakkarainen M, Raquez J-M, Mielewski DF, Narayan R, Netravali AN, Misra M (2022) Sustainable polymers. Nat Rev Methods Prim 2:1–27
Mohapatra S, Vishwakarma K, Joshi NC, Maity S, Kumar R, Ramchander M, Pattnaik S, Samantaray DP (2021) A review on PHAs: the future biopolymer. In: Environmental and agricultural microbiology, pp 83–100
Moradali MF, Rehm BHA (2020) Bacterial biopolymers: from pathogenesis to advanced materials. Nat Rev Microbiol 18:195–210. https://doi.org/10.1038/s41579-019-0313-3
More AS, Lebarbe T, Maisonneuve L, Gadenne B, Alfos C, Cramail H (2013) Novel fatty acid based di-isocyanates towards the synthesis of thermoplastic polyurethanes. Eur Polym J 49:823–833. https://doi.org/10.1016/j.eurpolymj.2012.12.013
Nanni A, Parisi M, Colonna M (2021) Wine by-products as raw materials for the production of biopolymers and of natural reinforcing fillers: a critical review. Polymers (Basel) 13:381
Nduko JM, Taguchi S (2021) Microbial production of biodegradable lactate-based polymers and oligomeric building blocks from renewable and waste resources. Front Bioeng Biotechnol 8
Niazi MBK, Zijlstra M, Broekhuis AA (2015) Influence of plasticizer with different functional groups on thermoplastic starch. J Appl Polym Sci 132:1–12. https://doi.org/10.1002/app.42012
Nurhamiyah Y, Amir A, Finnegan M, Themistou E, Edirisinghe M, Chen B (2021a) Wholly biobased, highly stretchable, hydrophobic, and self-healing thermoplastic elastomer. ACS Appl Mater Interfaces 13:6720–6730. https://doi.org/10.1021/acsami.0c23155
Nurhamiyah Y, Irvine G, Themistou E, Chen B (2021b) Novel biobased polyamide thermoplastic elastomer with medium hardness. Macromol Chem Phys 222:2100218. https://doi.org/10.1002/macp.202100218
Okkerse C, Van Bekkum H (1999) From fossil to green. Green Chem 1:107–114. https://doi.org/10.1039/a809539f
Ortelli S, Costa AL, Torri C, Samorì C, Galletti P, Vineis C, Varesano A, Bonura L, Bianchi G (2019) Innovative and sustainable production of biopolymers BT—Factories of the future: the Italian

Flagship initiative. In: Tolio T, Copani G, Terkaj W (eds). Springer International Publishing, Cham, pp 131–148
Padam BS, Chye FY (2020) Sustainable seaweed technologies. Elsevier Inc.
Paula GA, Benevides NMB, Cunha AP, Vit A, Morais S, Azeredo HMC, Pinto AMB (2015) Development and characterization of edible fi lms from mixtures of k-carrageenan, i-carrageenan, and alginate 47:140–145. https://doi.org/10.1016/j.foodhyd.2015.01.004
Perera KY, Sharma S, Pradhan D, Jaiswal AK, Jaiswal S (2021) Seaweed polysaccharide in food contact materials (Active packaging, intelligent packaging, edible films, and coatings). Foods 10:1–22. https://doi.org/10.3390/foods10092088
Pokhrel S (2015) A review on introduction and applications of starch and its biodegradable polymers. Int J Environ 4:114–125. https://doi.org/10.3126/ije.v4i4.14108
Pond CM (1998) Fats of life. Cambridge University Press, Cambridge
Popa VI (2018) Biomass for fuels and biomaterials. In: Biomass as renewable raw material to obtain bioproducts of high-tech value. Elsevier, pp 1–37
Prachayawarakorn J, Sangnitidej P, Boonpasith P (2010) Properties of thermoplastic rice starch composites reinforced by cotton fiber or low-density polyethylene. Carbohydr Polym 81:425–433. https://doi.org/10.1016/j.carbpol.2010.02.041
Quinzler D, Mecking S (2010) Linear semicrystalline polyesters from fatty acids by complete feedstock molecule utilization. Angew Chemie Int Ed 49:4306–4308. https://doi.org/10.1002/anie.201001510
Rai R, Tallawi M, Grigore A, Boccaccini AR (2012) Synthesis, properties and biomedical applications of poly(glycerol sebacate) (PGS): a review. Prog Polym Sci 37:1051–1078. https://doi.org/10.1016/j.progpolymsci.2012.02.001
Rajeswari A, Stobel Christy EJ, Pius A (2021) Chapter 5—Biopolymer blends and composites: processing technologies and their properties for industrial applications. In: Thomas S, Gopi S, Amalraj ABT-B, IA (eds). Elsevier, pp 105–147
Ranakoti L, Gangil B, Mishra SK, Singh T, Sharma S, Ilyas RA, El-Khatib S (2022) Critical review on polylactic acid: properties, structure, processing, biocomposites, and nanocomposites. Materials (Basel) 15
Ranganathan S, Dutta S, Moses JA, Anandharamakrishnan C (2020) Utilization of food waste streams for the production of biopolymers. Heliyon 6:e04891
Revati R, Majid MSA, Normahira M (2016) Biodegradable poly(lactic acid) scaffold for tissue engineering: a brief review
Rix E, Grau E, Chollet G, Cramail H (2016) Synthesis of fatty acid-based non-isocyanate polyurethanes, NIPUs, in bulk and mini-emulsion. Eur Polym J 84:863–872. https://doi.org/10.1016/j.eurpolymj.2016.07.006
Santoro O, Zhang X, Redshaw C (2020) Synthesis of biodegradable polymers: a review on the use of schiff-base metal complexes as catalysts for the ring opening polymerization (ROP) of cyclic esters. Catalysts 10
Sharma M, Dhingra HK (2021) An overview of microbial derived polyhydroxybutyrate (PHB): production and characterization. Microb Polym 143–176
Sharma V, Sehgal R, Gupta R (2021) Polyhydroxyalkanoate (PHA): properties and modifications. Polymer (Guildf) 212:123161. https://doi.org/10.1016/j.polymer.2020.123161
Sjoo M, Nillsson L (2018) Starch in food (structure, function and applications), 2nd edn. Woodhead Publishing
Sriroth K, Sangseethong K (2006) Biodegradable plastics from cassava starch. Acta Hortic 703:145–151. https://doi.org/10.17660/actahortic.2006.703.16
Stempfle F, Ortmann P, Mecking S (2016) Long-chain aliphatic polymers to bridge the gap between semicrystalline polyolefins and traditional polycondensates. Chem Rev 116:4597–4641. https://doi.org/10.1021/acs.chemrev.5b00705
Stempfle F, Quinzler D, Heckler I, Mecking S (2011) Long-chain linear C_{19} and C_{23} monomers and polycondensates from unsaturated fatty acid esters. Macromolecules 44:4159–4166. https://doi.org/10.1021/ma200627e

Subramanian V, Varade D (2017) Thermoelectric properties of biopolymer composites. Elsevier Inc.

Tan HW, Abdul Aziz AR, Aroua MK (2013) Glycerol production and its applications as a raw material: a review. Renew Sustain Energy Rev 27:118–127. https://doi.org/10.1016/j.rser.2013.06.035

Testud B, Pintori D, Grau E, Taton D, Cramail H (2017) Hyperbranched polyesters by polycondensation of fatty acid-based AB_n-type monomers. Green Chem 19:259–269. https://doi.org/10.1039/C6GC02294D

Thi H, Boonyaritthongchai P, Buanong M (2021) Scientia Horticulturae Chitosan- and κ-carrageenan-based composite coating on dragon fruit (Hylocereus undatus) pretreated with plant growth regulators maintains bract chlorophyll and fruit edibility. Sci Hortic (amsterdam) 281:109916. https://doi.org/10.1016/j.scienta.2021.109916

Tran TTB, Roach P, Nguyen MH, Pristijono P, Vuong QV (2020) Development of biodegradable films based on seaweed polysaccharides and Gac pulp (Momordica cochinchinensis), the waste generated from Gac oil production. Food Hydrocoll 99:105322. https://doi.org/10.1016/j.foodhyd.2019.105322

Trzaskowski J, Quinzler D, Bährle C, Mecking S (2011) Aliphatic long-chain C20 polyesters from olefin metathesis. Macromol Rapid Commun 32:1352–1356. https://doi.org/10.1002/marc.201100319

Türünç O, Firdaus M, Klein G, Meier MAR (2012) Fatty acid derived renewable polyamides via thiol–ene additions. Green Chem 14:2577. https://doi.org/10.1039/c2gc35982k

USDA Foreign Agricultural Services (2021) Oilseeds: world markets and trade

Vilela C, Silvestre AJD, Meier MAR (2012) Plant oil-based long-chain C 26 monomers and their polymers. Macromol Chem Phys 213:2220–2227. https://doi.org/10.1002/macp.201200332

Wang Y, Ameer GA, Sheppard BJ, Langer R (2002) A tough biodegradable elastomer. Nat Biotechnol 20:602–606. https://doi.org/10.1038/nbt0602-602

Yildirim S, Röcker B, Pettersen MK, Nilsen-Nygaard J, Ayhan Z, Rutkaite R, Radusin T, Suminska P, Marcos B, Coma V (2018) Active packaging applications for food. Compr Rev Food Sci Food Saf 17:165–199. https://doi.org/10.1111/1541-4337.12322

Young RJ, Lovell PA (2011) Introduction to polymers. CRC Press

Zhao H (2018) Enzymatic ring-opening polymerization (ROP) of polylactones: roles of non-aqueous solvents. J Chem Technol Biotechnol 93:9–19. https://doi.org/10.1002/jctb.5444

Potential Application of Agro-Industrial Byproduct for Bacterial Cellulose Production; Its Challenges and Emerging Trends for Food Packaging

Khatarina Meldawati Pasaribu, Nanang Masruchin, and Myrtha Karina

Abstract Most packaging used today is made of plastic, which is produced from fossil-based polymers. In terms of its ease of processing and cheapness, plastic is non-biodegradable. Apart from being a plastic substitute, cellulose-based packaging is bio-based and sustainable. Cellulose is commonly generated from vascular plants. However, numerous chemicals are required for cellulose isolation and purification. For plant cellulose replacement, bacterial cellulose is considered as the favorable resources. Bacterial cellulose, also well known as microbial cellulose, is the cellulose produced by the activity of non-pathogen gram-positive or gram-negative bacteria in the substrate containing carbon and nitrogen. Possessing a three-dimensional nanostructure, high reactive functional groups, high mechanical strength properties, and bacterial cellulose attracts much attention for research work or commercial purposes. However, Hestrin-Schramm, the synthetic or considered as standard medium for bacterial cellulose production, is expensive. Recently, there has been a lot of interest in searching for carbon and nitrogen sources as an alternative to synthetic bacterial growth media. Agro-industrial byproducts are derived from agriculture and food industry processing. Rich in carbohydrates and protein, these resources are suitable for bacterial cellulose production. This chapter aims to describe the agro-industrial residues for bacterial cellulose production and their recent possible application for food packaging.

Keywords Agro-industrial byproduct · Bacterial cellulose · Food packaging

K. M. Pasaribu · N. Masruchin · M. Karina (✉)
Research Center for Biomass and Bioproducts, National Research and Innovation Agency (BRIN), Cibinong 16911, Indonesia
e-mail: myrt001@brin.go.id

Research Collaboration Center for Biomass and Biorefinery, BRIN – UNPAD, Jatinangor 45363, Indonesia

N. Masruchin · M. Karina
Research Collaboration Center for Nanocellulose, BRIN - Andalas University, Padang 25163, Indonesia

M. A. R. Lubis et al. (eds.), *Biomass Conversion and Sustainable Biorefinery*, Green Energy and Technology, https://doi.org/10.1007/978-981-99-7769-7_3

1 Introduction

Due to its excellent properties such as high strength, hardiness, ease of processing, and cost-effectiveness, plastic has become a popular and important material for food packaging (Andrady et al. 2009). The common commercial plastics for food packaging are generally derived from petrochemical-based polymers, specifically polypropylene (PP), polystyrene (PS), polyvinyl chloride (PVC), high-density polyethylene (HDPE), low-density polyethylene (LDPE), and polyethylene terephthalate (PET) of which they account for almost 90% of total polymers manufacture (Napper and Thompson 2019). These plastics are non-biodegradable, and when their lifetime is over, the improper disposal of plastic waste would become a severe environmental problem. Due to poor management, plastic wastes are found in rivers. It was reported that 80% of plastic waste is distributed by more than 1,000 rivers globally (Meijer et al. 2022) which slowly goes to the ocean, becomes a **marine pollutant**, and gradually becomes a global oceanic serious problem (Long et al. 2022). Bulky plastics, often known as macro-plastics, have been found in the ocean, and their accumulation has significantly increased. As a result, marine animals were suffered and killed by plastic entanglement (Dasgupta et al. 2022). In the ocean, plastic gradually degrades into micro-plastic, whose accumulation is hazardous since it is easily swallowed by crustaceans whose digestive tract is usually consumed by humans (Rainieri and Barranco 2019). Besides causing the accumulation of plastic wastes, the manufacture of petroleum-based plastic triggers the reduction of its non-renewable resources. Nowadays, petrochemical industries are the source of more than 99% of the global plastics raw materials (British Plastics Federation 2019). Consequently, sustainable, renewable, and biodegradable-based biopolymers, specifically those originating from organic resources, are important as alternative resources for not only improving food quality but also producing environmentally-friendly food packaging.

The alternative source for petroleum-based polymer as a plastic replacement for various materials is addressed to cellulose, the most prevalent macromolecule on earth, a renewable and low-cost natural polymer. In general, cellulose is obtained from vascular plants, but it can also be found in tunicin, typically from *Microcosmus sulcatus*, *Halocynthia roretzi*, *Ciona intestinalis*, *Styela plicata*, and *Ascidia sp.* in the form of rod-like crystals (Zhao and Li 2014). Cellulose is also synthesized from green algae of *Cladophorales* (*Cladophora, Chaetomorpha,* and *Rhizoclonium*) and *Siphonocladales* (*Valonia*, *Dictyosphaeria*, *Siphonocladus,* and *Boergesenia*) (Mihranyan 2011). Additionally, cellulose is also found in the cell wall of *Saprolegnia monoica* (Fèvre et al. 1990).

Cellulose has been used in various fields for a long time for chemical-based materials, fabric supplies, and pulp paper production. For a wide range of applications, particularly when high purity and white color are considered necessary, cellulose should be isolated from lignin, hemicellulose, and pectin as the plant cell wall components through the pulping and bleaching process, which requires various synthetic chemicals. In addition, cellulose isolation and purification require a large amount of energy as well as the price of waste effluent and toxic material treatment.

Therefore, highly pure and environmentally friendly cellulose is an important topic of research subject. For this purpose, bacterial cellulose offers an alternative and interesting role as cellulose resources. In terms of cellulose isolation, bacterial cellulose is more beneficial than plant cellulose because it is free of lignin and hemicellulose, preventing the need for a pulping and bleaching process and making it more environmentally-friendly and energy efficient. Thus, bacterial cellulose is a source of pure cellulose as well as indicates higher water absorptivity, higher crystallinity, and a higher degree of polymerization (Salari et al. 2018).

Although BC possesses remarkable characteristics, it is costly to fabricate. The synthetic medium Hestrin–Schramm (HS) medium is one of the main factors causing to high production cost of BC. This synthetic medium is an indispensable material for bacteria to produce cellulose (Lahiri et al. 2021). Therefore, there is a need to explore the potential and low cost of natural resources for bacterial cellulose production. In this chapter, the production of bacterial cellulose by using abundant and low-cost agro-industrial byproducts and the possibility for food packaging application are presented.

2 Agro-Industrial Wastes for Bacterial Cellulose Production

Bacterial cellulose production depends on various basic factors such as oxygen supply (Wu and Li 2015), temperature (Lee et al. 2014), reactor design (Islam et al. 2017), and optimum pH (Reiniati et al. 2017). However, carbon and nitrogen play an important role in the synthesis of bacterial cellulose (Rajwade et al. 2015). Typically, the carbon and nitrogen sources were obtained from the Hestrin-Schramm culture medium. This expensive medium mostly contains synthetic glucose, peptone, yeast extracts, and various minerals. Therefore, exploring the low-cost carbon and nitrogen resources from agro-industrial byproducts for alternative resources is the right option.

Agro-industrial byproducts are defined as various wastes from the food and agriculture industries (Madeira et al. 2017). It is available in a huge amount every year. In terms of their abundance and sustainability, the agro-industrial byproducts can be deduced from the total volume produced from the plant waste from crops, approximately 250 million tons per year (FAO 2013). It was also informed that 1/3 of total human food production, or around 1.3 billion tons annually, is discarded worldwide (Duque-Acevedo et al. 2020). The use of agro-industrial residues for innovative products not only solves waste removal issues but also reduces pollution, reduces adverse effects on human and animal health (Zihare et al. 2018), and raises its value.

Agricultural commodities such as fruits, vegetables, legumes, and cereals and their processing generate unclear value byproducts (Almaraz-sánchez et al. 2022). Other agricultural primary activities such as harvesting, pruning, and collecting in the field crops generate residues in the form of logs, straws, leaves, husks, roots, and seeded pods from crops (Hiloidhari et al. 2020) as well as animal residues

(Forster-Carneiro et al. 2013). These underutilized products are important resources for new materials, chemicals (Madeira et al. 2017), and energy (Vandamme 2009). This is due to the important and valuable existing ingredients of the residues such as carbohydrates, protein, fibers, minerals, and vitamins (Lopes and Ligabue-Braun 2021). Carbohydrate and protein will be valuable resources and mediums for bacteria to propagate in the cellulose gel.

Depending on the material target, various pre-treatments of agro-industrial residue are commonly carried out. Chemical, biological, enzymatic, and physical pre-treatments such as milling, steam explosion/steam treatment, hydrothermal, irradiation, and chemical treatment such as alkaline hydrolysis are the most common pre-treatments (Singh nee' Nigam et al. 2009). These pre-treatments aim to reach the reactive molecule followed by procedures to obtain the desired raw material (De Corato et al. 2018). The fermentation, hydrolysis, precipitation, and filtration aim to eliminate the poisonous and low-value chemicals (Araújo et al. 2020), so that the existing organic substances could be changed into diverse products involving carbon and nitrogen as the intermediate compounds. The fine and highly de-crystalized structure was obtained by the milling process, whereas steam treatment/steam explosion increased the pore size of the fiber. Biological pre-treatment aims to degrade lignin; it generally involves the activity of white-rod fungi such as *Phanerochaete chrysosporium* (Singh nee' Nigam et al. 2009).

In agro-industrial industries, carbon and nitrogen could be found in the form of cellulose, hemicellulose, lignin, carbohydrate, or proteins (Urbina et al. 2021). The composition depends on the origin, type of raw material resources, and method of analysis. Based on these descriptions, they are particularly potential for BC production. The high carbohydrate and nitrogen contents of 56.9 and 28.5%, respectively, are found in oil palm frond (Rhaman and Naher 2021). Paddy straw also shows a high carbohydrate and nitrogen content of 50.2 and 84.9%, respectively.

The carbon and nitrogen generated from numerous agro-food industries is presented in Table 1.

2.1 Agro-Wastes

Available in a huge amount worldwide annually, agro-wastes are the most promising resource for the production of BC. Sugarcane straw is generated from sugarcane production, which plays a role in the daily required nutrient. Recently, sugarcane straw is utilized as bio-ethanol production, but its valorization is widely open to realization. It contains 35–45% of cellulose, which can be used for carbon resources. Sugarcane straw, which was previously boiled in water, was used as the medium for *Komagataeibacter xylinus ATCC 11142* and incubated statistically at 30 °C for 15 days. The dry pellicle of BC was weighed as 1.06 g/L (Dhar et al. 2019). A quite similar amount of BC yield was obtained from the root, stalk, and leaf parts of sweet sorghum, as of 2.28 g/L, stalk 1.82 g/L, and leaf 2.54 g/L, respectively. Commonly considered and used as an energy crop, sweet sorghum is an important resource

Table 1 Carbon and nitrogen content from agro-industrial byproducts

Resources	Carbon content (from)	Nitrogen content (from)	References
Cassava wastewater	Carbohydrates 58.11%	Total nitrogen 1.94%	Ribeiro et al. (2019)
Cheese whey	Lactose, 77%	Protein, 13%	Lopes et al. (2013)
Coconut oil cake	48.16%	1.69%	Sathish and Shetty (2013)
Coffee husk	18%	13%	Mussatto et al. (2011)
Corn cob	53.61%	1.9%	Sathish and Shetty (2013)
Grape pulp	Carbohydrate, 6.53%	1.96%	Ekanah et al. (2017)
Oat straw	Cellulose 40%, hemicellulose 27%, lignin 18%	NA	Singh nee' Nigam et al. (2009)
Oil palm empty fruit bunch	43.8–54.7%	0.25–1.21%	Chang (2014)
Oil palm frond	56.9%	28.5%	Rhaman and Naher (2021)
Orange peel	Cellulose (71.2 g/kg), hemicellulose (128 g/kg)	Crude protein (57.2 g/kg)	Ahmad et al. (2012)
Orange peel	Carbohydrate 52.90%	Crude protein 12.3%	Gotmare and Gade (2018)
Paddy straw	50.2%	84.9%	Rhaman and Naher (2021)
Pineapple waste	45.68%	0.61%	Sathish and Shetty (2013)
Rice bran extract	Glucose 38.3% cellulose 7.8%	13.2%	Choi (2020)
Rice husk	Cellulose 22%, hemicellulose 23%, lignin 15%	NA	Megawati et al. (2011)
Rice straw	Cellulose 32%, hemicellulose 24%, lignin 18%	NA	Limayem and Ricke (2012)
Rice washed	Carbohydrate, 90%	Protein, 8%	Srikandace et al. (2022)
Sawdust	55.2%	34.3%	Rhaman and Naher (2021)
Sugarcane bagasse	48.32%	0.2%	Sathish and Shetty (2013)
Sugarcane straw	Cellulose 36%, hemicellulose 21%, lignin 16%	NA	Saad et al. (2008)

(continued)

Table 1 (continued)

Resources	Carbon content (from)	Nitrogen content (from)	References
Tofu liquid waste	Carbohydrate, 25%	Protein, 65%	Srikandace et al. (2022)
Tomato juice	Carbohydrate, 2.52%	Protein, 1%	Ismail Abdullahi et al. (2016)
Wheat straw	Cellulose 27%, hemicellulose 21%, lignin 23%	NA	Adapa et al. (2011)

NA: Not available

for BC production, which will be possible for building block material application (Wang et al. 2021). A higher BC production of 2.86 g/L was reported from corn stalk, containing 3.87 g/L of glucose and glucan (35%), which were previously treated with acetic acid for *Acetobacter xylinum* ATCC 23767 (Cheng et al. 2017). Another corn residue can be obtained from the corn stover. Mainly contains glucose and xylose as well as available abundantly as agricultural residues, corn stover can be used as a low-cost feedstock in the manufacturing of BC. *Enterobacter sp.* FY-07 (CGMCC No. 6103) was used during the fermentation and incubated under static conditions at 30 °C for 24 h. The productivity was 14.35 g/L/ day but with the addition of xanthan gum, the productivity increased significantly up to 17.13 g/L/day. Interestingly, the pilot scale for BC has been reported by using oat hulls. Previously chemically treated with HNO_3, followed by enzymatic saccharification and the addition of sodium hydroxide, oat hulls were performed in a 100 L fermentor. After purification, 80.5 tons of 98%-wet BC gel per 100 tons of oat hulls were obtained with a 93% crystallinity index and composed of 100% cellulose I_α-allomorph (Skiba et al. 2020). Another approach to scale-up BC production has been explored. In a 30 L working volume, *Acetobacter xylinum* KJ1 was used, and the BC yield was achieved at 5.6 g/L using saccharified food wastes (Song et al. 2009). DHU-ATCC-1 strain, a mutant of *Komagataeibacter xylinus* ATCC 23770, was employed in a 75 L stirred-tank reactor to scale up BC production with the final yield of 17.3 gr/L using overripe bananas (Molina-Ramírez et al. 2020). Other possible agro-wastes for BC production are presented in Table 2.

2.2 *Fruit-Food Wastes*

Various studies have been reported regarding the use of agro-waste as a source of carbon and nitrogen for BC production. Green waste generated from fruits, vegetables, and food wastes is the potential resources due to its high glucose and fructose content. Citrus peel and pomace enzymolysis from beverage industrial waste were successfully used for BC by using *Komagataeibacter xylinus* CICC 10529 with a yield of 5.7 ± 0.7 g/L higher than from HS medium with 50 nm for its average diameter. The entire results confirmed the role of citrus peel and pomace enzymolysis

Table 2 Agro-industrial waste for BC production

Resource	Bacteria strain	Yield (gr/L)	References
Cacao mucilage exudate	*G. xylinus*	13.13	Saavedra-Sanabria et al. (2021)
Cashew tree	*K. rhaeticus*	2.3–6.0	Pacheco et al. (2017)
Cashew tree exudate	*K. rhaeticus*	2.8	Silva et al. (2010)
Coffee cherry husk	*G. hansenii UAC09*	6.24	Usha et al. (2011)
Corn stalk	*K. xylinum, ATCC 23767*	2.86	Cheng et al. (2017)
Corn stover	*Enterobacter sp. FY-07*	2.08	Gao et al. (2021)
Oat hulls	*Medusomyces gisevii Sa-12*	2.2	Skiba et al. (2020)
Pecan nutshell	*G. entanii*	2.8	Dórame-Miranda et al. (2019)
Prickly pear peels	*Lactiplantibacillus plantarum strain AS.6*	2.94	El-Gendi et al. (2023)
Sugarcane straw	*K. xylinus ATCC 11142*	1.06	Dhar et al. (2019)
Sweet sorghum	*Acetobacter xylinum ATCC 23767*	Root 2.28, stalk 1.82, leaf 2.54	Wang et al. (2021)
Wheat thin stillage	*G. sucrofermentans B-11267*	6.19	Revin et al. (2018)
Wheat straw	*K. xylinus* ATCC 23770	8.3	Chen et al. (2013)

as potential sources for BC production with similar characteristics to HS medium, being more environmentally-friendly and less expensive to produce (Fan et al. 2016). Other mango peel waste was also developed as an alternative culture medium for *Komagataeibacter xylinus DSMZ 2004*. The yield of BC was 6.32 g/L after 16 days of fermentation by the static culture technique. Structural analysis showed the diameter of BC from mango waste peel was 98.8 nm and showed a similar chemical structure to BC synthesized from pure sugar. This resulted BC was proposed for biomedical and pharmaceutical applications (García-Sánchez et al. 2020). Additionally, pineapple peels as an alternative medium were used for *Komagataeibacter xylinus* IITR DKH20 which was incubated for 384 h resulting in 11.44 g/L dried. The resulted BC revealed similar physicochemical properties to the BC produced using HS medium and was proposed for biomedical application (Khan et al. 2021). When wasted rotten tomato media was used as a substitute medium for *Gluconacetobacter hansenii* and cultivated for 7 days, the yield BC was 3.71 g/L. BC was produced after 7 days and aimed for medical and pharmaceutical fields (Fatima et al. 2021). However, a higher dry BC of 7.8 gr/L was obtained from the tomato juice when used as an optimization process for the 10 L production fermentation medium for *Acetobacter pasteurianus* RSV-4 after 7 days of incubation. Furthermore, *Komagataeibacter xylinus* DSM 6513 was successfully grown in a medium generated from red and white grape bagasse from the wine industry. It was reported that the white grape bagasse was a better

Table 3 Fruit wastes

Resource	Bacteria strain	Yield (gr/L)	References
Citrus peel pomace	*K. xylinus* CICC 10529	5.7	Fan et al. (2016)
Grape bagasse	*G. xylinus NRRL-B42*	8.0	Vazquez et al. (2013)
Kitchen waste	*K. rhaeticus K15*	4.76	Li et al. (2021)
Litchi extract	*K. xylinus CH001*	2.5	Yang et al. (2016)
Mango peels	*K. xylinus DSMZ200*	6.32	Sanchez et al. (2020)
Musk melon	*K. persimmonis GH-2*	8.08	Hungund et al. (2013)
Orange pulp	*A. pasteurianus RSV-4*	2.8	Kumar et al. (2019)
Pineapple peels	*K. xylinus IITR DKH20*	2.57	Khan et al. (2021)
Pineapple residue	*G. medellinensis*	3.24	Algar et al. (2015)
Rotten fruits	*G. xylinus*	0.06	Jozala et al. (2015)
Rotten tomato	*G. hansenii PJK KCTC 10505BP*	3.83	Fatima et al. (2021)
Tomato juice	*A. pasteurianus RSV-4*	7.8	Kumar et al. (2019)
Various fruit juice	*A. xylinum NBRC 13693*	0.2–2.1	Kurosumi et al. (2009)

substitution as a low-cost medium resource than the red grape. White grape bagasse also produced a five times higher yield, five times higher water holding capacity as well as greater flexibility than the HS medium. Red grape bagasse-based BC is suitable for the food industry, whereas white grape BC is appropriate for the textile and biomedical industries (Ogrizek et al. 2021). Another study reported that dried BC of 8.08 g/L was obtained from musk melon as a natural, cheaper carbon source medium for *Gluconeacetobacter persimmonis,* which was incubated at 30 °C for 14 days (Hungund et al. 2013). Kitchen wastes could also be the promising carbon and nitrogen resources for BC production, since it is composed of carbohydrates and protein. A new cellulose-producing bacteria, namely, *Komagataeibacter rhaeticus* K15, has been isolated from kombucha tea and shown the capability to use kitchen wastes as a carbon source for cellulose production of as much as 4.76 g/L (Li et al. 2021). The detailed lists of fruit wastes are shown in Table 3.

2.3 *Food and Beverage Industrial Wastes*

Based on daily activity, food-beverage industries generate large amounts of waste. Proper waste management can create economic benefits as well as provide free-contamination caused by their accumulation. On the other side, the wastes are rich in carbohydrates and protein thus promising to be a low-cost resource of media used for BC production. Beer manufacturing is a significant economic activity. Modern brewing is commonly a big industry that generates large quantities of byproducts nowadays. Waste beer yeasts are the second most common byproduct of the brewing industry that is discarded or fed to livestock. The waste beer yeast hydrolysates with 3% sugar concentration when it was treated by ultra-sonication, resulted BC yield of 7.02 g/L, nearly 6 times compared with the untreated waste beer yeast at 1.21 g/L (Lin et al. 2014). Cheese whey is today recognized as a source of functional and bioactive compounds, especially proteins and peptides but a significant amount of the whey produced globally is still not valorized whereas it contains rich nutrient components (Pires et al. 2021). The dry BC of 6.77 g/L was synthesized from the *K. xylinus* when enriched with β-galactosidase and proposed as a food packaging application. Corn steep liquor is a byproduct of the corn wet-milling production which generally consists mainly of water and other ingredients such as sugar and protein. Therefore, it is potential as nutrient medium for bacteria. A strain of *G. hansenii* UCP1619 was incubated in corn steep liquor at 30 °C for 10 days. A dry of 7.02 gr/L BC was resulted from this fermentation process and showed the future applications in the textile field (Costa et al. 2017). Another promising carbon source is thin stillage from rice wine distilleries. The strain of *G. xylinus* was incubated in the rice wine distillery by static cultivation for 7 days. A dry BC of 6.26 g/l was obtained which was reported almost 50% higher than produced in an HS-only medium with slightly denser reticulated structures and higher crystallinity (Wu and Liu 2013). Utilization of crude confectionery waste hydrolysates for *K. sucrofermentans* was reported for BC production. The waste contained 28.3% g/g free sugars, 28.4% g/g starch, 7.1% g/g protein, and 24.9% g/g fats was used in order to explore the lower cost alternative medium. BC was produced with a yield of 5.7 g/L and the potential as a bio-based packaging reinforcing agent (Efthymiou et al. 2022a, b). Other possible resources obtained from food and beverage industrial waste are presented in Table 4.

Table 4 Food and beverage industrial wastes

Resource	Bacteria strain	Yield (gr/L)	References
Cheese whey	*K. xylinus* DSM 2325	6.77	Rollini et al. (2020)
Confectionery wastes	*K. sucrofermentans*	5.7	Efthymiou et al. (2022a, b)
Corn steep liquor	*G. hansenii UCP1619*	7.02	Costa et al. (2017)
Jujube-processing industry	*K. xylinum CGMCC 2955*	2.2	Li et al. (2015)
Maple syrup	*A. xylinum BPR 2001*	1.51	Zeng et al. (2011)
Rice wine distillery	*G. xylinus*	6.26	Wu and Liu (2013)
Sugar cane molasses	*K. rhaeticus*	2.23–2.58	Machado et al. (2018)
Tofu liquid waste	*K. xylinum*	3.8	Srikandace et al. (2022)
Waste beer yeast	*K. hansenii CGMCC 3917*	7	Lin et al. (2014)
Whey	*G. sucrofermentans B-11267*	5.45	Revin et al. (2018)

2.4 Others

Other industrial wastes (Table 5) such as biodiesel wastes, Chinese medicinal herbs, cotton-based waste textiles, distillery effluent, dry olive mill residues, and tobacco extract wastes also show the potential resources for bacterial cellulose production.

3 Bacterial Cellulose and Its Properties

Bacterial cellulose, also known as bio-cellulose or microbial cellulose, is cellulose produced by the activity of non-pathogen, either positive or negative bacteria in a medium containing nitrogen and carbon as nutrient resources. The acetic bacterial that plays a role in the cellulose formation is commonly known as *Acetobacter xylinum* (Yamada et al. 1997), which is re-classified as *Gluconeacetobacter xylinum* and recently known as *Komagataeibacter xylinum.* This non-phatogen-aerobic bacteria can convert 108 glucose molecules per hour into cellulose and is considered as the most effective strain for bacterial cellulose production commercially due to its high productivity (Wang et al. 2019). Other producing bacteria are *Pseudomonas* (Ude et al. 2006), *Rhizobium* (Robledo et al. 2012), *Sarcina* (Yang et al. 2013a, b), *Agrobacterium* (Barnhart et al. 2013), and *Lactobacillus* (Khan et al. 2020). It was reported that Rhizobium and Agrobacterium produced cellulose in Rhizobium and Agrobacterium reportedly produced cellulose in exceedingly low yields. On the other hand, from the family of *Acetobacteriaceae* such as *Komagataeibacter*, *Acetobacter*, *Gluconacetobacter*, *Gluconobacter*, and *Asaia* together with the *Bacillus*, Leifsonia, *Salmonella*, *Erwinia*, *Enterobacter*, *Pseudomonas*, and *Shewanella* in non *Acetobacteriaceae* produce high cellulose (Li et al. 2022a, b). Generally bacterial cellulose is produced purposely, but it was hypothesized that cellulose is formed to protect bacteria from unfavorable factors such as UV radiation, harsh chemicals, and accessibility to oxygen (Retegi et al. 2010).

BC is illustrated by an ultrafine network structure, with emerging chains combining to generate sub-fibrils with a width of 1.5 nm (Ross et al. 1991). The spatial configuration of the pre-microfibril accumulation results in crystallinity of

Table 5 Other industrial wastes

Biodiesel waste	*G. xylinus NRRL-B42*	10	Vazquez et al. (2013)
Chinese medicinal herb	*Taonella mepensis*	0.54	Wu et al. (2021)
Cotton-based waste textiles	*G. xylinus ATCC 23770*	10.8	Hong et al. (2012)
Distillery effluent	*G. oboediens*	8.1	Jahan et al. (2018)
Dry olive mill residue	*K. sacchari sp.*	0.85	Gomes et al. (2013)
Elephant grass	*G. xylinus CH001*	6.4	Yang et al. (2013a, b)
Tobacco extract waste	*K. xylinum ATCC 23767*	5.2	Ye et al. (2019)

up to 84–89% (Czaja et al. 2004). The crystallinity highly relates to mechanical strength (Nishiyama et al. 2002) such as Young's modulus of BC, approximately in the range of 15–35 GPa, and the tensile strength typically in the range 200–300 MPa, respectively (Brown et al. 1976). The sub-fibrils are subsequently self-assembled to generate microfibrils, resulting in a fibrillar ribbon then tightly aggregating each other with a width of 50–80 nm. The resulted fibrillary ribbon is 200 times finer than cotton fiber with an extremely large surface area (Vitta and Thiruvengadam 2002). With a high surface area to mass ratio, bacterial cellulose shows high-water retention capability. The high-water content of bacterial celluloses is also due to the hydrophilicity property caused by the pore structure. The relative hydrophilicity was approximately around 40–50% (Bishop 2007). Additionally, BC shows the degree of polymerization around 14.000—16.000 at pH 4 but its polymerization lowered when pH increased to 5 (Tahara et al. 1997). A low degree of polymerization was also found when bacterial cellulose was synthesized spherical-type bubble column bioreactor (Choi et al. 2009).

BC is usually produced in a simple static method in which the container is filled with the acidic medium containing carbon–nitrogen where the bacteria strain is inoculated at room temperature for a certain time, usually from 1 to 2 weeks. The static method results in a thick pellicle at the top of the medium. The thickness increases with the increase of fermentation time (Fig. 1). High crystallinity, strong tensile strength, dense network structure, high-temperature resistance, and good flame retardancy were generated from bacterial cellulose in static culture. Additionally, under static culture conditions, bacterial cellulose was uniform in film shape as well as showed good biocompatibility and biodegradability (Gao et al. 2020).

Fig. 1 Hestrin-Schramm-based bacterial cellulose

On the other hand, agitated cultivation is simply carried out by growing the bacteria in a container containing growth nutrients and agitated for several days. The agitation technique aims to increase the oxygen supply for the bacteria which finally resulted in pellet-like aggregates. In the agitation culture, the crystallinity and cellulose Iα are lower than from the static culture. However, the degree of polymerization is higher than the static culture (Watanabe et al. 1998). Furthermore, through the agitation culture, the smaller particle size of bacterial cellulose resulted in the high-water holding capacity, compared to the static culture. Additionally, the lower Young's modulus and higher suspension viscosity were shown from the agitation technique (Ougiya et al. 1997). Additionally, more porous bacterial cellulose was produced by the agitation culture (Gao et al. 2020). So far it is considered that agitation is the most proper method for cost-effective BC production (Hu et al. 2013). The selection of these two methods is based on the BC application target with its various property considerations. The pellicle-type BC was developed for a plasmonic paper sensor (Purwidyantri et al. 2020) while the hollow-type spherical BC was proposed as a seamless capsule for drug delivery applications (Hoshi et al. 2018). In addition to the culture method, additional elements such as nutrients, type of bacterial strains, oxygen availability, and the alignment of its three-dimensional network the environment of fermentation also have an impact on the bacterial cellulose properties (Kim et al. 2019).

Generally, bacterial cellulose has far better properties than plant cellulose such as purity higher than 99% (Klemm et al. 2005), total surface area of more than 150 m^2/g (Ul-Islam et al. 2012), water holding capacity of more than 95% (Rebelo et al. 2018), and tensile strength 20–300 MPa (Feng et al. 2015). Crystallinity is also an important properties for evaluation since it relates to mechanical strength. A high crystallinity of up to 98.4% was observed in bacterial cellulose produced from confectionery using *Komagataeibacter sucrofermentans* after a 24 h HCl treatment (Efthymiou et al. 2022a, b). Additionally, a crystallinity of 75.37% was obtained from bacterial cellulose in sweet sorghum by *Acetobacter xylinum* ATCC 23767 (Wang et al. 2021). By using a laboratory-scale bioreactor with a 41 cm^2 cross-sectional area, overripe banana-based bacterial cellulose cultivated by *Komagataeibacter medellinensis* showed a crystallinity of 82.93% (Molina-ramírez et al. 2020). The high crystallinity highly corresponds with mechanical performance The tensile strength and Young's modulus of BC are 200–300 MPa and 15–35 GPa which are higher than synthetic polymer (Cacicedo et al. 2016). These values usually vary depending on the bacterial strain, cultivation method, culture nutrient as well as drying method. A tensile strength of around 27.3–37.2 MPa was achieved when bacterial cellulose was oven-dried (Illa et al. 2019). The tensile strength of bacterial cellulose-based stalk and leaf of sweet sorghum when prepared in the medium of *Acetobacter xylinum* ATCC 23767 was 8.24 MPa and 4.83 MPa, respectively. When bacterial cellulose was produced in rotten guava mixed with cheese whey by using *Komagataeibacter intermedius MO*, the tensile of 30 MPa was achieved (Lotfy et al. 2021). Furthermore, rotten banana-based bacterial cellulose showed tensile strength and Young's modulus of 280.6 MPa and 9.4 MPa, respectively, much higher than those synthesized by Hestrin-Schramm medium (Molina-Ramírez et al. 2018). Young's modulus of 8.7

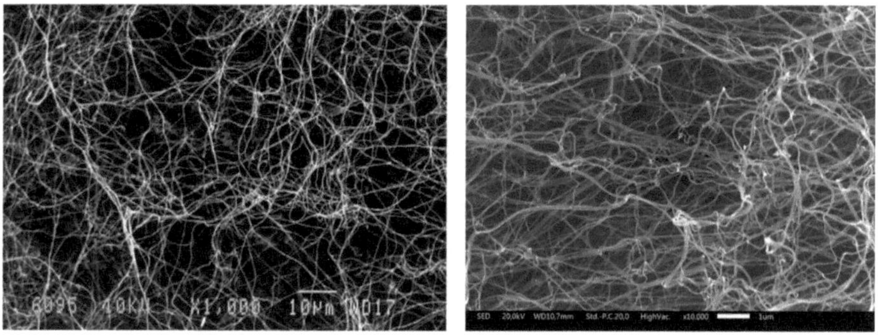

Fig. 2 Purification by 2% NaOH (left) (Skiba et al. 2020), purification by boiling water (right) (Srikandace et al. 2022)

GPa was achieved from bacterial cellulose synthesized using tofu liquid water, higher than from bacterial cellulose cultivated in the Hestrin-Schramm medium. This bacterial cellulose also revealed the same irregular three-dimensional network made of disordered dense fibrils arrangement with that produced from the synthetic medium (Srikandace et al. 2022).

The resulted bacterial cellulose requires purification by removing the remaining bacterial cell or nutrients in the medium. Different from plant cellulose which is chemically linked with hemicellulose and lignin and thus makes it difficult to remove impurities, purification of bacterial cellulose is much easier to carry out. Purification is easily carried out by boiling dilute sodium hydroxide followed by rinsing it with water (Revin et al. 2018). However, it has been proved that by boiling in water for 10 min with 2–3 replication after the water is decanted, pure bacterial cellulose was obtained (Srikandace et al. 2022). Figure 2 shows the morphology of bacterial cellulose purified by sodium hydroxide and boiling water, respectively.

3.1 Drying of Bacterial Cellulose

The drying method plays a role in the performance and properties of bacterial cellulose. As bacterial cellulose is too thick and slippery, the dry state is preferable for its wider application and it is more easily handled with stable properties. Various drying method has been reported for bacterial cellulose treatment, such as oven drying (Illa et al. 2019), microwave heating and air convection heating (Gao et al. 2020), and evaporation (Zeng et al. 2014). These drying techniques provide various performance alterations. When compared with the freeze-drying technique at −84 °C for 24 h, oven drying of bacterial cellulose resulted in higher crystallinity, decreased fiber diameter, narrowed size distribution, and increased mechanical properties (Illa et al. 2019). However, the swelling ability of the bacterial cellulose gel was reduced through freeze drying (Clasen et al. 2006). Additionally, whitish BC with higher porosity was shown by the freeze-dried method whereas transparent and film volume reduction was deduced from bacterial cellulose by oven drying (Vasconcellos and Farinas 2018). Furthermore, a long drying time of 120 h at 100 °C provided tensile strength of 250.7 MPa and a tensile modulus of 18.6 GPa (Abral et al. 2021). The supercritical drying technique provided mechanically robust and extremely light films of bacterial cellulose (Zeng et al. 2014) whereas freeze drying at −30 °C resulted in transparent film with higher porosity (Urbina et al. 2019a, b). It was reported that the lyophilizer technique employed at −50 to 20 °C for 36 h yielded a loose reticulated porous structure with a high-water absorption capacity (Feng et al. 2015). Microwave heating was carried out in a short time but it provided bacterial cellulose with slightly lower crystallinity and a higher swelling degree with the wrinkled surface (Indriyati and Puspitasari 2019).

3.2 *Bacterial Cellulose-Based Food Packaging*

There have been various studies were carried out to use bacterial cellulose for food packaging thus the evaluation of its properties for that purpose is indispensable. Bacterial cellulose is hydrophilic due to its rich hydroxyl group content, therefore bacterial cellulose has low barrier properties. Water vapor permeability (WVP) of bacterial cellulose is frequently studied for food packaging applications. It appraises the amount of water vapor that can pass through the package layer from the inner or outer environment, which possibly leads to unfavorable alterations in the product's characteristics. For this purpose, bacterial cellulose produced by *Gluconeacetobacter hansenii* CGMCC3917 was used as a reinforcing agent and it was incorporated with agar for edible packaging. The concentration of 3–5% of bacterial cellulose showed favorable WVP properties. The application of BC up to 10% decreased WVP up to 25.7% (Wang et al. 2018). The WVP ranged from 1.87×10^{-11} to 2.04×10^{-10} g/m s Pa was obtained from a composite film containing bacterial cellulose, glycerol, and polyvinyl alcohol. The film is the potential for food packaging to keep the quality of food as well as increase the shelf-life (Cazón et al. 2020).

Another important characteristic consideration for food packaging is mechanical strength which plays a role during production, storage, application, transportation, and distribution. Mechanical properties such as ultimate tensile strength, elongation at break, tensile Young's modulus, tensile toughness to break, ultimate puncture strength, puncture deformation, puncture Young's modulus, and puncture toughness to break were evaluated for bacterial nanocomposite film incorporated with polyvinyl alcohol, glycerol and boric acid. Due to its suitable mechanical properties, the resulted nanocomposites are suitable for disposable packaging (Rouhi et al. 2017). Improvement of mechanical properties as well as possess good antibacterial activity and antioxidant capacity was shown from alternative edible and environment-friendly sheets for food packaging made of bacterial cellulose, curdlan, and cinnamon essential oil (Zhou et al. 2022). Other enhancements of mechanical properties together with barrier and antibacterial properties were reported from a sheet prepared from bacterial cellulose with the addition of konjac glucomannan and cucurmin. The resulted film was claimed to support beef freshness (Li et al. 2022a, b). In addition, the wrapping nanopaper generated from bacterial cellulose and *Lactobacillus plantarum* was efficient against *Lactobacillus monocytogenes* in freshly ground beef (Shafipour Yordshahi et al. 2020).

When bacterial cellulose was combined with cyanidin-3-glucoside, it resulted in a smart pH-sensitive sheet that possessed an antioxidant characteristic and was applicable for tilapia filet freshness non-destructive packaging indicator (Shi et al. 2022). Another smart film based on bacterial cellulose was developed by 2,2,6,6-tetramethylpiperidine-1-oxyl radical (TEMPO)-oxidation containing thymol and anthocyanin-rich purple potato extract. It was reported that the film showed improved thermal stability, UV protection, and water vapor barrier characteristics but somewhat decreased tensile strength. With real-time assessment of freshness, these particular characteristics of composite film illustrate the prospective tool for commercial shrimp packaging (Wen et al. 2021). Interestingly, isolates of sunflower protein and bacterial cellulose with improved mechanical properties, water vapor permeability, and solubility were developed for food packaging materials, specifically for fresh fruit preservation (Efthymiou et al. 2022a, b).

Transparency as another important characteristic should be taken into account not only for the product performance but also for consumer satisfaction. Transparency and hydrophobicity improvement as well as antioxidant capacity was obtained when bacterial cellulose-apple pomace-based nanopapers were combined with hydrophobic medium-chain-length polyhydroxyalkanoate as a coating agent. The film was developed for active packaging application (Urbina et al. 2019a, b). Additionally, bacterial cellulose from sago liquid waste was developed for meat sausage packaging. Its transparency was improved by the addition of carboxymethyl cellulose into the bacterial cellulose. This treatment improved mechanical characteristics as well as kept sausage quality for 6 days at room temperature (Yanti et al. 2021). Cheese whey permeates as a by-product of whey ultrafiltration, as a cheap substrate for bacterial cellulose production by *Komagataeibacter xylinus*, and conjugated with Sakacin-A, produced by *Lactobacillus sakei* was reported as the potential antimicrobial packaging material (Rollini et al. 2020).

4 Conclusion

As petroleum-based plastic supplies continue to decline the price rises whereas its demand increases in line with the population growth as well as the awareness of environmental rules have prompted an exploration for low-cost bacterial cellulose production for environmentally-friendly food packaging. Even though the resources are varied based on the type of activity, it has been reported that agricultural wastes show potential as an alternative source of carbohydrates and nitrogen for bacterial cellulose production. The studies contributed to investigating the applicable technique for food packaging application. If this biomass is used to produce bacterial cellulose massively or on a large scale, this not only increases the value added of the residues but also supports the waste management from the related agro-industrial operation as well as the possibility of creating economic growth.

Acknowledgements The authors are grateful for any support provided by the National Research and Innovation Agency, BRIN-Indonesia.

References

Abdullahi I, Abdullahi N, Abdu AM, Ibrahim AS (2016) Proximate, mineral and vitamin analysis of fresh and canned tomato. Biosci Biotechnol Res Asia 13(2):1163–1169

Abral H, Chairani MK, Rizki MD, Mahardika M, Handayani D, Sugiarti E, Muslimin AN, Sapuan SM, Ilyas RA (2021) Characterization of compressed bacterial cellulose nanopaper film after exposure to dry and humid conditions. J Market Res 11:896–904. https://doi.org/10.1016/j.jmrt.2021.01.057

Adapa PK, Tabil LG, Schoenau GJ, Canam T, Dumonceaux T (2011) Quantitative analysis of lignocellulosic components of non-treated and steam exploded barley, canola, oat and wheat straw using fourier transform infrared spectroscopy. J Agricul Sci Technol B 1:177–188

Ahmad WA, Yusof NZ, Nordin N, Zakaria ZA, Rezali MF (2012) Production and characterization of violacein by locally isolated chromobacterium violaceum grown in agricultural wastes. Appl Biochem Biotechnol 167(5):1220–1234. https://doi.org/10.1007/s12010-012-9553-7

Algar I, Fernandes SCM, Mondragon G, Castro C, Garcia-astrain C, Gabilondo N (2015) Pineapple agroindustrial residues for the production of high value bacterial cellulose with different morphologies. J Appl Polym Sci 41237:1–8. https://doi.org/10.1002/app.41237

Almaraz-sánchez I, Amaro-reyes A, Acosta-gallegos JA, Mendoza-Sánchez M (2022) Processing agroindustry by-products for obtaining value-added products and reducing environmental impact. J Chem 1–13

Andrady AL, Neal MA, Andrady AL, Neal MA (2009) Applications and societal benefits of plastics Applications and societal benefits of plastics. Philos Trans R Soc B Biol Sci 364(June):1977–1984. https://doi.org/10.1098/rstb.2008.0304

Araújo D, Castro MCR, Figueiredo A, Vilarinho M, Machado A (2020) Green synthesis of cellulose acetate from corncob: physicochemical properties and assessment of environmental impacts. J Clean Prod 260:120865. https://doi.org/10.1016/j.jclepro.2020.120865

Barnhart DM, Su S, Baccaro BE, Banta LM, Farrand SK (2013) CelR, an Ortholog of the Diguanylate cyclase PleD of Caulobacter, regulates cellulose synthesis in agrobacterium tumefaciens. Appl Environ Microbiol 79(23):7188–7202

Bishop C (2007) Vacuum deposition onto webs, films, and foils. In: Engineering (General), 3rd ed., pp 165–190

British Plastics Federation (2019) Oil consumption. Oil Consumption. https://www.bpf.co.uk/press/oil_consumption

Brown RM, Willlison JHM, Richardson CL (1976) Cellulose biosynthesis in Acetobacter xylinum: visualization of the site of synthesis and direct measurement of the in vivo process. Proc Natl Acad Sci 73(12):4565–4569. https://doi.org/10.1073/pnas.73.12.4565

Cacicedo ML, Castro MC, Servetas I, Bosnea L, Boura K, Tsafrakidou P, Dima A, Terpou A, Koutinas A, Castro GR (2016) Progress in bacterial cellulose matrices for biotechnological applications. Bioresour Technol 213:172–180. https://doi.org/10.1016/j.biortech.2016.02.071

Cazón P, Velazquez G, Vázquez M (2020) Characterization of mechanical and barrier properties of bacterial cellulose, glycerol and polyvinyl alcohol (PVOH) composite films with eco-friendly UV-protective properties. Food Hydrocolloids 99:105323. https://doi.org/10.1016/j.foodhyd.2019.105323

Chang SH (2014) An overview of empty fruit bunch from oil palm as feedstock for bio-oil production. Biomass Bioenergy 62:174–181. https://doi.org/10.1016/j.biombioe.2014.01.002

Chen L, Hong F, Yang X, Han S (2013) Biotranformation of wheat straw to bacterial cellulose and its mechanism. Biores Technol 135:464–468

Cheng Z, Yang R, Liu X, Liu X, Chen H (2017) Green synthesis of bacterial cellulose via acetic acid pre-hydrolysis liquor of agricultural corn stalk used as carbon source. Biores Technol 234:8–14

Choi CN, Song HJ, Kim MJ, Chang MH, Kim SJ (2009) Properties of bacterial cellulose produced in a pilot-scale spherical type bubble column bioreactor. Korean J Chem Eng 26(1):136–140. https://doi.org/10.1007/s11814-009-0021-1

Choi HJ (2020) Agricultural biowaste, rice bran, as carbon source to enhance biomass and lipid production: analysis with various growth rate models. Water Sci Technol 82(6):1120–1130. https://doi.org/10.2166/wst.2020.342

Clasen C, Sultanova B, Wilhelms T, Heisig P, Kulicke WM (2006) Effects of different drying processes on the material properties of bacterial cellulose membranes. Macromol Symp 244:48–58. https://doi.org/10.1002/masy.200651204

Costa AFS, Almeida FCG, Vinhas GM, Sarubbo LA (2017) Production of bacterial cellulose by Gluconacetobacter Hansenii using corn steep liquor as nutrient sources. Front Microbiol 8(October):1–12. https://doi.org/10.3389/fmicb.2017.02027

Czaja W, Romanovicz D, Brown R (2004) Structural investigations of microbial cellulose produced in stationary and agitated culture. Cellulose 11(3–4):403–411

Dasgupta S, Sarraf M, Wheeler D (2022) Plastic waste cleanup priorities to reduce marine pollution: a spatiotemporal analysis for Accra and Lagos with satellite data ☆. Sci Total Environ 839(May):156319. https://doi.org/10.1016/j.scitotenv.2022.156319

De Corato U, De Bari I, Viola E, Pugliese M (2018) Assessing the main opportunities of integrated biorefining from agro-bioenergy co/by-products and agroindustrial residues into high-value added products associated to some emerging markets: a review. Renew Sustain Energy Rev 88:326–346. https://doi.org/10.1016/j.rser.2018.02.041

Dhar P, Pratto B, Gonçalves Cruz AJ, Bankar S (2019) Valorization of sugarcane straw to produce highly conductive bacterial cellulose/graphene nanocomposite films through in situ fermentation: Kinetic analysis and property evaluation. J Clean Prod 238:117859. https://doi.org/10.1016/j.jclepro.2019.117859

Dórame-Miranda RF, Gámez-Meza N, Medina-Juárez LÁ, Ezquerra-Brauer JM (2019) Bacterial cellulose production by Gluconacetobacter Entanii using pecan nutshell as carbon source and its chemical functionalization. Carbohydr Polym 207(November 2018):91–99. https://doi.org/10.1016/j.carbpol.2018.11.067

Duque-Acevedo M, Belmonte-Ureña LJ, Cortés-García FJ, Camacho-Ferre F (2020) Agricultural waste: review of the evolution, approaches and perspectives on alternative uses. Glob Ecol Conserv 22:e00902. https://doi.org/10.1016/j.gecco.2020.e00902

Efthymiou M, Tsouko E, Pateraki C (2022a) Property evaluation of bacterial cellulose nanostructures produced from confectionery wastes. Biochem Eng J 186(August):108575. https://doi.org/10.1016/j.bej.2022.108575

Efthymiou M-N, Tsouko E, Papagiannopoulos A, Athanasoulia I-G, Georgiadou M, Pispas S, Briassoulis D, Tsironi T, Koutinas A (2022b) Development of biodegradable films using sunflower protein isolates and bacterial nanocellulose as innovative food packaging materials for fresh fruit preservation. Sci Rep 12(1):6935. https://doi.org/10.1038/s41598-022-10913-6

Ekanah KS, State E, Obueh HO (2017) Nutritional and antinutritional evaluation of grape fruit (Citrus paradisi) juice using different extraction methods. J Adv Food Sci Technol 4(2):84–90

El-Gendi H, Salama A, El-Fakharany EM, Saleh AK (2023) Optimization of bacterial cellulose production from prickly pear peels and its ex situ impregnation with fruit byproducts for antimicrobial and strawberry packaging applications. Carbohydr Polym 302:120383. https://doi.org/10.1016/j.carbpol.2022.120383

Fan X, Gao Y, He W, Hu H, Tian M, Wang K, Pan S (2016) Production of nano bacterial cellulose from beverage industrial waste of citrus peel and pomace using Komagataeibacter xylinus. Carbohydr Polym 151:1068–1072. https://doi.org/10.1016/j.carbpol.2016.06.062

FAO R (2013) Food wastage footprint: impacts on natural resources

Fatima A, Yasir S, Khan, Mohd. S., Manan, S., Ullah, M. W., & Ul-Islam, M. (2021). Plant extract-loaded bacterial cellulose composite membrane for potential biomedical applications. J Bioresour Bioprod 6(1):26–32. https://doi.org/10.1016/j.jobab.2020.11.002

Feng X, Ullah N, Wang X, Sun X, Li C, Bai Y, Chen L, Li Z (2015) Characterization of bacterial cellulose by Gluconacetobacter Hansenii CGMCC 3917. J Food Sci 80(10):E2217–E2227. https://doi.org/10.1111/1750-3841.13010

Fèvre M, Girard V, Nodet P (1990) Cellulose and β-Glucan synthesis in Saprolegnia. In: Kuhn PJ, Trinci APJ, Jung MJ, Goosey MW, Copping LG (eds) Biochemistry of cell walls and membranes in Fungi. Springer, Berlin Heidelberg, pp 97–107

Forster-Carneiro T, Berni MD, Dorileo IL, Rostagno MA (2013) Biorefinery study of availability of agriculture residues and wastes for integrated biorefineries in Brazil. Resour Conserv Recycl 77:78–88. https://doi.org/10.1016/j.resconrec.2013.05.007

Gao H, Sun Q, Han Z, Li J, Liao B, Hu L, Huang J, Zou C, Jia C, Huang J, Chang Z, Jiang D, Jin M (2020) Comparison of bacterial nanocellulose produced by different strains under static and agitated culture conditions. Carbohydr Polym 227. https://doi.org/10.1016/j.carbpol.2019.115323

Gao G, Liao Z, Cao Y, Zhang Y, Zhang Y, Wu M (2021) Highly efficient production of bacterial cellulose from corn stover total hydrolysate by Enterobacter sp. FY-07. Bioresour Technol 341(August 2021). 125781. https://doi.org/10.1016/j.biortech.2021.125781

García-Sánchez ME, Robledo-Ortíz JR, Jiménez-Palomar I, González-Reynoso O, González-García Y (2020) Production of bacterial cellulose by Komagataeibacter xylinus using mango waste as alternative culture medium Producción. Revista Mexicana De Ingeniería Química Production 2:851–865

Gomes FP, Silva NH, Trovatti E, Serafim LS, Fuarte MF, SIlvestre AJ, Neto CP, Freira CS (2013) Production of bacterial cellulose by Gluconeacetobacter sacchari using dry olive mill residue. Biomass Bioenergy 55:205–211

Gotmare S, Gade J (2018) Orange peel: a potential source of phytochemical compounds. Int J ChemTech Res 11(02):240–243

Hiloidhari M, Bhuyan N, Gogoi N, Seth D, Garg A, Singh A, Prasad S, Kataki R (2020) Agroindustry wastes: biofuels and biomaterials feedstocks for sustainable rural development. In: Kumar RP, Gnansounou E, Raman JK, G. B. T.-R. B. R. for S. E, Baskar B (eds) Refining biomass residues for sustainable energy and bioproducts. Academic Press, pp 357–388. https://doi.org/10.1016/B978-0-12-818996-2.00016-8

Hong F, Guo X, Zhang S, Han S, Yang G, Jönsson LJ (2012) Bacterial cellulose production from cotton-based waste textiles: enzymatic saccharification enhanced by ionic liquid pretreatment. Biores Technol 104:503–508. https://doi.org/10.1016/j.biortech.2011.11.028

Hoshi T, Yamazaki K, Sato Y, Shida T, Aoyagi T (2018) Production of hollow-type spherical bacterial cellulose as a controlled release device by newly designed floating cultivation. Heliyon 4(10):e00873. https://doi.org/10.1016/j.heliyon.2018.e00873

Hu Y, Catchmark JM, Vogler EA (2013) Factors impacting the formation of sphere-like bacterial cellulose particles and their biocompatibility for human osteoblast growth. Biomacromolecules 14(10):3444–3452. https://doi.org/10.1021/bm400744a

Hungund B, Prabhu S, Shetty C, Acharya S, Prabhu V, Sg G (2013) Production of bacterial cellulose from Gluconacetobacter persimmonis GH-2 using dual and cheaper carbon sources. Microb Biochem Technol 5(2):31–33. https://doi.org/10.4172/1948-5948.1000095

Illa MP, Sharma CS, Khandelwal M (2019) Tuning the physiochemical properties of bacterial cellulose: effect of drying conditions. J Mater Sci 54(18):12024–12035. https://doi.org/10.1007/s10853-019-03737-9
Indriyati YI, Puspitasari T (2019) Comparative study of bacterial cellulose film dried using microwave and air convection heating. J Eng Technol Sci 51(1):121–132
Islam MU, Ullah MW, Khan S, Shah N, Park JK (2017) Strategies for cost-effective and enhanced production of bacterial cellulose. Int J Biol Macromol 102:1166–1173. https://doi.org/10.1016/j.ijbiomac.2017.04.110
Jahan F, Kumar V, Saxena RK (2018) Distillery effluent as a potential medium for bacterial cellulose production: a biopolymer of great commercial importance. Biores Technol 250:922–926
Jozala AF, Pértile RAN, dos Santos CA, de Carvalho Santos-Ebinuma V, Seckler MM, Gama FM, Pessoa A (2015) Bacterial cellulose production by Gluconacetobacter xylinus by employing alternative culture media. Appl Microbiol Biotechnol 99(3):1181–1190. https://doi.org/10.1007/s00253-014-6232-3
Khan H, Kadam A, Dutt D (2020) Studies on bacterial cellulose produced by a novel strain of Lactobacillus genus. Carbohydr Polym 229:115513. https://doi.org/10.1016/j.carbpol.2019.115513
Khan H, Saroha V, Raghuvanshi S, Bharti AK, Dutt D (2021) Valorization of fruit processing waste to produce high value-added bacterial nanocellulose by a novel strain Komagataeibacter xylinus IITR DKH20. Carbohydr Polym 260:117807. https://doi.org/10.1016/j.carbpol.2021.117807
Kim Y, Ullah MW, Ul-Islam M, Khan S, Jang JH, Park JK (2019) Self-assembly of bio-cellulose nanofibrils through intermediate phase in a cell-free enzyme system. Biochem Eng J 142:135–144. https://doi.org/10.1016/j.bej.2018.11.017
Klemm D, Heublein B, Fink H-P, Bohn A (2005) Cellulose: fascinating biopolymer and sustainable raw material. Angew Chemie Int Edn 44(22):3358–3393. https://doi.org/10.1002/anie.200460587
Kumar V, Sharma DK, Bansal V, Mehta D, Sangwan R, Yadav SK (2019) Efficient and economic process for the production of bacterial cellulose from isolated strain of Acetobacter pasteurianus of RSV-4 bacterium. Bioresour Technol 275:430–433. https://doi.org/10.1016/j.biortech.2018.12.042
Kurosumi A, Sasaki C, Yamashita Y, Nakamura Y (2009) Utilization of various fruit juices as carbon source for production of bacterial cellulose by Acetobacter xylinum NBRC 13693. Carbohydr Polym 76(2):333–335. https://doi.org/10.1016/j.carbpol.2008.11.009
Lahiri D, Nag M, Dutta B, Dey A, Sarkar T, Pati S (2021) Bacterial cellulose: production, characterization, and application as antimicrobial agent. Int J Mol Sci 22:12984
Lee KY, Buldum G, Mantalaris A, Bismarck A (2014) More than meets the eye in bacterial cellulose: biosynthesis, bioprocessing, and applications in advanced fiber composites. Macromol Biosci 14:10–32
Li Z, Wang L, Hua J, Jia S, Zhang J, Liu H (2015) Production of nano bacterial cellulose from waste water of candied jujube-processing industry using Acetobacter xylinum. Carbohydr Polym 120:115–119. https://doi.org/10.1016/j.carbpol.2014.11.061
Li Z, Azi F, Ge Z, Liu Y, Yin X, Dong M (2021) Bio-conversion of kitchen waste into bacterial cellulose using a new multiple carbon utilizing Komagataeibacter rhaeticus: fermentation profiles and genome-wide analysis. Int J Biol Macromol 191(May):211–221. https://doi.org/10.1016/j.ijbiomac.2021.09.077
Li G, Wang L, Deng Y, Wei Q (2022a). Research progress of the biosynthetic strains and pathways of bacterial cellulose. J Ind Microbiol Biotechnol 49(1) (Oxford University Press). https://doi.org/10.1093/jimb/kuab071
Li N, Yang X, Lin D (2022b) Development of bacterial cellulose nanofibers/konjac glucomannan-based intelligent films loaded with curcumin for the fresh-keeping and freshness monitoring of fresh beef. Food Packag Shelf Life 34:100989. https://doi.org/10.1016/j.fpsl.2022.100989

Limayem A, Ricke SC (2012) Lignocellulosic biomass for bioethanol production: current perspectives, potential issues and future prospects. Progr Energy Combust Sci 38(4):449–467. https://doi.org/10.1016/j.pecs.2012.03.002
Lin D, Lopez-Sanchez P, Li R, Li Z (2014) Production of bacterial cellulose by Gluconacetobacter Hansenii CGMCC 3917 using only waste beer yeast as nutrient source. Biores Technol 151:113–119
Long Z, Pan Z, Jin X, Zou Q, He J, Li W, Waters CN, Turner SD, do Sul JAI, Yu X, Chen J, Lin H, Ren J (2022) Anthropocene microplastic stratigraphy of Xiamen Bay, China: a history of plastic production and waste management. Water Res 226:119215. https://doi.org/10.1016/j.watres.2022.119215
Lopes FC, Tichota DM, Pereira JQ, Segalin J, de Oliveira Rios A, Brandelli A (2013) Pigment production by filamentous fungi on agro-industrial byproducts: an eco-friendly alternative. Appl Biochem Biotechnol 171(3):616–625. https://doi.org/10.1007/s12010-013-0392-y
Lopes FC, Ligabue-Braun R (2021) Agro-industrial residues: eco-friendly and inexpensive substrates for microbial pigments production. Front Sustain Food Syst 5:589414
Lotfy VF, Basta AH, Abdel-monem MO, Abdel-hamed GZ (2021) Utilization of bacteria in rotten Guava for production of bacterial cellulose from isolated and protein waste. Carbohydr Polym Technol Appl 2(February):100076. https://doi.org/10.1016/j.carpta.2021.100076
Machado RTA, Bagliotti A, Miguel R, Faza D, Antonio SG, Gutierrez J, Tercjak A, Berretta AA, Ribeiro SJL, Lazarini SC, Lustri WR, Barud HS (2018) Komagataeibacter rhaeticus grown in sugarcane molasses-supplemented culture medium as a strategy for enhancing bacterial cellulose production. Ind Crops Prod 122(May):637–646. https://doi.org/10.1016/j.indcrop.2018.06.048
Madeira JV, Contesini FJ, Calzado F, Rubio MV, Zubieta MP, Lopes DB, de Melo RR (2017) Agro-industrial residues and microbial enzymes: an overview on the eco-friendly bioconversion into high value-added products. In: G. B. T.-B. of M. E. Brahmachari (ed) Biotechnology of microbial enzymes: production, biocatalysis and industrial applications. Academic Press, pp 475–511. https://doi.org/10.1016/B978-0-12-803725-6.00018-2
Megawati, Sediawan WB, Sulistyo H, Hidayat M (2011) Kinetics of sequential reaction of hydrolysis and sugar degradation of rice husk in ethanol production: effect of catalyst concentration. Bioresour Technol 102(2):2062–2067
Meijer LJJ, van Emmerik T, van der Ent R, Schmidt C, Lebreton L (2022) More than 1000 rivers account for 80% of global riverine plastic emissions into the ocean. Sci Adv 7(18):eaaz5803. https://doi.org/10.1126/sciadv.aaz5803
Mihranyan A (2011) Cellulose from cladophorales green algae: from environmental problem to high-tech composite materials. J Appl Polym Sci 119(4):2449–2460. https://doi.org/10.1002/app.32959
Molina-Ramírez C, Castro C, Zuluaga R, Gañán P (2018) Physical characterization of bacterial cellulose produced by Komagataeibacter medellinensis using food supply chain waste and agricultural by-products as alternative low-cost feedstocks. J Polym Environ 26(2):830–837. https://doi.org/10.1007/s10924-017-0993-6
Molina-ramírez C, Cañas-gutiérrez A, Castro C, Zuluaga R (2020) Effect of production process scale-up on the characteristics and properties of bacterial nanocellulose obtained from overripe Banana culture medium. Carbohydr Polym 240(April):116341. https://doi.org/10.1016/j.carbpol.2020.116341
Mussatto SI, Machado EMS, Martins S, Teixeira JA (2011) Production, composition, and application of coffee and its industrial residues. Food Bioprocess Technol 4(5):661–672. https://doi.org/10.1007/s11947-011-0565-z
Napper IE, Thompson RC (2019) Marine plastic pollution: other than microplastic. In: Letcher TM, D. A. B. T.-W., Vallero E (eds) Waste. Academic Press, pp 425–442. https://doi.org/10.1016/B978-0-12-815060-3.00022-0
Nishiyama Y, Langan P, Chanzy H (2002) Crystal structure and hydrogen-bonding system in cellulose Iβ from synchrotron X-ray and neutron fiber diffraction. J Am Chem Soc 124(31):9074–9082. https://doi.org/10.1021/ja0257319

Ogrizek L, Lamovšek J, Cuš F, Leskovšek M, Gorjanc M (2021) Properties of bacterial cellulose produced using white and red grape bagasse as a nutrient source. Processes 9:1–15
Ougiya H, Watanabe K, Morinaga Y, Yoshinaga F (1997) Emulsion-stabilizing effect of bacterial cellulose. Biosci Biotechnol Biochem 61(9):1541–1545. https://doi.org/10.1271/bbb.61.1541
Pacheco G, Nogueira CR, Meneguin AB, Trovatti E, Silva MC, Machado RT, Ribeiro SJ, da Silva Filho EC, Barud HD (2017) Development and characterization of bacterial cellulose produced by cashew tree residues as alternative carbon source. Ind Crops Prod107:13–19
Pires AF, Marnotes NG, Rubio OD, Garcia AC, Pereira CD (2021) Dairy by-products: a review on the valorization of whey and second cheese whey. Foods 12(10):1067
Purwidyantri A, Karina M, Hsu C, Srikandace Y, Prabowo BA, Lai C (2020) Facile bacterial cellulose nano fibrillation for the development of a plasmonic paper sensor. ACS Biomater 6:3122–3131. https://doi.org/10.1021/acsbiomaterials.9b01890
Rainieri S, Barranco A (2019) Microplastics, a food safety issue? Trends Food Sci Technol 84:55–57. https://doi.org/10.1016/j.tifs.2018.12.009
Rajwade JM, Paknikar KM, Kumbhar JV (2015) Applications of bacterial cellulose and its composites in biomedicine. Appl Microbiol Biotechnol 99(6):2491–2511. https://doi.org/10.1007/s00253-015-6426-3
Rebelo RA, Archer AJ, Chen X, Liu C, Yang G, Liu Y (2018) Dehydration of bacterial cellulose and the water content effects on its viscoelastic and electrochemical properties. Sci Technol Adv Mater 19(1):203–211. https://doi.org/10.1080/14686996.2018.1430981
Reiniati I, Hrymak AN, Margaritis A (2017) Recent developments in the production and applications of bacterial cellulose fibers and nanocrystals. Crit Rev Biotechnol 37(4):510–524. https://doi.org/10.1080/07388551.2016.1189871
Retegi A, Gabilondo N, Peña C, Zuluaga R, Castro C, Gañan P, de la Caba K, Mondragon I (2010) Bacterial cellulose films with controlled microstructure–mechanical property relationships. Cellulose 17(3):661–669. https://doi.org/10.1007/s10570-009-9389-7
Revin V, Liyaskina E, Nazarkina M, Bogatyreva A, Shchankin M (2018) Cost-effective production of bacterial cellulose using acidic food industry by-products. Braz J Microbiol 49:151–159. https://doi.org/10.1016/j.bjm.2017.12.012
Rhaman SMSA, Naher L (2021) Evaluating carbon, nitrogen and heavy metal content in different agriculture biomass for mushroom substrate. In: International conference on bioengineering and technology, pp 1–7
Ribeiro JE, da Silva Sant'Ana AM, Martini M, Sorce C, Andreucci A, de Melo DJ, da Silva FL (2019) Rhodotorula glutinis cultivation on cassava wastewater for carotenoids and fatty acids generation. Biocatal Agricul Biotechnol 22:101419. https://doi.org/10.1016/j.bcab.2019.101419
Robledo M, Rivera L, Jiménez-Zurdo JI, Rivas R, Dazzo F, Velázquez E, Martínez-Molina E, Hirsch AM, Mateos PF (2012) Role of Rhizobium endoglucanase CelC2 in cellulose biosynthesis and biofilm formation on plant roots and abiotic surfaces. Microb Cell Fact 11(1):125. https://doi.org/10.1186/1475-2859-11-125
Rollini M, Musatti A, Cavicchioli D, Bussini D, Farris S, Rovera C, Romano D, De Benedetti S, Barbiroli A (2020) From cheese whey permeate to Sakacin-A/bacterial cellulose nanocrystal conjugates for antimicrobial food packaging applications: a circular economy case study. Sci Rep 10(1):21358. https://doi.org/10.1038/s41598-020-78430-y
Ross P, Mayer R, Benziman M (1991) Cellulose biosynthesis and function in bacteria. Microbiol Rev 55(1):35–58
Rouhi M, Razavi SH, Mousavi SM (2017) Optimization of crosslinked poly(vinyl alcohol) nanocomposite films for mechanical properties. Mater Sci Eng: C 71:1052–1063. https://doi.org/10.1016/j.msec.2016.11.135
Saad MBW, Oliveira LRM, Cândido RG, Quintana G, Rocha GJM, Gonçalves AR (2008) Preliminary studies on fungal treatment of sugarcane straw for organosolv pulping. Enzyme Microb Technol 43(2):220–225. https://doi.org/10.1016/j.enzmictec.2008.03.006

Saavedra-Sanabria OL, Duran D, Cabezas J, Hernandez I, Blanco-tirado C, Combariza MY (2021) Cellulose biosynthesis using simple sugars available in residual cacao mucilage exudate. Carbohyd Polym 274(April):118645. https://doi.org/10.1016/j.carbpol.2021.118645

Salari M, Sowti Khiabani M, Rezaei Mokarram R, Ghanbarzadeh B, Samadi Kafil H (2018) Development and evaluation of chitosan based active nanocomposite films containing bacterial cellulose nanocrystals and silver nanoparticles. Food Hydrocoll 84:414–423. https://doi.org/10.1016/j.foodhyd.2018.05.037

Sanchez ME, Robledo-Ortíz JR, Jimenez-Palomar I, González-Reynoso O (2020) Production of bacterial cellulose by Komagataeibacter xylinus using mango waste as alternative culture medium. Revista Mexicana De Ingeniería Química 19(2):851–865

Sathish T, Shetty R (2013) Intensification of fructosyltransferases and fructo-oligosaccharides production in solid state fermentation by aspergillus awamori GHRTS. Indian J Microbiol 53(3):337–342. https://doi.org/10.1007/s12088-013-0380-5

Shafipour Yordshahi A, Moradi M, Tajik H, Molaei R (2020) Design and preparation of antimicrobial meat wrapping nanopaper with bacterial cellulose and postbiotics of lactic acid bacteria. Int J Food Microbiol 321:108561. https://doi.org/10.1016/j.ijfoodmicro.2020.108561

Shi C, Ji Z, Zhang J, Jia Z, Yang X (2022) Preparation and characterization of intelligent packaging film for visual inspection of tilapia fillets freshness using cyanidin and bacterial cellulose. Int J Biol Macromol 205:357–365. https://doi.org/10.1016/j.ijbiomac.2022.02.072

Silva TM, Santiago PO, Purcena LLA, Fernandes KF (2010) Study of the cashew gum polysaccharide for the horseradish peroxidase immobilization—Structural characteristics, stability and recovery. Mater Sci Eng C 30:526–530. https://doi.org/10.1016/j.msec.2010.01.016

Singh nee' Nigam P, Gupta N, Anthwal A (2009) Pre-treatment of agro-industrial residues. In: Singh nee' Nigam P, Pandey A (eds) Biotechnology for agro-industrial residues utilisation: utilisation of agro-residues. Springer Netherlands, pp 13–33. https://doi.org/10.1007/978-1-4020-9942-7_2

Skiba EA, Budaeva VV, Ovchinnikova EV, Gladysheva EK, Kashcheyeva EI, Pavlov IN, Sakovich GV (2020) A technology for pilot production of bacterial cellulose from oat hulls. Chem Eng J 383. https://doi.org/10.1016/j.cej.2019.123128

Song H-J, Li H, Seo J-H, Kim M-J, Kim S-J (2009) Pilot-scale production of bacterial cellulose by a spherical type bubble column bioreactor using saccharified food wastes. Korean J Chem Eng 26(1):141–146. https://doi.org/10.1007/s11814-009-0022-0

Srikandace Y, Apriyana AY, Zahrad SA, Ramdhani W, Asri PPP, Andriani D, Abdullah AHD, Syampurwadi A, Satoto R, Karina M (2022) Bacterial cellulose production by Komagataeibacter xylinus using rice-washed water and tofu processing wastewater with the addition of sodium glutamate. Fibers Polym 23(5):1190–1196. https://doi.org/10.1007/s12221-022-4729-4

Tahara N, Tabuchi M, Watanabe K, Yano H, Morinaga Y, Yoshinaga F (1997) Degree of polymerization of cellulose from acetobacter xylinum BPR2001 decreased by cellulase produced by the strain. Biosci Biotechnol Biochem 61(11):1862–1865. https://doi.org/10.1271/bbb.61.1862

Ude S, Arnold DL, Moon CD, Timms-Wilson T, Spiers AJ (2006) Biofilm formation and cellulose expression among diverse environmental pseudomonas isolates. Environ Biol 8(11):1997–2011

Ul-Islam M, Khan T, Park JK (2012) Water holding and release properties of bacterial cellulose obtained by in situ and ex situ modification. Carbohydr Polym 88(2):569–603

Urbina L, Alonso-varona A, Saralegi A, Palomares T, Eceiza A (2019a) Hybrid and biocompatible cellulose/polyurethane nanocomposites with water-activated shape memory properties. Carbohyd Polym 216(March):86–96. https://doi.org/10.1016/j.carbpol.2019.04.010

Urbina L, Eceiza A, Gabilondo N, Corcuera MÁ, Retegi A (2019b). Valorization of apple waste for active packaging: multicomponent polyhydroxyalkanoate coated nanopapers with improved hydrophobicity and antioxidant capacity. Food Packag Shelf Life 21:100356. https://doi.org/10.1016/j.fpsl.2019.100356

Urbina L, Corcuera MA, Gabilondo N, Eceiza A, Retegi A (2021) A review of bacterial cellulose: sustainable production from agricultural waste and applications in various fields. Cellulose 4:8229–8253. https://doi.org/10.1007/s10570-021-04020-4

Usha M, Rastogi NK, Appaiah KAA (2011) Statistical optimization of medium composition for bacterial cellulose production by Gluconacetobacter Hansenii UAC09 using coffee cherry husk extract—An agro-industry waste. J Microbiol Biotechnol Biotechnol 21(7):739–745. https://doi.org/10.4014/jmb.1012.12026
Vandamme EJ (2009) Agro-industrial residue utilization for industrial biotechnology products. In: Singh nee' Nigam P, Pandey A (eds) Biotechnology for agro-industrial residues utilisation. Springer Netherlands, pp 3–11. https://doi.org/10.1007/978-1-4020-9942-7_1
Vasconcellos VM, Farinas CS (2018) The effect of the drying process on the properties of bacterial cellulose films from Gluconacetobacter Hansenii. Chem Eng Trans 64:145–150. https://doi.org/10.3303/CET1864025
Vazquez A, Foresti ML, Cerrutti P, Galvagno M (2013) Bacterial cellulose from simple and low cost production media by Gluconacetobacter xylinus. J Polym Environ 21:545–554. https://doi.org/10.1007/s10924-012-0541-3
Vitta S, Thiruvengadam V (2002) Multifunctional bacterial cellulose and nanoparticle-embedded composites. Curr Sci 102(10):1398–1405
Wang X, Guo C, Hao W, Ullah N, Chen L, Li Z, Feng X (2018) Development and characterization of agar-based edible films reinforced with nano-bacterial cellulose. Int J Biol Macromol 118:722–730. https://doi.org/10.1016/j.ijbiomac.2018.06.089
Wang J, Tavakoli J, Tang Y (2019) Bacterial cellulose production, properties and applications with different culture methods—A review. Carbohydr Polym 219:63–76. https://doi.org/10.1016/j.carbpol.2019.05.008
Wang Q, Nnanna PC, Shen F, Huang M, Tian D, Hu J, Zeng Y, Yang G, Deng S (2021) Full utilization of sweet sorghum for bacterial cellulose production: a concept of material crop. Ind Crops Prod 162:113256. https://doi.org/10.1016/j.indcrop.2021.113256
Watanabe K, Tabuchi M, Morinaga Y, Yoshinaga F (1998) Structural features and properties of bacterial cellulose produced in agitated culture. Cellulose 5(3):187–200. https://doi.org/10.1023/A:1009272904582
Wen Y, Liu J, Jiang L, Zhu Z, He S, He S, Shao W (2021) Development of intelligent/active food packaging film based on TEMPO-oxidized bacterial cellulose containing thymol and anthocyanin-rich purple potato extract for shelf life extension of shrimp. Food Packag Shelf Life 29:100709. https://doi.org/10.1016/j.fpsl.2021.100709
Wu J, Liu R (2013) Cost-effective production of bacterial cellulose in static cultures using distillery wastewater. J Biosci Bioeng 115(3):284–290
Wu SC, Li MH (2015) Production of bacterial cellulose membranes in a modified airlift bioreactor by Gluconacetobacter xylinus. J Biosc Bioeng 120(4):444–449. https://doi.org/10.1016/j.jbiosc.2015.02.018
Wu Y, Huang T, Li Z, Huang Z, Lu Y, Gao J, Hu Y, Huang C (2021). In-situ fermentation with gellan gum adding to produce bacterial cellulose from traditional Chinese medicinal herb residues hydrolysate. Carbohydr Polym 270(April). https://doi.org/10.1016/j.carbpol.2021.118350
Yamada Y, Hoshino K, Ishikawa T (1997) The phylogeny of acetic acid bacteria based on the partial sequences of 16S ribosomal RNA: the elevation of the subgenus Gluconoacetobacter to the generic level. Biosci Biotechnol Biochem 61(8):1244–1251
Yang XY, Huang C, Guo HJ, Xiong L, Li YY, Zhang HR, Chen XD (2013a) Bioconversion of elephant grass (Pennisetum purpureum) acid hydrolysate to bacterial cellulose by Gluconacetobacter xylinus. J Appl Microbiol 115:995–1002. https://doi.org/10.1111/jam.12255
Yang Y, Jia J, Xing J, Chen J, Lu S (2013b) Isolation and characteristics analysis of a novel high bacterial cellulose producing strain Gluconacetobacter intermedius CIs26. Carbohydr Polym 92(2):2012–2017. https://doi.org/10.1016/j.carbpol.2012.11.065
Yang X-Y, Huang C, Guo H-J, Xiong L, Luo J, Wang B, Lin XQ, Chen X-F, Chen X-D (2016) Bacterial cellulose production from the litchi extract by Gluconacetobacter Xylinu. Prep Biochem Biotechnol 46(1):37–41. https://doi.org/10.1080/10826068.2014.958163

Yanti NA, Ahmad SW, Ramadhan LOAN, Jamili M, Walhidayah T, Mamangkey J (2021) Properties and application of edible modified bacterial cellulose film based sago liquid waste as food packaging. Polymers 13(20):3570. https://doi.org/10.3390/polym13203570

Ye J, Zheng S, Zhang Z, Yang F, Ma K, Feng Y, Zheng J (2019) Bacterial cellulose production by Acetobacter xylinum ATCC 23767 using tobacco waste extract as culture medium. Bioresour Technol 274:518–524. https://doi.org/10.1016/j.biortech.2018.12.028

Zeng X, Small DP, Wan W (2011) Statistical optimization of culture conditions for bacterial cellulose production by Acetobacter xylinum BPR 2001 from maple syrup. Carbohyd Polym 85:506–513. https://doi.org/10.1016/j.carbpol.2011.02.034

Zeng M, Laromaine A, Roig A (2014) Bacterial cellulose films: influence of bacterial strain and drying route on film properties. Cellulose 21(6):4455–4469. https://doi.org/10.1007/s10570-014-0408-y

Zhao Y, Li J (2014) Excellent chemical and material cellulose from tunicates: diversity in cellulose production yield and chemical and morphological structures from different tunicate species. Cellulose 21(5):3427–3441. https://doi.org/10.1007/s10570-014-0348-6

Zhou L, Fu J, Bian L, Chang T, Zhang C (2022) Preparation of a novel curdlan/bacterial cellulose/cinnamon essential oil blending film for food packaging application. Int J Biol Macromol 212:211–219. https://doi.org/10.1016/j.ijbiomac.2022.05.137

Zihare L, Spalvins K, Blumberga D (2018) Multi criteria analysis for products derived from agro-industrial by-products. In: International scientific conference environmental and climate technologies, vol 147, pp 452–457. https://doi.org/10.1016/j.egypro.2018.07.045

Biomass Valorization for Bioenergy Production

Sri Suhartini, Nimas Mayang S. Sunyoto, Hendrix Y. Setyawan, Novita Ainur Rohma, Elviliana, Dinda Agustin Pratiwi, Fitria, Efri Mardawati, Kasbawati, and Lynsey Melville

Abstract Biomass is currently seen as a potential to be used as bioenergy resources. Its high availability and renewability generate extensive interest for further valorization. In Indonesia, research and development of transforming biomass into bioenergy via different pathways is expanding. Conversion of biomass via physical/mechanical, biochemical, and thermochemical offers produces bioenergy in the form of liquid (i.e., biodiesel, bioethanol, and bio-oil), gasses (i.e., biogas and syngas), and solid (i.e., biopellets, biochars, and briquettes). These types of bioenergy are essential for substituting fossil-based fuels, hence have positive impacts on reducing carbon emissions and climate change. Different mechanisms of process occur during the conversion. Specific measures to the influencing factors are crucial to ensure the optimum performance efficacy. This chapter discusses various bioenergy routes from

S. Suhartini (✉) · N. M. S. Sunyoto · H. Y. Setyawan · N. A. Rohma · Elviliana · D. A. Pratiwi
Department of Agro-Industrial Technology, Faculty of Agricultural Technology, Universitas Brawijaya, Malang, East Java 65145, Indonesia
e-mail: ssuhartini@ub.ac.id

Fitria
Research Centre for Biomass and Bioproducts, National Research and Innovation Agency (BRIN), Cibinong, West Java, Indonesia

Department of Biological Systems Engineering, Washington State University, Richland, Washington, USA

E. Mardawati
Department of Agro-Industrial Technology, Faculty of Agro-Industrial Technology, Universitas Padjadjaran, Sumedang, West Java, Indonesia

Kasbawati
Applied Mathematics Laboratory, Department of Mathematics, Hasanuddin University, Makassar, Sulawesi, Indonesia

L. Melville
Bioresource and Bioeconomy Research Group, Faculty of Computing, Engineering and Built Environment, Birmingham City University, Birmingham, West Midlands, UK

S. Suhartini
Centre of Excellence in Bioenergy and Biorefinery, Faculty of Agricultural Technology, Universitas Brawijaya, 65145 Malang, Indonesia

M. A. R. Lubis et al. (eds.), *Biomass Conversion and Sustainable Biorefinery*, Green Energy and Technology, https://doi.org/10.1007/978-981-99-7769-7_4

biomass substrates from the process' mechanisms to examples, in particular anaerobic digestion, transesterification, fermentation, densification, and thermochemical pathways.

Keywords Anaerobic digestion · Transesterification · Fermentation · Pelletizing · Pyrolysis

1 Introduction

Energy demand continues to increase in line with the increasing rate of population growth and economic growth. For example, according to the Ministry of Energy and Mineral Resources of the Republic of Indonesia (Kementerian ESDM 2021), energy demand in Indonesia is expected to continue to increase until 2030 with an average annual increase of 1.6%. Energy is used in various fields, one of which is as fuel for industrial, transportation, and household activities. The most widely used energy source to date comes from fossil fuels (Ulfa et al. 2021). Fossil fuels are non-renewable fuels and their availability is limited, thus over time there will be scarcity. In addition, fossil fuels can produce emissions (such as NO_x and SO_2), when accumulated, will cause acid rain (Zikri et al. 2018). These problems have an impact on climate change (Istiani et al. 2021).

Also, along with the increasing demand for energy, it is necessary to use renewable energy as an alternative to fossil fuels. Renewable energy can be created from the utilization of biomass including agro-industrial, plantation, agricultural, and forestry wastes. Biomass is an organic material obtained from living things in nature (Calvo-Serrano et al. 2019). Usually, biomass comes from products that are renewable, urban waste, forests, and residues originating from the agricultural sector. Biomass-derived from plants is obtained by the reaction between carbon dioxide and water, air, and the sun through the process of photosynthesis to produce carbohydrates which form a group of biomasses. Biomass is composed of three main components, 40–60%(w), 10–30%(w), and 20–30%(w) (Reyes et al. 2021). Biomass-derived from forest and agricultural plants is generally composed of components of cellulose, hemicellulose, lignin, fat, protein, starch, and sugar. Biomass also contains water and organic components such as nitrogen, sulfur, alkali, alkaline earth, heavy metals, magnesium, chlorine, and potassium.

According to Kusumaningrum and Munawar (2014), biomass has the potential to be used as an energy source as it is abundant, cheap, environmentally friendly, and renewable. While Tursi (2019) defined biomass is organic material produced directly or indirectly by living organisms and is available in a renewable manner. Use of biomass as bioenergy sources can minimize the negative impacts on the environment (Febrianti et al. 2020). Also, the benefits of using biomass-based bioenergy are that it does not emit sulfur that causes acid rain, produces less amount of ash than coal, is abundant in availability, renewable, and relatively fast to produce (Munawar and Subiyanto 2014).

Sources of biomass include agricultural residues such as rice husks, corn cobs, and corn fiber, remaining forest products such as wood, agricultural and plantation industry residues such as palm oil waste including empty oil palm bunches (EFB), shells, fronds, palm tree trunks, wood processing industry sawdust, cocoa shells, sugarcane bagasse, pulp sludge, urban waste such as used paper, and dry leaves (Febrianti et al. 2020). Agricultural industry waste is another source of biomass (Wang et al. 2022). This waste can be detrimental to the environment, like other industrial wastes. The form of this waste can be solids, gases, or liquids. In handling agricultural and industrial waste, it is necessary to group it based on its form and on its raw components, such as carbohydrates, proteins, or fats. Indonesia has potential biomass resources of up to 50,000 MW which are dominated by plants and industrial organic waste. Installed biomass capacity in Indonesia reaches 312 MW (Yana et al. 2022). Biomass is formed by the main lignocellulosic compounds including cellulose, hemicellulose, and lignin. These three compounds function to form complex chemical bonds to become the basic material for plant cell walls. The cellulose content in the biomass varies and ranges from 40 to 50%. Hemicellulose in biomass has a percentage of about 15–35%. Lignin in biomass has a percentage of around 10–25% (Tursi 2019), and is considered as a complex structure that covers the cell walls (Hermiati et al. 2010). The lignocellulosic content contained in biomass can be used to produce energy through the biomass conversion process. Biomass is unique in nature and has different concentration of lignocellulosic compounds, as shown in Table 1 (Hermiati et al. 2010).

Biomass sustainability is critically important to be created in the supply of energy sources. The government's role as a policymaker is vital in developing and implementing renewable energy sources. The support from the Indonesian government is by releasing a Minister of Energy and Mineral Resources Regulation regarding the implementation of co-firing to increase biomass economies of scale and reduce dependence on coal.

Table 1 Lignocellulose content in various biomass

Biomass	Cellulose (%)	Hemicellulose (%)	Lignin (%)
Cocoa shell	36.47	18.90	60.67
Sugarcane bagasse	50	25	25
Corncob	41	36	16
Wheat straw	30	50	15
Empty fruit bunches of palm oil	44.21	16.68	35.51
Rice husk	34.4	24.3	19.2

Source Hermiati et al. (2010)

2 Bioenergy from Biomass Resources: Mechanisms and Example

Biomass as a fuel without going through the conversion process has bad properties due to its low density. Direct use of biomass also causes respiratory problems due to carbon monoxide (CO) and sulfur dioxide (SO_2) gases. To reduce this, biomass can be converted into energy such as heat, electricity, liquid, solid, and gaseous fuels (Moura et al. 2022). Various opportunities routes for biomass valorization to bioenergy, either to physical, biochemical, thermal, mechanical, thermal, and thermochemical pathways (Fig. 1). These can be done in several ways including densification, gasification, carbonization, pyrolysis, and anaerobic digestion processes. Each of these conversion principles consists of a specific process that is useful to convert biomass into bioenergy. These routes can produce bioenergy in different forms such as liquids (i.e., biodiesel, bioethanol), gases (biogas, syngas), and solids (biopellet, biochar, briquettes) (Parinduri and Parinduri 2020).

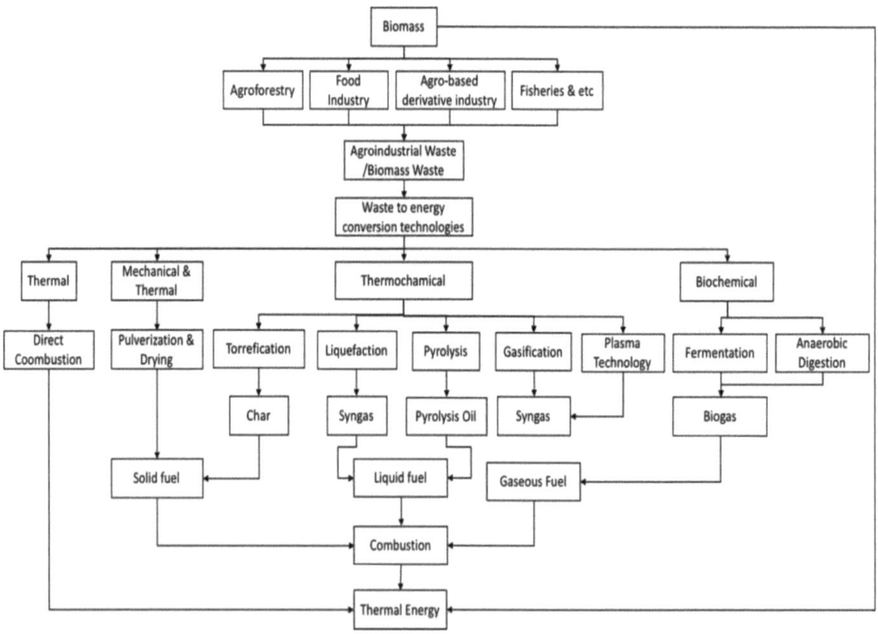

Fig. 1 Biomass conversion to energy pathways

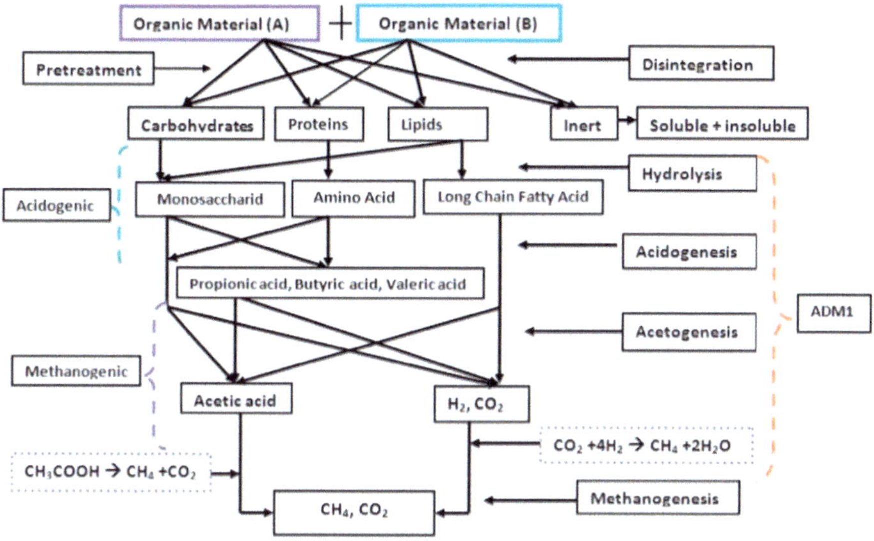

Fig. 2 Stages in the anaerobic digestion process (Hagos et al. 2017)

2.1 *Anaerobic Digestion*

2.1.1 Definition and Mechanisms

The anaerobic digestion (AD) process consists of four stages: hydrolysis, acidogenesis, acetogenesis, and methanogenesis, where the dominant biogas containing methane gas is produced in the methanogenesis process. The stages of the anaerobic digestion process are shown in Fig. 2. Biogas produced can later be converted into electrical and heat energy or developed into biomethane (Montingelli et al. 2016). Besides biogas, AD also generated organic residues, known as digestate. The AD system enhances the recovery of nutrients (N and P) in the final product. The resulting P composition has the potential to be reused as fertilizer (Li et al. 2022).

2.1.2 Hydrolysis

Hydrolysis as the initial stage of AD is a process of breaking down large and complex polymers such as fats, oils, starch, polysaccharides, and proteins into monomeric or oligomeric components such as amino acids, long-chain fatty acids, and simple sugars. This process is carried out by extracellular enzymes that are owned by microorganisms into the total volume of solution (Thanarasu et al. 2022). This initial stage, known as biological pretreatment, will also produce intermediate products such as ammonia and long-chain fatty acids that can affect the AD process. Thus, the adjustment process in the microbial consortium as well as the bioreactor needs to be carried out. Operational strategies include co-digestion, pretreatment with enzymes, chemically or mechanically, as well as dividing the AD stages into two major or other stages. Hydrolysis is also known as one of the rate-limiting reactions in which components of complex chemical compounds such as fats, polysaccharides, and proteins are deconstructed into monomers. Hydrolyses are included as individual monomers from suspended solids. The results of this modeling can also determine sufficient exposure time to provide pretreatment as an effort to increase biogas production (Hirmiz et al. 2019). Some preventive steps to avoid biogas production barriers are carried out in several ways, namely, by adding inoculum and adjusting the pH at the hydrolysis–acidogenesis stage. The result of this adjustment can increase bioenergy production two times higher than usual.

2.1.3 Acidogenesis

The acidogenesis stage begins with the absorption of monomer and polymer products into the cell membrane of acidogenic bacteria into volatile fatty acids (VFA), alcohol, and other inorganic compounds such as CO_2, H_2, H_2S, and NH_3 (Richard et al. 2019). VFA control is an important factor in biogas production, because the accumulation of these compounds will reduce the effectiveness of biogas production (Thanarasu et al. 2022). Microorganisms involved in this stage include *Acetivibrio, Bacteroides, Clostridium, Eubacterium, Lactobacillus,* and *Streptococcus.* One of the syntrophic microorganisms, namely, *C. Cloacimonas*, plays a key role at this stage. This microbial ability is related to the production of methane gas from the hydrogen chain. The coenzyme associated with it is Methanoculleus (Niu et al. 2022). The main challenge in this stage is to avoid VFA conversion to methane gas. The focus of research must emphasize this issue to avoid inhibiting biomethane production. The study of (Al-Sulaimi et al. 2022) showed a negative effect of VFA acidification on production efficiency and biogas biodegradation from waste sludge substrates, especially for bioreactor conditions that use thermophilic reactor conditions. The high concentration of VFA will cause the pH to drop and cause an unfavorable reactor environment for the methanogen process.

2.1.4 Acetogenesis

Substrate through subsequent breakdown which produces H_2 and CO_2 accompanied by acetate is referred to as the process of acetogenesis. Acetogens are an important step in the biodegradation process of organic matter to maintain the efficiency of the biogas production process. Acetogenic bacteria will use production from hydrolysis and oxidize pyruvate. Pyruvate is part of the intermediate product in the AD process and will be converted to acetate (Thanarasu et al. 2022). Other results such as the hydrogen gas produced will initiate syntrophic relationships in the AD-transfer system between hydrogen species (Meegoda et al. 2018). Acetogen species can be divided into several parts, namely, hydrogen generators and proton-to-hydrogen reducers. Most of the known homoacetogen species are from the genera *Acetobacterium, Acetoanaerobium, Acetogenium, Butyribacterium, Clostridium, Eubacterium,* and *Pelobacter* (Borja and Rincón 2017). In a single stage, microorganisms that play a role in acetogenesis and methanogenesis develop simultaneously. The two microorganisms require different environmental conditions. So, currently, another study is being carried out regarding the optimum conditions for both bacteria in single conditions or in two-stage conditions (Qian et al. 2019).

2.1.5 Methanogenesis

The end product of acetogenesis is acetate, H_2, and the methylated mixture is used by methanogenic bacteria for gas production. Acetate used by methanogens can become a methyl group and CO_2, and then this group is reduced to methane using electrons provided by the carboxyl group (Thanarasu et al. 2022). This methanogenesis takes place at the end and takes a long retention time between 15 days and 3 months (Tabatabaei et al. 2020). Microorganisms that play a role in this stage are methanogens which are a group of archaebacteria that are unicellular and very sensitive to the presence of oxygen (Borja and Rincón 2017). One of the inhibiting factors in this stage is the presence of excessive ammonia gas from a high nitrogen content substrate as the main inhibitor. The toxic level of this gas is caused by un-ionized ammonia, due to its ability to enter the bacterial cell membrane to disrupt the balance of potassium and methane protons. The handling practice is to provide additional water, but this also increases operational costs and is inefficient. Because of this, several research and industrial projects have begun to focus on adapting environmental conditions to these methanogenic microbes (Capson-Tojo et al. 2020). There are two pathways involved in this step involving acetic acid and carbon dioxide, the two main products from the previous step to produce methane gas (Kirk and Gould 2020):

$$\textit{Acetotrophic}\ \text{Reaction} \quad CH_3COOH \rightarrow CH_4 + CO_2 \Delta G0' = -30,9\ \text{kJ/mol}$$

$$\textit{Hydrogenotrophic}\ \text{Reaction} \quad 4H_2 + CO_2 \rightarrow CH_4 + 2H_2O\ \Delta G0' = -135,4\ \text{kJ/mol}$$

The main product resulted from AD is called biogas. Biogas is a mixture of gases produced by methanogenic bacteria through an anaerobic digestion process that transforms complex organic compounds with the help of microorganisms under anaerobic conditions (Gonzalez-Gil et al. 2018). Complex organic compounds can be found in biomass. However, biomass used as raw material for biogas generally has low economic value or is waste, such as industrial agriculture waste (agro-industry), forest product waste, and livestock waste. Many gases contained in biogas include methane (CH_4), carbon dioxide (CO_2), ammonia (NH_3), water vapor (H_2O), hydrogen sulfide (H_2S), methyl siloxane, nitrogen (N_2), oxygen (O_2), volatile components (VOC), carbon monoxide (C.O.), and hydrocarbons. The most extensive content in biogas is 50–80% CH_4 and 20–50% CO_2 from the amount of gas produced. Biogas' energy depends on methane concentration, where the higher the methane content, the greater the energy content or calorific value (Neves et al. 2004). Physically, methane is colorless, odorless, and flammable. Chemically, methane is composed of one C atom and four H atoms. This compound is very stable due to the presence of C–H bonds and requires an energy of 438.8 kJ/mol to break it down (Park and Lee 2013). A simple purification process can purify biogas by passing the gas through a NaOH solution to bind CO_2 (one impurity) and converting it to precipitate sodium bicarbonate (Na_2CO_3) H_2O. This process will increase the purity of methane (CH_4). Based on research conducted by Lasocki et al. (2015) using 1 M NaOH for 10 min can increase the concentration of CH_4 and remove H_2S and CO_2 altogether.

2.1.6 Enhancing ADPprocess: Pathways

As AD technology is primarily dependent on microbial activity, ensuring the feedstock has higher accessibility for the microorganisms is a must. Pretreatment on biomass especially with higher lignocellulosic content should be considered before treated with AD. There are common pretreatment methods, i.e., physical, mechanical, biological, and chemical. Ampese et al. (2022) used the dried apple pomace as AD feedstock VS that through physical pretreatment using a mill to reduce its size resulted in 36.61 L CH_4/kg VS that was predicted to generate electricity, heat, and carbon reduction in 1.92 kWh/ton, 8.63 MJ/ton, 0.62 kg, and CO_{2eq}/ton, respectively. Steam explosion (STEX), as part of physicochemical pretreatment, is being investigated on many energy crops and ecological plants. A lignin-rich macrophyte (*Potamogeton maackianus*) under semi-continuous AD is pretreated via STEX and resulted in enhancing the hydrolysis efficiency from 19% (108 $\pm$ 31 L/kg VS) up to 50% (200 $\pm$ 36 L/kg VS) (Akizuki et al. 2022). The STEX gives an easier breakdown for the microbial activity and enhances the biomethane yield.

Minimizing the energy input and faster reaction is part of merits of chemical pretreatment, but another consideration that should be noted is the toxicity profile from the materials that has to meet the principle of green chemistry. Lomwongsopon and Aramrueang (2022) seek the biogas potential from cassava pulp pretreated using mild-chemical with concentrations 0.1–0.3 w/v% of H_2SO_4, HCl, NaOH, and KOH. KOH under 2.0 w/v% showed the highest methane yield of 324 $\pm$ 4 L/kg VS. Biological pretreatment as one of the favorable methods to explore is generally utilized microorganism activity to degrade and reduce the thickness of cell wall of lignocellulose biomass. For instance, stover as the feedstock is being investigated via AD and given the biological pretreatment using microbial consortium to positively affect the activity of functional microorganisms. A 62.85% increment on the peak phase is shown than the untreated stover and significant methane production is resulted (Zhao et al. 2019).

Along with its development, AD with one feedstock produces low yields and is inefficient (Beniche et al. 2021). The application of Anaerobic co-digestion (AcoD) which is an AD system that uses two or more substrates to improve the process, stability, or production of biogas is preferred because it can increase biogas production by diluting toxic substances and provide a synergistic growing environment for microorganisms and can also handle multiple wastes at once (Taboada-Santos et al. 2019). For instance, the study of (Wang et al. 2022) the anaerobic co-digestion between excess sludge and chicken manure has effectively increased the methane yield in mesophilic and thermophilic temperatures up to 123.1 L/kg VS and 171.3 L/kg VS, respectively. Proper mixing ratio to have the balance support factor (i.e. pH, C/N) is the main key to enhance the biogas production.

2.1.7 Examples

Commodity of palm oil has been the major crops in some tropical countries, i.e., Indonesia, Thailand, and Malaysia as the three largest producers of biomass. From the industrial process, empty fruit bunches of palm oil (EFB) are the generated waste that could be valorized via AD route to produce biogas. (Suhartini et al. 2020) evaluated the untreated EFB biomass with batch AD under mesophilic temperature for 30 days and the specific methane potential was 0.110 $m^3/kgVS_{added.}$ Similarly, (Hidayat et al. 2020) investigated the biomass potential with fungal pretreatment to manage the higher content of lignin which is one of the barriers of AD steps. The pretreatment is capable to disrupt the lignocellulose content and enhance methane production than the untreated along with increment in the organic matter.

Interest in utilizing the macroalgae along with other feedstock has gained more studies with various perspectives either with co-digestion or additional of other bioactive compounds to obtain higher methane yield. In our previous studies, using G. verrucosa with the addition of tofu dregs substrate, a higher SMP was produced in a 20% tofu dregs mixture (compared to only 10%) with a value of 120 L CH_4/kgVS (Suhartini et al. 2022a).

In line with the synergetic effect from co-digestion, a study by Wickham et al. (2019) showed a positive effect of AcD, with a stable composition of methane gas in the biogas of 60–65%. In this case, AcD does not show any negative implications such as disrupting the stability of the reactor or other toxic gases. However, not all AcoD can provide optimal results, in the study (T.A.S et al. 2020) compared mono and co-digestion substrates and resulted in mono-digestion gave higher yields of 0.61 $m^3/KgVS_{reduced}$ compared to AcoD of 0.39 m^3 $/KgVS_{reduced}$. Therefore, AcoD needs to be carried out to see the relation impact on biomethane production either antagonistic or synergetic. Apart from the biomass potential on biogas production and pretreatment work, for the pilot and industrial scales of AD implementation, sometimes the foaming issues are the major concern. This condition could be due to the presence of unwanted microorganisms and other operational factors. Trace element supplementation, adjustment of organic loading rate, addition of anti-foams, and water dilution could be the solution for the foaming issue (Suhartini et al. 2019).

2.2 *Transesterification*

2.2.1 Definition and Mechanisms

Transesterification as the most common mechanism to produce biodiesel will be discussed in detail in this section. Figure 3 shows the main process that starts from long-chain fatty acids react with alcohol with catalyst and through three stepwise that ultimately produce methyl esters as biodiesel and glycerol as the co-product (Meher et al. 2006). In addition, this reaction is divided into two sections: catalytic or non-catalytic. As shown in Fig. 4, catalytic process could be conducted with homogeneous catalysis, heterogeneous catalysis, or combinatorial catalysis. Acids, alkalis, and enzymes are the materials to be utilized for synthesizing biodiesel under catalytic conditions. FAME produced from various substrates of waste is attracting more researchers to utilize the low cost of feedstock and minimize the waste presence. This study used chicken fat oil and waste cooking oil with acid catalyst and gained the FAME yield of 90.8% and 92.3%, respectively, and categorized as the suitable feedstock for biodiesel feedstock (Shatesh Kumar et al. 2020). Meanwhile, the non-catalytic is carried out with supercritical fluid. This process could be produced more than one product such as other value-added by-products, i.e., triacetin. Under the economic and environment assessment, the supercritical fluid, the progress to address the saponification issue and other unwanted waste generation through other steps, which provides more advantages will have more feasibility if using the glycerol-free process to gain high revenue (Ang et al. 2014).

Table 2 shows the advantages and disadvantages of each process of transesterification under catalytic or non-catalytic process. It should be noted that every process will work effectively under the specific feedstock. The heterogeneous catalytic process emerged to combat drawbacks of homogeneous catalyst role, but if homogenous catalyst coupling with other methods, i.e., ultrasound-assisted extraction using the *Annona squamosa seed oil* will result in rich oil content of 94.7% yield (Sundaramahalingam et al. 2021). The study revealed that the combined methods seem more feasible from energy balance and have a positive impact on the economic view.

Transesterification of biomass or organic waste results in biodiesel. Biodiesel as a green fuel produced from animal oil, vegetable oil, and algae oil has the physicochemical similarity with fuel diesel but has low energy density and generates lower

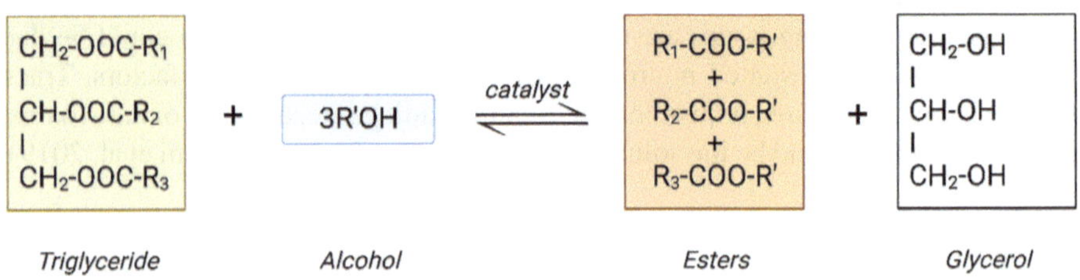

Fig. 3 Transesterification reaction (Meher et al. 2006)

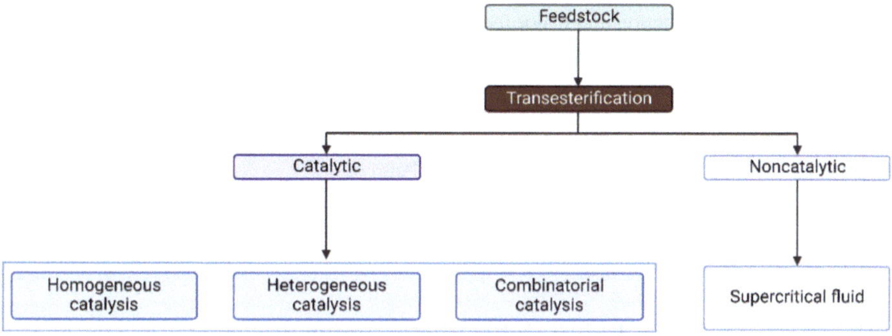

Fig. 4 Overview of transesterification in biodiesel production processes. Adapted from Nayab et al. (2022)

Table 2 Overview of transesterification in biodiesel production processes

Process	Materials	Advantages	Disadvantages	References
Homogeneous catalysis	H_2SO_4, NaOH, $NaOCH_3$, KOH	– Faster reaction rate – Required mild reaction condition – High corrosiveness – Non-reusability – Difficult to be separated	– Contaminated the outcome, affected the separation process – Soap formation – Low purity – Toxic waste generation	Sakai et al. (2009), Kaur and Ali (2011)
Heterogeneous catalysis	CaO, CaO–KOH, CaMgO, CaZnO	– Reusable – Easy separation – Reduces wastewater generation – Lower toxicity	– Partial recoverability and reusability – Energy and waste conservation problem – Longer retention time	Lee et al. (2014)
Combinatorial catalysis	Mixed catalysis	– Increase the yield – Decrease the toxic product	– High expenses	Kim et al. (2008)
Supercritical fluid	Variance in pressure, temperature, molar ration between alcohol and lipid, and retention time	– Shorter duration progress – Efficient conversion step – Catalyst-free operation – Lower quality of feedstock can be used	– Operational cost	Deshpande et al. (2017), Makareviciene and Sendzikiene (2021)

Table 3 Biodiesel production and major feedstock (OECD-FAO 2021)

Country	Production (%)	Major feedstock
European Union (EU)	32.3	Palm oil, rapeseed oil, used cooking oils
United States	18.1	Soybean oil, used cooking oils
Indonesia	15	Palm oil
Brazil	12.2	Soybean oil
Argentina	5	Soybean oil

GHG emissions (Okechukwu et al. 2022). This substitution could tackle the climate change issues due to reliance on fossil fuel-based sources. Biodiesel or fatty acid methyl ester (FAME) is generated by transesterification and esterification, and is produced of mixed fatty acid esters (biodiesel) and glycerol as the by-products. Global biodiesel production is projected to reach 50 billion L in 2029, which was 53% higher than in 2016 (Rezania et al. 2019). EU has the first position as the largest country in the production of biodiesel followed by the USA and Indonesia. The major feedstock in detail is shown in Table 3. Biomass and its waste utilization in biodiesel production could lead to bioeconomy concept which offers sustainability pathways to generate renewable energy sources.

2.2.2 Pretreatment for Biodiesel Production

In order to enhance the efficient and effective result, biomass is pretreated either with mechanical, physical, chemical, or biological. This pretreatment to assure the extraction will give the maximum result. For instance, the lignocellulose biomass is rich in lignin content, and high crystallinity which will hinder the process of biodiesel production, especially with tolerance to microbe and chemical conditions. Biomass pretreatment for biodiesel production has been investigated in several studies. Biological pretreatment is being sought from other studies. It has been reported that the corn straw is pretreated with *Mycobacterium smegmatis* LZ-K2 to prove the microorganism's capability in reducing the lignin content (Zhang et al. 2019) and the enzyme system shown the possibility further to treat such lignin-rich biomass. In other pretreatment, i.e., dilute acid could produce ethanol and biodiesel. This case of wheat straw valorization under moderate conditions with 140–160 °C and 0.3–0.6% of sulfuric acid concentration could degrade the lignin content more than the other parts. Meanwhile, the physical pretreatment has also been investigated to enhance the yield of FAME. Priyadarshi and Paul (2018) conducted the advanced microwave technology (AMT) pretreatment in regard to achieve high yield of FAME with source from kitchen waste. It resulted to have 96.89 wt% as the maximum FAME yield. A similar pretreatment was also identified from the study to increase the yield of FAME and reported that 93 wt% of the FFA conversion is achieved with AMT as the pretreatment method to reduce water and the viscosity reduction (Idowu et al. 2019).

2.2.3 Bioeconomy Concept on Biodiesel Production

Majority of the world is facing crisis in energy, water, and food nexus. Most of all technology still lack sustainability and viability from the economic and environment view. For instance, in order to produce one renewable energy product, the energy balance resulted in negative values which is not possible to proceed to the pilot scale or industrial scale. This issue is affected by many factors, including feedstock characteristic and technology itself. Hence, implementing bioeconomy concept into the conversion technology is urged to be adopted. Not only produce one product but also more value-added products to balance the energy balance or generate more energy output. Same condition applied for this biodiesel production. Most studies have investigated the chance of producing more products.

Angelaalincy et al. (2023) reported the investigation on managing the environmental contamination from arsenic with phytoremediation and nanopartible-mediated adsorption while co-produce the biodiesel and increase up to 125% the lipid content and incline the FAME production. This technology combats the pollution issue and provides the new renewable energy outcome which promotes the biorefinery system. Jeyakumar et al. (2022) studied the third generation of biofuel, *Padina tetrastromatica* and *Sargassum swartzii* macroalgae, to produce bioethanol and the solid waste generated is being sought the biodiesel production potential. The result shows that both marine biomasses are capable of extracting ethanol yield and FAME yield more than 80%. Another example of biomass valorization to promote the circular economy is converting *Macaúba* husk into biodiesel and biomethane. The generated waste from biodiesel production is firstly pretreated by subcritical water hydrolysis to breakdown the lignocellulose compound and via anaerobic digestion (AD) to have the clean energy, i.e., biogas (Ampese et al. 2021). The pretreatment successfully improves the biodegradability of feedstock and the biogas production could be further utilized in combined heat and power scenarios to enhance the functionality. Another feedstock being identified with its potential to obtain the same output (biomethane and biodiesel) is the energy crop, sunflower plant. The study result shown the biodiesel yield and biogas production in 96.2% and 342.7 N mL/g VS, respectively (Ebrahimian et al. 2022). This study explored hot water, sodium carbonate, and phosphoric acid as the three physiochemical pretreatments to increase the hydrolysis step that is beneficial for AD technology and showed a positive effect to maintain the recalcitrant content of its biomass.

2.2.4 Examples

A study from Halim et al. (2022) regarding the extraction of oil and protein from the species *Nannochloropsis sp.* (microalgae) developed a hypotonic autolytic and osmotic incubation pretreatment method for easy penetration of microalgae cell walls up to 4.9 times compared to only using conventional methods. After pretreatment, cell disruption (CD) was continued with the mechanical disruption (HPH) or chemical disruption (pH 12) method. From the experiments, the cell walls that did not undergo pretreatment only experienced 12% CD. The remaining more than 80% indicates that the defense wall of *Nannochloropsis* cells is very thick and difficult to penetrate, so it requires pretreatment to weaken this defense and also increase CD performance (the next process) to maximize liberation products. The results of the pretreatment of autolytic incubation made CD HPH and pH 12 increase, respectively, by 2.4x and 4.9x. The increase in CD performance is due to the high efficiency of disruption which is characterized by the thinning of the cell wall due to its weak defense and makes the biomass more sensitive to the treatment it receives. Other pretreatment results from hypotonic osmotic shock also provide good output. The CD of HPH and PH12 increased by 1.6x and 1.4x, respectively. The low result of this osmotic performance is because this method further increases membrane permeability.

The capability of fixing CO_2 by microalgae is very high and this process can also synergize with each other to produce other energy products such as omega-3, biohydrogen, bioethanol, and biodiesel. The study of Srinuanpan et al. (2018) showed the potential for synergy in the production of biogas and also biodiesel by cultivating the microalgae *Scnedesmus sp.* CO_2 gas reduction of >96% and CH_4 levels obtained >98% are obtained and shown the effective of elaboration of converstion technologies. Another study examining the addition of bacterial culture in microalgae cultivation to increase the FAME yield was carried out by Kumsiri et al. (2021). The use of *Picicocus intestinalis* WA3 bacteria in the cultivation of microalgae *Tetradesmus obliquus* gives a 1.3x increase in biomass production, 1.39x chlorophyll content, and 1.55x lipid productivity. Another way to enhance the biodiesel yield is feedstock valorization. The mixed culture of microalgae (MC) and food waste have been used as a feeding source for the black soldier fly larvae (BSFL) to enrich its nutrients and the BSFL is utilized as the medium to extract biodiesel from it. The result shown that FAME yield of BFSL with MC feeding is higher up to 140% comparing to the food waste feeding (Mahmoud et al. 2022).

2.3 *Densification*

2.3.1 Introduction and Mechanisms

Densification, which defined as a way of developing material functions to increase the energy content per unit volume, heat capacity, and reduce combustion ash (Qadry et al. 2018). The densification process is the process of compacting biomass by pressing or pressing thus increasing the potential energy and product mass density (Pradhan et al. 2018). The densification process is mostly carried out on bulk materials or those that have irregular physical forms such as biomass. This is because biomass has a low natural specific energy content. The densification process has several advantages, including facilitating the handling, storage, and transportation processes, controlled particle size distribution for product uniformity and density, improved product composition quality, increased product calorific value per unit volume, and increased energy content (Gong et al. 2021).

The densification is divided into extruding, briquetting, and pelleting. Extruding process involves compressing the material on a screw (screw) or piston to produce a solid and compact product (Istiani et al. 2021). Briquetting process produces a product shaped like a tube with varying dimensions as needed. Pelleting process involves the flow of material from a rotating roll accompanied by pressure into the holes for the biopellets (dies) (Widjaya et al. 2019). According to Winata (2013), the pelletization process is drying and forming biomass using high pressure to produce cylindrical solid biomass with a maximum diameter of 2.5 cm. Pelletization is carried out to produce denser biomass with a smaller volume and higher density. The quality of solid biomass is influenced by the strength and durability of the particle bonds, die diameter, die temperature, pressure, and preheating of the biomass mixture. The pelleting process can be carried out using the cold method (or without a heating process).

One of the products that resulted from densification is biopellet. Biopellets are a form of renewable energy (Ulfa et al. 2021); in a pellet-shaped made from compressed biomass materials (Munawar and Subiyanto 2014). Biopellets have a cylindrical shape resembling a pipe and have a solid texture, but have a smaller size than briquettes, and generally used for residential stoves, boilers, gasifiers, etc. (Rusdianto and Choiron 2015). Biopellets are widely used by people in American and European countries for space heating in winter and boiler fuel in industry. The size of the biopellets is about 6–10 mm in diameter and about 10–25 mm in length. The physical characteristics of biopellets such as color depend on the raw materials used (Hadiyane et al. 2021). An example of biopellets from coffee waste is shown in Fig. 5.

The advantages of transforming biomass into biopellets include high fuel efficiency and consistency (i.e., reduce NO_x emissions, particulate matters, and volatile organic matters), produce fewer emissions than wood, relatively easy to manufacture, no risk of explosion (Rusdianto and Choiron 2015). Biopellets are a superior fuel due to their higher energy content than the original biomass material. Also, biopellets

Fig. 5 Biopellet from coffee waste (Hadiyane et al. 2021)

have a denser structure, are easy to handle, environmentally friendly, and easy to distribute (Qadry et al. 2018).

According to Frodeson et al. (2019), the two methods commonly used in the manufacture of biopellets are pelletization using screw pressing and hydraulic pressing. Threaded pressing can be done using a single-or twin-screw press. Pressing using a threaded press has several advantages including greater production capacity because the process takes place continuously and saves production time. The way a threaded press works is by applying the screw principle. The material is pressed using the thrust of a rotating screw, then it will be pushed out as a result of the applied pressure. The more the material goes to the end of the screw, the greater the pressure experienced by the material. According to Damayanti et al. (2017), hydraulic pressing is done by applying hydraulic pressure to the material of 2000 lb/in^2 to allow the material to be molded.

The process of making biopellets generally consists of several stages including preparation of biomass raw materials, pretreatment in the form of initial drying of raw materials and size reduction, densification process (pelleting), final drying of biopellets, and packaging (Istiani et al. 2021). According to Raudhatul Jannah et al. (2022), the size reduction is carried out with a chopping machine to make the size of the raw material range from 3 to 5 mm and then flouring is carried out to transform biomass material into powder. The pelletization process is carried out using a ring die pellet mill. The powdered biomass is put into the machine to be molded into biopellets according to the desired size. Drying of biopellets aimed to reduce the water content to 10–15% before packaging (Damayanti et al. 2017). Factors that affect the characteristics of pellets include characteristics and composition of biomass, adhesive, particle size, moisture content, densification equipment, pressing conditions, and post-production handling (Zikri et al. 2018).

The main parameter in determining the quality of fuel products is the calorific value. The higher the calorific value, the better the quality of the biopellets (Ulfa et al. 2021). High-quality pellets are generally obtained by densification under high pressure with the addition of an adhesive. In the pelleting process in medium capacity, effective adhesives are derived from natural and cheap ingredients such as sago flour or tapioca flour (Damayanti et al. 2017). The physical criteria for good biopellets include surface structure are smooth dense and solid, no moldy during long-term storage, and do not emit excessive smoke (Lubis et al. 2016). Also, biopellets with good quality must meet predetermined standards as regulated in SNI 8675:2018 (NSA 2018) about biomass pellets for energy, as can be seen in Table 4.

2.3.2 Examples

There are several studies that have examined the manufacture of biopellets from various types of biomass, whether with the addition of adhesive or not. An example is the research by (Damayanti et al. 2017), regarding the effect of sieve size and the addition of tapioca flour adhesive on the characteristics of biopellets from cocoa shells. The research design used was RAK with a factor of sieve size variation and

Table 4 Quality standard for biopellets in Indonesia based on SNI 8675:2018

No	Parameter	Unit	Requirement
1	Density	g/cm^3	Min. 0.8
2	Moisture	%	Max. 12
3	Ash	%	Max. 5
4	Volatile matters	%	Max. 80
5	Bound carbon content	%	Min. 14
6	Calorific value	MJ/kg	Min. 16.5

Source National Standardization Agency (2018)

the percentage of adhesive addition of tapioca starch. Sieve sizes used were 20, 40, 60, and 80 mesh. The percentage of tapioca flour adhesive addition used was 0, 10, and 20%. From the research, it was found that the optimal results were found in the treatment of a 20 mesh sieve size and the addition of 20% adhesive. The resulting quality for the parameters of moisture content and density complies with SNI 8675: 2018, while the parameters of ash content and calorific value do not comply with SNI 8675: 2018 concerning biomass pellets for energy.

Qadry et al. (2018) study about the characteristics of biopellets from a mixture of palm shells and sawdust. The method in this study is experimental. The composition of the mixed ingredients is 100% palm shells, 100% sawdust, 30% palm shells and 70% sawdust, 70% palm shells and 30% sawdust, 50% palm shells, and 50% sawdust. The sieve size used is 80 mesh. From the research, it was found that the best quality biopellets were found in the combined treatment of a mixture of raw materials 70% palm shell and 30% sawdust. Testing the characteristics of the resulting biopellets complied with SNI 8675:2018 for the parameters of density, moisture content, ash content, volatile matter content, and bound carbon content. Meanwhile, the calorific value parameter does not meet SNI 8675:2018 in several treatments.

Istiani et al. (2021) studied biopellet production from candlenut shell with a different mixture ratio of sago stem and sawdust. The treatment used was the composition of the ingredients including 65% candlenut shell, 5% sago stem, 5% sawdust, 25% tapioca adhesive; 55% candlenut shell, 10% sago stem, 10% sawdust, 25% tapioca adhesive; 45% candlenut shell, 15% sago stem, 15% sawdust, 25% tapioca adhesive; 35% candlenut shell, 20% sago stem, 20% sawdust, 25% tapioca adhesive. From the research results, it was found that the average density of biopellets was 0.29 g/cm^3, the average moisture content was 10.31%, the average calorific value was 4181 cal/g, the average ash content was 10.25%, and the average -the average volatile matter content is 71.31%. The best treatment was obtained from a mixture of 55% hazelnut shells, 10% sago bark, 10% sawdust, and 25% tapioca adhesive. The parameters of density and ash content obtained did not meet SNI 8675:2018 for several treatments.

Ulfa et al. (2021) studied the quality of biopellets from rice husk waste. This study used a factorial RAL model with particle size factors and adhesive variations.

Variations in particle size were 40, 60, and 80 mesh, with different adhesive concentrations at 25% and 30% tapioca flour. The results reported that the powder size and the amount of added adhesive and the interaction between the two did not significantly affect the density, moisture content, ash content, volatile matter content, and bound carbon content. However, the powder size, the adhesive concentration, and their interaction had a significant effect on the calorific value. The resulting density values ranged from 0.7 to 0.95 g/cm^3, with an average value of ash content (16.5–19.9%), moisture content (13.1–14.5%), volatile matter (57.3–63.6%), bound carbon content (4.7–9.8%), and calorific value (2781–3378 cal/g). The calorific value from the resulting biopellets still did not meet the parameter of SNI 8675:2018. Hence, in-depth studies to improve the quality are required.

A study by Raudhatul Jannah et al. (2022) on biopellets from rattan shavings and mixed sawdust biomass using sago adhesive. The research method uses RAL with factors including powder size and powder composition. The powder sizes used were 20, 40, and 60 mesh. The powder composition (on a wet weight basis) used was rattan (100), sawdust (100), rattan:sawdust (70:30), rattan:sawdust (50:50), and rattan:sawdust (30:70). The best biopellet results were obtained from the treatment with a composition of 100% sawdust and 40 mesh size. The quality of the biopellets produced for the parameters of density, volatile matter content, moisture content, and calorific value complies with SNI 8021:2014, while the parameters of ash content and bound carbon content do not comply with SNI 8021:2014.

2.4 *Fermentation*

2.4.1 Definition and Mechanisms

Fermentation is one of the main steps critical in the conversion of lignocellulosic or sugars into bioethanol. There are several research methods and conversion design configurations that are economical and environmentally friendly. The fermentation steps of second-generation bioethanol can be carried out in four ways, namely, separate hydrolysis and fermentation (Separate Hydrolysis and Fermentation or SHF), saccharification and simultaneous fermentation (Simultaneous Saccharification and Fermentation or SSF), saccharification and simultaneous co-fermentation (Simultaneous Saccharification and Co-Fermentation or SSCF) and consolidated bioprocessing (CBP) (Cardona and Sánchez 2007; Aditiya et al. 2016).

The Separate Hydrolysis and Fermentation (SHF) method is a method for making bioethanol in which the cellulose hydrolysis and fermentation stages take place in different reactors (Saini et al. 2015). Raw materials containing cellulose are hydrolyzed or saccharified and then followed by pentose fermentation (C5). Ethanol is then distilled and the remaining hydrolysate is flowed into the second reactor to get the hexose component (C6) fermented. Bioethanol is also distilled after hexose fermentation (Balat et al. 2008). This method has the advantage of optimizing the operating conditions at each stage. The Simultaneous Saccharification and Fermentation (SSF) method is a method for making bioethanol in which the cellulose hydrolysis stage and the pentose fermentation (C5) stage are carried out simultaneously in one reactor. Next, hexose (C6) fermentation will be carried out in another reactor (Cardona and Sánchez 2007). In addition to the lower operating costs of this method compared to the SHF method, the possibility of inhibition of enzyme-related products can also be avoided so that the bioethanol produced is higher (Saini et al. 2015). In this method, the fermentation of pentoses (C5) and hexoses (C6) is separated because microorganisms that utilize both sugars (pentoses (C5) and hexoses (C6)) are slower than microorganisms that only assimilate hexoses (C6). This method also prevents the reduction of monomers formed after the hydrolysis process, so SSF is claimed to produce higher bioethanol yields (Cardona and Sánchez 2007).

In the Simultaneous Sacharification and Co-Fermentation (SSCF) method, different microorganisms are mixed for pentose (C5) and hexose (C6) fermentation in one reactor. This method allows mixed culture microbes to carry out saccharification continuously without separation which is continued by the fermentation process into bioethanol (Cardona and Sánchez 2007). However, the ability to ferment pentoses (C6) together with hexoses (C5) is not widespread among microorganisms and is one of the biggest obstacles in the industrial production of ethanol using this method (Talebnia et al. 2010). The Consolidated Bioprocessing (CBP) method is the conversion of lignocellulose using one or more microorganisms into bioethanol in one unit operation without additional enzymes (Branco et al. 2018). The microorganisms used can produce saccharolytic enzymes (cellulases and hemicellulases) themselves, hydrolyze substrates, and are capable of fermenting both pentoses (C5) and hexoses (C6) so that they are considered the most cost-effective method (Aditiya et al. 2016). Despite its advantages that are more efficient in conversion and require less energy, this method has disadvantages, namely, longer conversion times and lower productivity (Miskat et al. 2020).

Bioethanol is a colorless liquid made through a fermentation process from carbohydrates which involves biological processes. As much as 95% of ethanol in the world is made with vegetable-based ingredients and the rest is made synthetically (Lane and Morrissey 2010). In industry, bioethanol is used as a raw material for industrial alcohol derivatives, mixtures of alcohol, basic ingredients for the pharmaceutical industry, and mixtures of vehicle fuels (Ishola et al. 2014). In developed countries, bioethanol is widely used as an ingredient in gasoline mixtures to increase the octane rating of fuel. Bioethanol is mixed with gasoline at volume fractions of 5, 10, and 85% (fuel name E5–E85). E85 fuel is used for vehicles with flexible fuels (flexible fuel vehicles or FFV), while the E5 and E10 fractions can only be used for vehicles without engine modifications (Bušić et al. 2018) According to data from FAO Agricultural Outlook 2015, the two highest bioethanol producers are Brazil and the United States. In Brazil, sales of FFV vehicles are sold consistently every year, indicating that the demand for bioethanol in Brazil is always there.

The production of bioethanol from lignocellulosic biomass is a development of second-generation biofuels. The use of lignocellulosic biomass has advantages and disadvantages. The advantage is that the substrate is not competitive with food ingredients while the disadvantage is that it requires high efficiency for breaking down lignin as a wrapper for cellulose and hemicellulose. Cellulose is the component most responsible for the production of bioethanol from lignocellulosic biomass (Prasad et al. 2019). In general, there are three main stages in bioethanol production, including pretreatment, hydrolysis, and fermentation. However, the production of bioethanol from biomass, one of which is OPEFB, requires longer stages, namely, pretreatment, hydrolysis, filtration, fermentation, and purification. The stages of the OPEFB conversion process into bioethanol can be seen in Fig. 6. Pretreatment is used to produce monomer from OPEFB which then has the potential to be used as raw material for bioethanol fermentation (Suhartini et al. 2022b). The choice of the pretreatment method can greatly affect the economy because increasing conversion efficiency tends to add significant overall costs (Rabemanolontsoa and Saka 2016). Less significant lignin removal can reduce the rate of hydrolysis and decrease the efficiency of the conversion process, therefore removal of lignin prior to hydrolysis is essential to ensure higher production of C5 and C6 sugars (Wan Azelee et al. 2014).

2.4.2 Pretreatment for Bioethanol Production

Lignin as an aromatic phenolic compound can be attacked by laccase enzymes so that its molecular structure is easily degraded. Laccase enzymes can attack phenolic lignin to form phenoxy radicals but cannot oxidize non-phenolic lignin directly because of its low redox potential. To this end, the laccase enzyme can oxidize low molecular weight mediators which act as electron carriers and diffuse into the insoluble lignin structure to oxidize them. Enzymatic pretreatment of biomass generally uses specific enzymes isolated from certain microbes. This pretreatment tends to be used to degrade lignin which can be carried out by complex lignin enzymes, consisting of lignin-peroxidase (Li-P), manganese-peroxidase (Mn-P), and laccase (Lac). These enzymes are capable of converting and breaking down lignin compounds into their constituent components, namely, conipheryl alcohol, p-cumaryl alcohol, and synapcyl alcohol. After lignin is degraded into its constituents, cellulose, and hemicellulose become more accessible (Ravindran and Jaiswal 2016).

Hendriks and Zeeman (2009) stated that pretreatment of lignocellulosic biomass can also be carried out using combination, thermal, steam, and chemical methods which include alkaline, acidic, and organosolv liquids. Physical pretreatment, i.e., cutting biomass into smaller sizes is widely applied to increase the surface area and reduce the degree of polymerization (Taherzadeh and Karimi 2008). On the other hand, chemical pretreatment, namely, the use of sulfuric acid (H_2SO_4) and sodium hydroxide (NaOH), showed excellent results for lignin degradation, reduced crystallinity, and degree of polymerization so as to increase the digestibility of cellulose (Tsabitah et al. 2014). In its application, H_2SO_4 and NaOH are widely applied in pretreatment of biomass to produce biogas and bioethanol although there are drawbacks, namely, the emergence of chemical waste that requires further handling. Biological pretreatments, such as ensiling using EM4 and ensiling using molasses have also been widely studied as alternative options for producing biogas and

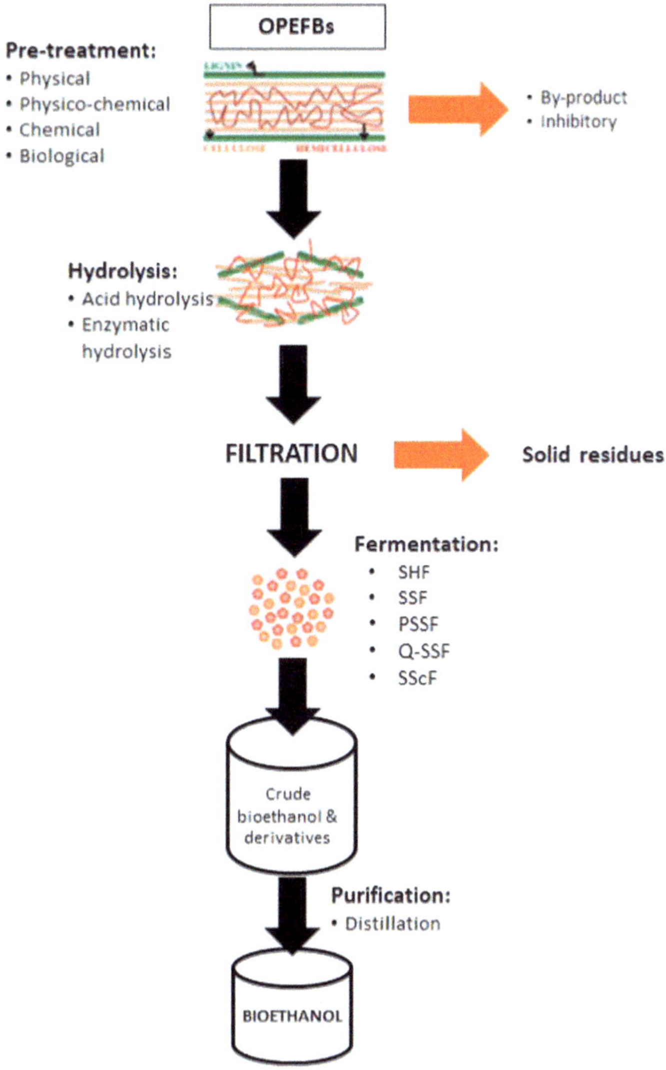

Fig. 6 Stages of the conversion process of OPEFB into bioethanol (Suhartini et al. 2022b)

bioethanol because of their economical and environmentally friendly advantages (Lunsin et al. 2018). The relatively slow rate of biological pretreatment is the main weakness of biological pretreatment. As for the use of the laccase enzyme, as a branch of biological pretreatment, it is proven to be able to degrade lignin in lignocellulosic biomass more effectively (Schroyen et al. 2015). The effectiveness and efficiency of pretreatment in degrading lignin content is highly dependent on the composition of the biomass and operating conditions.

2.4.3 Examples

Releasing the sugar content from its biomass from the complex structure via pretreatment should be facilitated with careful consideration into the economic viability and sustainability regarding the chemical or materials used. Sugarcane bagasse (SB) has been investigated as the feedstock for bioethanol production. Couple pretreatment has been applied with chemical and biological to assure the delignification could be higher. Imidazole, a green solvent, was used to take account for disruption in the complex of cell wall structure of SB. Total bioethanol production of 30%, with ethanol volume of 110.3 L/ton sugarcane has been reached and shows a great correlation between the methods to manage its lignin content (Valladares-Diestra et al. 2022). Following the physicochemical pretreatment, other lignocellulosic biomass, *Miscanthus* as part of energy crops is being exposed to some methods for having much lower lignin content. After pretreated with a novel chemical surfactant, mild chemical (H_2SO_4 or NaOH) in steam explosion was used as the pretreatment for *mischantus* and these combined methods have demonstrated an increment in lignocellulose accessibility and will promote the enzymatic saccharification in its biomass (Sun et al. 2020). Similarly, a study by Gao et al. (2021) coupling the steam explosion with green-liquor pretreatment and leading to the highest bioethanol yield of 20.3% (%dry biomass) compared with other reported bamboo processes. Conclusively, pretreatment is becoming one of the great strategies to enhance bioethanol production by giving more accessibility to the biomass structure and increasing the hydrolysis activity.

2.5 *Thermochemical*

2.5.1 Direct Combustion

Combustion is a chemical reaction between flammable matter and oxygen, resulting in the release of heat (Glassman 1977). Direct combustion is one of the oldest methods that is quite simple and is commonly used in thermochemical energy production. Direct combustion is an energy conversion carried out by burning sufficient agromaterial in the air to produce heat, steam, or electricity (Mandø 2013). The hydrogen and carbon contained in the fuel will react with oxygen and release energy. During combustion, there are substances that vaporize together with a portion of the carbon in the form of flammable gaseous hydrocarbons and the release of carbon monoxide with thermal degradation. Carbon monoxide is formed due to the reduction reaction of CO_2 with C, with the following chemical formula:

$$C + O_2 \rightarrow CO_2 \quad + 8084 \text{ kCals/kg of Carbon}$$
$$2C + O_2 \rightarrow 2\,CO \quad + 2430 \text{ kCals/kg of Carbon}$$
$$2H_2 + O_2 \rightarrow 2H_2O \quad + 28{,}922 \text{ kCals/kg of Hydrogen}$$
$$S + O_2 \rightarrow SO_2 \quad +2{,}224 \text{ kCals/kg of Sulphur}$$

In general, combustion is carried out in a furnace, steam turbine, or boiler (Ye et al. 2022) with a temperature range of 800–1000 °C. Three stages in the direct combustion of agro-materials are evaporation of the water, then distillation and combustion of volatiles occur after which carbon is bound and oxygen reacts at high temperatures (Lackner 2013). Energy is produced in the last two stages; this process is suitable for the conversion of all types of biomass that have a low moisture content (<50%). Several factors affect the direct combustion process, such as the volatile content, water, ash, and tar of the material (Lackner 2013). Direct combustion is considered carbon neutral, although it produces pollutants in the form of nitrogen oxides, sulfur oxides, carbon dioxide, and dioxins.

The heat generated from the direct combustion process is difficult to store, so it must be used immediately. The production cost of the direct combustion process is slightly higher than the pyrolysis and gasification processes. This is because the preparation of raw materials for direct combustion requires drying, cutting, and crushing before being fed into the boiler.

2.5.2 Torrefaction

Torrefaction is basically an improvement process from the direct combustion process on a large scale. Torrefaction is the heat degradation of biomass in inert or nitrogen (Wang et al. 2023). The temperature used for the torrefaction process ranges from 200 to 300 °C without oxygen within a few hours (depending on the condition of the biomass) in non-oxidized conditions with the aim of improving the physicochemical properties of biomass as a solid fuel. Torrefaction is a mild pyrolysis of the fibrous structure of crushed biomass (Han et al. 2022). Its heating value and hydrophobicity are increased to increase the stability of the biomass during storage. The main product produced in this process is charcoal.

Torrefaction can improve the characteristics of the resulting fuel by reducing the water content, increasing the heating value, reducing the oxygen/carbon ratio, and increasing the energy density. There are two kinds of methods used in torrefaction, the wet process and the dry process (Akbari et al. 2021). In wet torrefaction, agro-materials are treated with compressed hot water and produce solid fuel products, aqueous compounds, and gases. However, wet torrefaction has the disadvantage that it is still necessary to separate the excess water contained in the biomass. Dry torrefaction requires intensive drying of agro-materials at higher temperatures. The temperature used for dry torrefaction ranges from 230 to 300 °C in the absence of oxygen.

2.5.3 Pulverization and Drying

In the pulverization and drying process, the biomass is reduced in size and dried with the aim of increasing the quality of the raw material so that it is easy to burn (Sarnavi et al. 2023). Besides that, reducing the size can make it easier for the biomass to be put into the reactor/furnace, and drying can help the next process to be more efficient. The result of this process is solid fuel.

Based on the crushing principle, the biomass pulverizer is classified into hammer, blade, and combined types (Wei et al. 2014). According to the pulverization method and approach, the biomass pulverizer is classified into chopping, rolling cutting, and combined pulverizing types. According to the purpose and particle size of the crushed material, it is classified into coarse, twisted, and fine types.

Drying is a process of removing the water content contained in a material using evaporation to obtain a dry solid product (El-Mesery and El-khawaga 2022). There is a process of transfer of heat and water vapor from the surface of the material to the air without changing the shape of the material. There are two types of drying, namely, natural and artificial drying. Natural drying can be done by placing the material in the free air, and under the heat of the sun, the air is allowed to remove the moisture contained in the biomass, while the sun's heat can help evaporate water from the material. Artificial drying is done by utilizing the heat generated from combustion. This drying can be done using a dryer, such as using a tunnel dryer or oven. The advantage of artificial drying over natural drying is that the temperature and drying rate can be controlled according to needs and will not be affected by climate change. However, artificial drying also has the disadvantage of requiring a higher cost when compared to natural drying.

2.5.4 Liquefaction

Liquefaction is included in the thermochemical agro-material conversion technology (Kavitha et al. 2023). In the process, agro-materials are converted to liquid fuel products through a complex sequence of physical and chemical changes. Agro-materials are decomposed into small molecules. These small molecules are unstable and reactive and can be polymerized into oily compounds. Liquefaction can be done directly or indirectly. Direct liquefaction involves fast pyrolysis to produce tar and pyrolysis oil (Folkedahl et al. 2011). Indirect liquefaction involves the use of a catalyst to convert non-condensable gaseous products from pyrolysis or gasification into liquid products.

In direct liquefaction (Folkedahl et al. 2011), the biomass is directly converted into products without drying. Under these conditions, water remains in a liquid state and has a low viscosity and high capacity to dissolve inorganic compounds. Indirect liquefaction involves low temperature (300–350 °C) and high pressure (5–20 MPa) with a residence time of about 30 min and is often carried out using a catalyst in the presence of hydrogen, so this process can also be called catalytic liquefaction. The composition of the resulting pyrolysis oil and char is influenced by the liquefaction method used. The difference between liquefaction and pyrolysis is that pyrolysis usually occurs at higher temperatures and lower pressures, and it is necessary to dry the raw material first. This liquefaction technology is rarely used because it requires a more expensive reactor. From the liquefaction process, liquid fuels, commonly known as biofuels, such as ethanol and methanol, will be obtained. The most frequently used biofuel is ethanol which can be obtained from sugarcane, corn, and other grains.

2.5.5 Pyrolysis

Pyrolysis is a process in which organic materials containing carbon are decomposed thermochemically (Liu et al. 2022). The pyrolysis process involves little or no oxygen in a closed vessel. Oxygen is only required in small quantities for combustion to occur in order to produce enough heat to start the endothermic pyrolysis process. Then slowly, the oxygen in the vessel will run out because there is no oxygen supply from outside during the process, resulting in incomplete combustion in the vessel, and charcoal is produced. The mechanism of the pyrolysis process of biomass generally consists of hydrolysis reactions, primary reactions, and secondary reactions. Biomass will decompose thermally without oxygen and involves many reactions of biomass biomolecules.

Pyrolysis of biomass can be influenced by four factors, including water content, decomposition of cellulose, lignin, and decomposition of hemicellulose (Vuppaladadiyam et al. 2022). Pyrolysis can be carried out at high temperatures up to 800 °C and low pressures up to 700 kPa. In extreme pyrolysis, carbonization will occur, which will only produce charcoal. If the pyrolysis reactor is equipped with a condenser (Fig. 7), there will be a change in the form of the resulting gas to liquid commonly known as pyrolysis oil. The products produced during the pyrolysis are charcoal, pyrolysis oil, tar, and non-condensable and flammable gases, such as CO, CO_2, CH_4, and H_2 (Samer 2017).

A pyrolysis reactor is an equipment to decompose organic compounds by the heating process without direct contact with outside air at a temperature of 300–600 °C. This reactor is sealed to prevent excessive heat from escaping. The pyrolysis reactor is made of a cylindrical tank made of stainless steel, with rock walls installed between the reactor covers to minimize heat escape. The heat source used is a gas stove with LPG fuel, and there is a thermocouple connected to a pipe in the reactor so that the temperature can be measured and adjusted as desired. The large volume of the reactor will affect the amount of input that will be processed into pyrolysis oil.

The working principle of the pyrolysis reactor is to carry out the process of burning biomass without air (O_2). The decomposition of organic matter will occur as the temperature used increases. There are several types of reactors used in the pyrolysis process, namely, batch/semi-batch, fluidized bed reactors, fixed bed reactors, screw kilns, and spouted beds. Batch/semi-batch-type reactors are the type of reactor that is widely used by researchers because the design is simpler and easier to operate.

There are four stages of the pyrolysis mechanism based on the lignocellulose contained in the material, including removal of water content, decomposition of

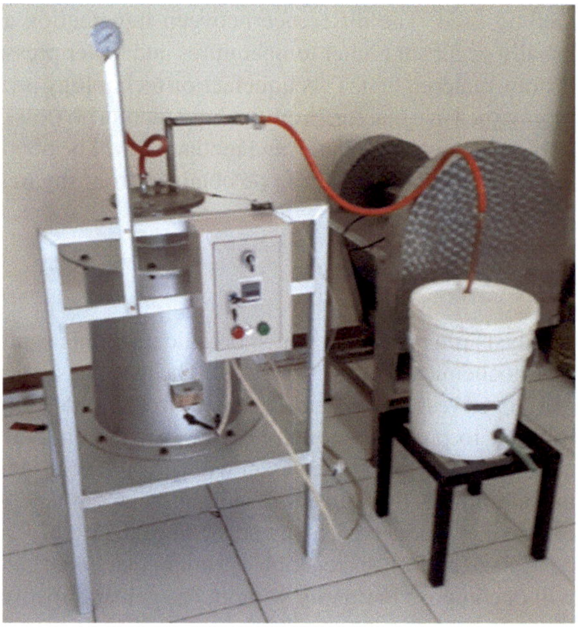

Fig. 7 A small pyrolysis reactor

hemicellulose, decomposition of cellulose, and decomposition of lignin. The dewatering step occurs at a temperature of less than 200 °C. If the water content in the material is too high, there will be a risk of producing large amounts of ash. The hemicellulose decomposition stage occurs at a temperature of 200–280 °C at this stage to produce syngas and minor pyrolysis oil. The cellulose decomposition stage occurs at temperatures above 240–350 °C. At this stage, syngas, pyrolysis oil, and minor biochar products are produced. Furthermore, the last stage is the decomposition of lignin which occurs at a temperature of 280–500 °C (Samer 2017). At this stage, the products of pyrolysis oil and biochar are produced.

The mechanism of the pyrolysis process can be seen in Fig. 8. Cellulose and hemicellulose mainly form volatile products on heating due to the thermal splitting of the sugar units. Lignin mainly forms char because it is not easily broken down into smaller molecules.

Some factors that can affect pyrolysis products include temperature, type of agro-material, heating rate, the proportion of oxygen, and equipment design. In general, the resulting charcoal is up to 50% of the dry agro material, the gas product ranges from 5 to 20% depending on the temperature used, while the tar product can be up to 25% by weight percent of the dry agro material. Charcoal has low sulfur and nitrogen content, making it much easier to store and transport than the agro-raw materials used. The resulting tar still contains water and can be separated by distillation or extraction. Pyrolysis oil obtained from the condensation of the resulting gas needs to be acid-neutralized so that it can be used as fuel. The quality and quantity of the product produced are also influenced by the temperature used during the pyrolysis

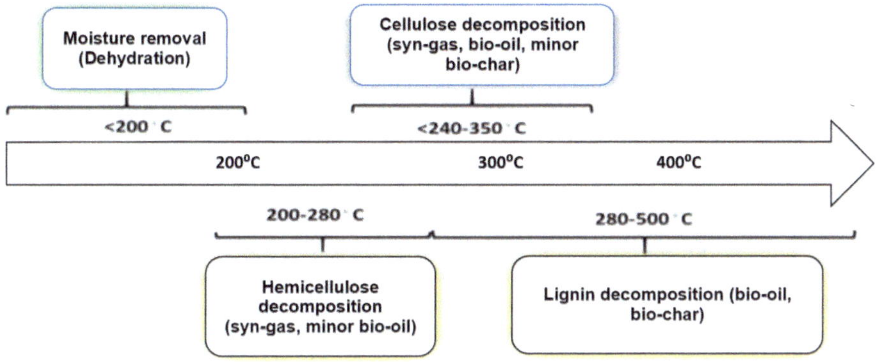

Fig. 7 Biomass degradation temperature (Samer 2017)

process. There are three types of pyrolysis based on temperature levels (Weir et al. 2022), as seen in Table 5.

Slow pyrolysis is a process in which biomass is heated at a slow temperature rate in an inert atmosphere to a maximum temperature (de Almeida et al. 2022). Slow pyrolysis takes place at low heating rates (0.1–0.8 °C/s), longer times (5–30 min or even 25–35 h), and temperatures around 300–550 °C. This pyrolysis process aims to obtain maximum bio-charcoal, syngas, and pyrolysis oil simultaneously.

Table 5 Characteristics of biomass thermochemical conversion technologies

		Pyrolysis		Liquefaction	Gasification
		Slow	Fast		
Feedstock					
Feed Size		Any	Small	Very small	Mixed, large
Moisture content		Low	Very low	Very low	50% max
Parameters					
Temperature °C		400–600	450–900	250–400	1000–1500
Pressure, bar		0.01–1	1	100–200	Up to 20
Maximum throughput achieved to date, dry/year		5	0.05	0.1	40
Product (dry basis on dry feed)					
Gas	Yield, %wt	Up to 40	Up to 70	20	100–250
	HHV, MJ/NM3	5–10	10–20	2–6	May 15
Liquid	Yield, %wt	Up to 30	Up to 70	Up to 50	Up to 3
	HHV, MJ/kg	23	23	30	23
Solid	Yield, %wt	30	Up to 15	Up to 25	Nil (ash)
	HHV, MJ/kg	30	30	30	–

This process can also eliminate the content of smoke production and the formation of solid products. This method can produce a solid uniform product with low water content and higher energy content than the initial biomass. Fast pyrolysis takes place at higher heating rates (10–1000 °C), and very short residence times (0.5–2 s), with temperatures around 850–1250 °C (Hu et al. 2022). This pyrolysis aims to get more pyrolysis oil products than bio-charcoal and syngas. In general, fast pyrolysis produces 60–75% liquid, 15–25% solid, and 10–20% gas. Flash pyrolysis is almost the same as fast pyrolysis, with some modifications and improvements (Cornelissen et al. 2009). This pyrolysis takes place at a very high heating rate of around 1000 °C, a residence time of 0.1–1 s, with a temperature of 800–1000 °C. This pyrolysis aims to get the maximum syngas product.

Pyrolysis technology has several advantages, including, namely, the process is simple, so it does not require high expertise to operate. This technology can eliminate carcinogenic compounds such as PAHs (Polycyclic Aromatic Hydrocarbons). This technology is quite efficient. Namely, it can produce three products with only one process required. The resulting pyrolysis oil product is able to overcome the problem of traditional smoking of fish/meat. The main components of pyrolysis oil are organic compounds and water (15–30%). Therefore, pyrolysis oil has the opportunity to be converted into fuel and chemical sources. The biggest component of bio-oil is hydrocarbons which are traditionally produced from petroleum. Given the dwindling reserves of petroleum, pyrolysis oil has the potential to be developed as a substitute fuel for petroleum. Pyrolysis oil can be used as fuel in furnaces or boilers. However, the high oxygen content in pyrolysis oil components can cause undesirable fuel characteristics, such as increased viscosity, corrosion, low heating value, and unstable during storage (can form a precipitate). Several organic compounds of high economic value found in pyrolysis oil are levoglucosan, toluene, xylene, limonene, and phenol. The constituent components of pyrolysis oil are highly dependent on the composition of the raw materials, the pyrolysis process, and the reaction conditions. Therefore, research on raw materials and pyrolysis process variations is important to extract chemical components with high economic value.

2.5.6 Gasification

Gasification is the partial (imperfect) combustion of solid fuels, which can produce flammable gas and ash (Valizadeh et al. 2022). Gasification of agro-materials belongs to the thermochemical conversion technology, in which solid agro-materials are converted into gaseous fuels. Under controlled conditions, characterized by low oxygen supply and high temperatures, most of the agro materials can be converted into producer gas fuels consisting of carbon monoxide (CO), hydrogen (H_2), carbon dioxide, nitrogen, and methane (CH_4). Producer gas is a mixture of flammable and non-flammable gases. Basically, the gasification process begins with a pyrolysis process at a temperature of around 150–900 °C, followed by an oxidation process of the pyrolysis gas at a temperature of 900–1,400 °C, then a reduction process at a temperature of 600–900 °C (Valizadeh et al. 2022).

The gasification process can produce relatively higher gas, which is around 85%, compared to pyrolysis, which can only produce around 35% gas. Gas fuel from biomass gasification can be used for cooking, both at the household level and in small industries in undeveloped areas. In addition, the gas is used to drive turbines and combustion engines, as fuel in steam boilers, and for lighting. The application of biomass thermal gas is a more efficient alternative and produces lower pollution than direct combustion, although it has the disadvantage of higher investment costs.

There are four different process stages during fuel gasification (Akhtar et al. 2018), among them are:

a. Fuel drying

Agrofuel enters through the top of the gasifier and then moves down during the process. Heat radiation from the combustion zone helps reduce fuel moisture entering the drying zone. Temperatures in this zone range less than 120 °C.

b. Pyrolysis

Dry fuel moves downward and is exposed to heating at higher temperatures, namely, above 200 °C. At this temperature, the fuel begins to lose its volatile content. In this zone, no air is allowed to enter. When the temperature reaches 400 °C, an exothermic reaction occurs, where the structure of the wood material or other organic solids begins to break down.

c. Combustion/Oxidation

The combustion reaction is an exothermic reaction resulting in a theoretical oxidation temperature of up to 1450 °C. The combustible substance of the fuel usually consists of carbon, hydrogen, and oxygen. In complete combustion, carbon dioxide is obtained from carbon in the fuel and water is obtained from hydrogen, usually in the form of water vapour. This heating causes some of the charcoal to be oxidized, and the rest undergoes a reduction process.

d. Reduction

Partial combustion products, such as water, carbon dioxide, and partially cracked pyrolysis products, move through the hot charcoal bed, where subsequent reduction reactions occur. The temperature in the reduction zone ranges from 800 to 1000 °C. The lower the temperature in the reduction zone, the lower the heating value of the gas.

2.5.7 Plasma Technology

Plasma is a gas that undergoes ionization in an electric discharge or a mixture of electrons, radicals, and negative and positive ions. Plasma technology can be used to decompose organic compounds from solid, liquid, or gas waste (Dimitrakellis et al. 2022). This technology is carried out by heating the waste using a device similar to an oven/microwave. The heating principle used in plasma technology is that in the combustion chamber, two high-voltage electrodes (about 10,000 V) are given, and then the waste is put into the furnace. The electrode is given an electric voltage so that plasma will form, which will decompose the waste. Plasma technology has the advantage that it does not require a large area, takes a short time, and does not involve chemicals.

In wastewater treatment, plasma technology plays a role in various oxidizing processes of the compounds (Dimitrakellis et al. 2022). A reaction occurs on ions and electrons in the plasma in the liquid waste, and then UV light and shockwave are formed. These ions and electrons contain very high energy causing the decomposition of water (H_2O) and producing active species such as OH, O, H, and H_2O_2. Active specifics are strong oxidants that can oxidize organic compounds and kill bacteria in liquid waste.

3 Conclusion

Biomass can be derived from any organic material including agricultural crop residues, agro-industrial waste, household waste, forestry waste, etc. Different types of biomass has also unique characteristics, which may play a key factor in the selection of the best conversion routes. Lignocellulosic biomass, considered as third-generation sources of bioenergy, is potential to be directly valorized into bioenergy, either via physical/mechanical, biochemical, and thermochemical. However, the results may not be optimum. Hence, pretreatment or other measures are often required to improve the conversion process' efficacy. Specific to anaerobic technology, it is widely adopted in the global world from small to commercial scale. The resulted biogas can be converted into electricity (and heat) or biomethane and the digestate into biofertilizer. Transesterification is the main process to transform biomass into biodiesel, mostly suitable for biomass with high in lipid or oil contents. While fermentation of lignocellulosic biomass to bioethanol remains a challenge, therefore integration with pretreatment and an improved hydrolysis step are necessary. Densification of biomass into biopellets can be one of the alternatives with lower operating cost, however, the risk of carbon emission from burning the fuels need to be considered. Thermochemical pathways offer greater and faster conversion to bioenergy, yet a high cost for the capital investment and operation may hinder the technology for commercial application in Indonesia.

Acknowledgements The authors would like to thank the Department of Agro-industrial Faculty of Agricultural Technology, Universitas Brawijaya, for the support in writing this book. We would like also to thank *Pusat Kolaborasi Riset Biomasa dan Biorefinery* for the opportunity to join in this publication.

References

Abou El-Nour KMM, Eftaiha A, Al-Warthan A, Ammar RAA (2010) Synthesis and applications of silver nanoparticles. Arab J Chem 3:135–140

Ahamed M, AlSaalhi MS, Siddiqui MKJ (2010) Silver nanoparticle applications and human health. Clin Chim Acta 411:1841–1848

Akhavan O, Ghaderi E (2012) Cu and CuO nanoparticles immobilized by silica thin films as antibacterial materials and photocatalysts. Surf Coat Technol 205:219–223

Ali SM, Yousef NMH, Nafady NA (2015) Application of biosynthesized silver nanoparticles for the control of land snail *Eobania vermiculata* and some plant pathogenic fungi. J Nanomater 218904

Ali A, Zafar H, Zia M, Ul Haq I, Phull AR, Ali JS, Hussain A (2016) Synthesis, characterization, applications, and challenges of iron oxide nanoparticles. Nanotechnol Sci Appl 9:49–67

Ali K, Dwivedi S, Azam A, Saquib Q, Al-Said MS, Alkhedhairy AA, Musarrat J (2017) Aloe vera extract functionalized zinc oxide nanoparticles as nanoantibiotics against multidrug resistant clinical bacterial isolates. J Colloid Interface Sci 472(2016):145–156

Anand R, Kulothungan S (2014) Silver mediated bacterial nanoparticles as seed dressing against crown rot pathogen of groundnut. Arch Appl Sci Res 6:109–113

Anjum NA, Adam V, Kizek R, Duarte AC, Pereira E, Iqbal M, Lukatkin AS, Ahmad I (2015) Nanoscale copper in the soil-plant system: toxicity and underlying potential mechanisms. Environ Res 138:306–325

Aqel A, El-Nour KMMA, Ammar RAA, Al-Warthan A (2012) Carbon nanotubes, science and technology part (I) structure, synthesis and characterization. Arab J Chem 5:1–23

Arokiyaraj S, Saravanan M, Udaya Prakash NK (2013) Enhanced antibacterial activity of iron oxide magnetic nanoparticles treated with *Argemone mexicana* L. leaf extract: an in vitro study. Mat Res Bull 48:3323–3327

Asanithi P, Chaiyakun S, Limsuwan P (2012) Growth of silver nanoparticles by DC magnetron sputtering. J Nanomater 2012:963609

Asha-Rani PV, Mun GLK, Hande MP, Valiyaveettil S (2009) Cytotoxicity and genotoxicity of silver nanoparticles in human cells. ACS Nano 3:279–290

Astefanei A, Núñez O, Galceran MT (2015) Characterisation and determination of fullerenes: a critical review. Anal Chim Acta 882:1–21

Barrena R, Casals E, Colon J, Font X, Sanchez A, Puntes V (2009) Evaluation of the ecotoxicity of model nanoparticles. Chemosphere 75:850–857

Barrett CS, Cohen JB, Faber J, Jenkins R, Leyden DE, Russ JC, Predecki PK (1986) Advances in X-ray analysis, vol 29. Plenum Press, New York, NY, USA

Bhuyan T, Mishra K, Khanuja M, Prasad R, Varma A (2015) Biosynthesis of zinc oxide nanoparticles from *Azadirachta indica* for antibacterial and photocatalytic applications. Mater Sci Semicond Process 32:55–61

Boro RC, Kaushal J, Nangia Y, Wangoo N, Bhasinc A, Suri CR (2011) Gold nanoparticles catalyzed chemiluminescence immunoassay for detection of herbicide 2,4-dichlorophenoxyacetic acid. Analyst 136:2125–2130

Brice-Profeta S, Arrio MA, Troncetal E (2005) Magneticorderin γ-Fe2O3 nanoparticles: a XMCD study. J Magn Magn Mat 288:354–365

Cao YC (2002) Nanoparticles with Raman spectroscopic fingerprints for DNA and RNA detection. Science 80:1536–1540

Chauhan RPS, Gupta C, Prakash D (2012) Methodological advancements in green nanotechnology and their applications in biological synthesis of herbal nanoparticles. Int J Bioassays 7:6–10

Chen Z, Meng HA, Xing GM, Chen C, Zhao Y, Jia G, Wang T, Yuan H, Ye C, Zhao F, Chai Z, Zhu C, Fang X, Ma B, Wan L (2006) Acute toxicological effects of copper nanoparticles in vivo. Toxicol Lett 163:109–120

Chen X, Schluesener HJ (2008) Nanosilver: a nanoproduct in medical application. Toxicol Lett 176:1–12

Chen T, Wong YS, Zheng W, Bai Y, Huang L (2008) Selenium nanoparticles fabricated in *Undaria pinnatifida* polysaccharide solutions induce mitochondria-mediated apoptosis in A375 human melanoma cells. Coll Surf B 67:26–31

Corredor E, Testillano PS, Coronado MJ, González-Melendi P, Fernández-Pacheco R, Marquina C, Ibarra MR, de la Fuente JM, Rubiales D, Pérez-de-Luque A, Risueño MC (2009) Nanoparticle penetration and transport in living pumpkin plants: in situ subcellular identification. BMC Plant Biol 9:45

De Jaeger N, Demeyere H, Finsy R, Sneyers R, Vanderdeelen J, van der Meeren P, van Laethem M (1991) Particle sizing by photon correlation spectroscopy. Part I. Monodisperse latices. Influence of scattering angle and concentration of dispersed material. Part Part Sys Charac 8:179–186

de Oliveira JL, Campos EVR, Bakshi M, Abhilash PC, Fraceto LF (2014) Application of nanotechnology for the encapsulation of botanical insecticides for sustainable agriculture: prospects and promises. Biotechnol Adv 32:1550–1561

Doshi R, Braida W, Christodoulatos C, Wazne M, O'Connor G (2008) Nanoaluminum: transport through sand columns and environmental effects on plant and soil communication. Environ Res 106:296–303

Dubas ST, Pimpan V (2008) Humic acid assisted synthesis of silver nanoparticles and its application to herbicide detection. Mater Lett 62:2661–2663

Duman F, Ismail Ocsoy I, Kup FO (2016) Chamomile flower extract-directed CuO nanoparticle formation for its antioxidant and DNA cleavage properties. Mat Sci Eng C 60:333–338

Ekimov AI, Onuschchenko AA (1981) Quantum size effect in three-dimensional microscopic semiconductor crystals. JETP Lett 34:345–348

El-Rahman AFA, Mohammad TGM (2013) Green synthesis of silver nanoparticle using *Eucalyptus globulus* leaf extract and its antibacterial activity. J Appl Sci Res 9:6437–6440

Embiale A, Hussein M, Husen A, Sahile S, Mohammed K (2016) Differential sensitivity of *Pisum sativum* L. cultivars to water-deficit stress: changes in growth, water status, chlorophyll fluorescence and gas exchange attributes. J Agron 15:45–57

Eric D (1987) Engines of creation: the coming area of nanotechnology. Anchor Library of Science

Eustis S, El-Sayed MA (2006) Why gold nanoparticles are more precious than pretty gold: noble metal surface plasmon resonance and its enhancement of the radiative and nonradiative properties of nanocrystals of different shapes. Chem Soc Rev 35:209–217

FAO (2009) How to feed the world in 2050. FAO Report Oct. 2009:1–35

Faraji M, Yamini Y, Rezaee M (2010) Magnetic nanoparticles: synthesis, stabilization, functionalization, characterization, and applications. J Iran Chem Soc 7:1–37

Feynman RP (1960) There's plenty of room at the bottom. Eng Sci 23:22–36

Freestone I, Meeks N, Sax M, Higgitt C (2007) The lycurgus cup—A Roman nanotechnology. Gold Bull 40:270–277

Getnet Z, Husen A, Fetene M, Yemata G (2015) Growth, water status, physiological, biochemical and yield response of stay green sorghum Sorghum bicolor (L.) Moench varieties-a field trial under drought-prone area in Amhara regional state. Ethiopia J Agron 14:188–202

Grillo R, Pereira AES, Nishisaka CS, Lima RD, Oehlke K, Greiner R, Leonardo F, Fraceto LF (2014) Chitosan/tripolyphosphate nanoparticles loaded with paraquat herbicide. An environmentally safer alternative for weed control. J Hazard Mater 278:163–171

Gujrati M, Malamas A, Shin T, Jin E, Sun Y, Lu ZR (2014) Multifunctional cationic lipid-based nanoparticles facilitate endosomal escape and reduction-triggered cytosolic siRNA release. Mol Pharm 11:2734–2744

Guo D, Xie G, Luo J (2014) Mechanical properties of nanoparticles: basics and applications. J Phys D Appl Phys 47:13001

Hafez EE, Hassan HS, Elkady M, Salama E (2014) Assessment of antibacterial activity for synthesized zinc oxide nanorods against plant pathogenic strains. Int J Sci Technol Res 3:318–324

Hassan MS, Amna T, Yang OB, El-Newehy MH, Al-Deyab SS, Khil MS (2012) Smart copper oxide nanocrystals: Synthesis, characterization, electrochemical and potent antibacterial activity. Coll Surf B Biointerf 97:201–206

Hirsch LR, Stafford RJ, Bankson JA, Sershen SR, Rivera B, Price RE, Hazle JD, Halas NJ, West JL (2003) Nanoshell-mediated near-infrared thermal therapy of tumours under magnetic resonance guidance. Proc Natl Acad Sci USA 100:13549–13554

Holgate ST (2010) Exposure, uptake, distribution and toxicity of nanomaterials in humans. J Biomed Nanotechnol 6:1–19

Holzinger M, Le Goff A, Cosnier S (2014) Nanomaterials for biosensing applications: a review. Front Chem 2:63

Hossain Z, Mustafa G, Komatsu S (2015) Plant responses to nanoparticle stress. Int J Mol Sci 16:26644–26653

Hou J, Shang J, Jiao C, Jiang P, Xiao H, Luo L, Liu T (2013) A core crosslinked polymeric micellar platium(IV) prodrug with enhanced anticancer efficiency. Macromol Biosci 13:954–965

Mironava T, Simon M, Rafailovich MH, Rigas B (2013) Platinum folate nanoparticles toxicity: cancer vs. normal cells. Toxicol in Vitro 27:882–889

Husen A, Iqbal M, Aref MI (2014) Growth, water status and leaf characteristics of *Brassica carinata* under drought stress and rehydration conditions. Braz J Bot 37:217–227

Husen A, Iqbal M, Aref IM (2016) IAA-induced alteration in growth and photosynthesis of pea (*Pisum sativum* L.) plants grown under salt stress. J Environ Biol 37:421–429

Husen A (2017) Gold nanoparticles from plant system: synthesis, characterization and application. In: Ghorbanpourn M, Manika K, Varma A (eds) Nanoscience and plant–soil systems, vol 48. Springer International Publication, pp 455–479

Husen A, Iqbal M, Aref IM (2017) Plant growth and foliar characteristics of faba bean (*Vicia faba* L.) as affected by indole-acetic acid under water-sufficient and water-deficient conditions. J Environ Biol 38:179–186

Husen A, Iqbal M, Sohrab SS, Ansari MKA (2018) Salicylic acid alleviates salinity-caused damage to foliar functions, plant growth and antioxidant system in ethiopian mustard (*Brassica carinata* A. Br.) Agri Food Sec 7:44

Husen A, Siddiqi KS (2014a) Carbon and fullerene nanomaterials in plant system. J Nanobiotechnol 12:16

Husen A, Siddiqi KS (2014b) Phytosynthesis of nanoparticles: concept, controversy and application. Nano Res Lett 9:229

Husen A, Siddiqi KS (2014c) Plants and microbes assisted selenium nanoparticles: characterization and application. J Nanobiotechnol 12:28

Hussein M, Embiale A, Husen A, Aref IM, Iqbal M (2017) Salinity-induced modulation of plant growth and photosynthetic parameters in faba bean (*Vicia faba*) cultivars. Pak J Bot 49:867–877

Istiani W, Sribudiani E, Somadona S (2021) Biopellet from waste candlenut shell (*Aleurites moluccana*) with mixture of sago stem biomass (*Metroxylon sago*) and sawdust as an alternative energy source. Wahana Forestra: Jurnal Kehutanan 67:170–180

Iavicoli I, Leso V, Beezhold DH, Shvedova AA (2017) Nanotechnology in agriculture: opportunities, toxicological implications, and occupational risks. Toxicol App Pharmacol 329:96–111

Iijima S (1991) Helical microtubules of graphitic carbon. Nature 354:56–58

Ibrahim KS (2013) Carbon nanotubes-properties and applications: a review. Carbon Lett 14:131–144

Iravani S (2011) Green synthesis of metal nanoparticles using plants. Green Chem 13:2638–2650

Jasim B, Thomas R, Mathew J, Radhakrishnan EK (2016) Plant growth and diosgenin enhancement effect of silver nanoparticles in Fenugreek (*Trigonella foenum-graecum* L.). Saudi Pharm J 25:443–447

Jasim B, Thomas R, Mathew J, Radhakrishnan EK (2017) Plant growth and diosgenin enhancement effect of silver nanoparticles in Fenugreek (*Trigonella foenum-graecum* L.) Saudi Pharma J 25:443–447

Jiang J, Oberdörster G, Biswas P (2009) Characterization of size, surface charge, and agglomeration state of nanoparticle dispersions for toxicological studies. J Nanopart Res 11:77–89
Jung JH, Cheol OH, Soo NH, Ji JH, Soo KS (2006) Metal nanoparticle generation using a small ceramic heater with a local heating area. J Aerosol Sci 37:1662–1670
Kaida T, Kobayashi K, Adachi M, Suzuki F (2004) Optical characteristics of titanium oxide interference film and the film laminated with oxides and their applications for cosmetics. J Cosmet Sci 55:219–220
Kang T, Wang F, Lu L, Zhang Y, Liu T (2010) Methyl parathion sensors based on gold nanoparticles and Nafion film modified glassy carbon electrodes. Sens Actuat B 145:104–109
Kanhed P, Birla S, Gaikwad S, Gade A, Seabra AB, Rubilar O, Duran N, Rai M (2014) In vitro antifungal efficacy of copper nanoparticles against selected crop pathogenic fungi. Mater Lett 115:13–17
Karnan T, Selvakumar SAS (2016) Biosynthesis of ZnO nanoparticles using rambutan (*Nephelium lappaceum* L.) peel extract and their photocatalytic activity on methyl orange dye. J Mol Struct 1125:358–365
Khadri H, Alzohairy M, Janardhan A, Kumar AP, Narasimha G (2013) Green synthesis of silver nanoparticles with high fungicidal activity from olive seed extract. Adv Nanopart 2:241–246
Khodakovskaya M, Dervishi E, Mahmood M, Xu Y, Li Z, Watanabe F, Biris AS (2009) Carbon nanotubes are able to penetrate plant seed coat and dramatically affect seed germination and plant growth. ACS Nano 3:3221–3227
Khot LR, Sankaran S, Maja JM, Ehsani R, Schuster EW (2012) Applications of nanomaterials in agricultural production and crop protection: a review. Crop Prot 35:64–70
Kim DY, Saratale RG, Shinde S, Syed A, Ameen F, Ghodake G (2018) Green synthesis of silver nanoparticles using *Laminaria japonica* extract: characterization and seedling growth assessment. J Clean Prod 172:2910–2918
Klug HP, Alexander LE (1974) X-ray diffraction procedures for poly-crystallite and amorphous materials, 2nd edn. Wiley, New York, NY, USA
Kneipp K, Wang Y, Kneipp H, Perelman LT, Itzkan I, Dasari RR, Fela MS (1997) Single molecule detection using surface-enhanced Raman scattering (SERS). Phys Rev Lett 78:1667–1670
Kot M, Major Ł, Lackner JM, Chronowska-Przywara K, Janusz M, Rakowski W (2016) Mechanical and tribological properties of carbon-based graded coatings. J Nanomater 2016:1–14
Laware S, Raskar S (2014) Influence of zinc oxide nanoparticles on growth, flowering and seed productivity in onion. Int J Curr Microbiol Appl Sci 3:874–881
Lee CW, Mahendra S, Zodrow K, Li D, Tsai YC, Braam J, Alvarez PJ (2010) Developmental phytotoxicity of metal oxide nanoparticles to *Arabidopsis thaliana*. Environ Toxicol Chem 29:669–675
Leonhardt U (2007) Optical metamaterials: Invisibility cup. Nat Photon 1:207–208
Lidén G (2011) The European Commission tries to define nanomaterials. Ann Occup Hyg 55:1–5
Lin D, Xing B (2007) Phytotoxicity of nanoparticles: inhibition of seed germination and root growth. Environ Pollut 150:243–250
Liu J, Qiao SZ, Hu QH (2011) Magnetic nanocomposites with mesoporous structures: synthesis and applications. Small 7:425–443
Liu R, Lal R (2015) Potentials of engineered nanoparticles as fertilizers for increasing agronomic productions. Sci Tot Environ 514:131–139
Luo M, Liu D, Zhao L, Han J, Liang Y, Wang P, Zhou Z (2014) A novel magnetic ionic liquid modified carbon nanotube for the simultaneous determination of aryloxyphenoxy-propionate herbicides and their metabolites in water. Anal Chim Acta 852:88–96
Mahmoud WMM, Rastogi T, Kümmerer K (2017) Application of titanium dioxide nanoparticles as a photocatalyst for the removal of micropollutants such as pharmaceuticals from water. Curr Opin Green Sust Chem 6:1–10
Mallakpour S, Sirous F (2015) Surface coating of α-Al_2O_3 nanoparticles with poly(vinyl alcohol) as biocompatible coupling agent for improving properties of bio-active poly(amide-imide) based nanocomposites having l-phenylalanine linkages. Prog Org Coat 85:138–145

McWilliams A (2018) Nanocomposites, nanoparticles, nanoclays and nanotubes: global markets to 2022. BCC Research Report Overview, pp 1–7

Min Y, Li J, Liu F, Yeow EK, Xing B (2014) NIR light mediated photoactivation pt based antitumor prodrug and simultaneous cellular apoptosis imaging via upconversion nanoparticles. Angew Chem Int Ed Engl 53:1012–1016

Mishra S, Keswani C, Abhilash PC, Fraceto LF, Singh HB (2017) Integrated approach of agri-nanotechnology: challenges and future trends. Front Plant Sci 8:471

Morsy MK, Khalaf HH, Sharoba AM, El-Tanahi HH, Cutter CN (2014) Incorporation of essential oils and nanoparticles in pullulan films to control foodborne pathogens on meat and poultry products. J Food Sci 79:M675–M682

Mucalo MR, Bullen CR, Manley-Harris M, McIntire TM (2002) Arabinogalactan from the Western larch tree: a new, purified and highly water-soluble polysaccharide-based protecting agent for maintaining precious metal nanoparticles in colloidal suspension. J Mat Sci 37:493–504

Murugan K, Labeeba MA, Panneerselvam C, Dinesh D, Suresh U, Subramaniam J, Madhiyazhagan P, Hwang J, Wang L, Nicoletti M, Benelli G (2015) *Aristolochia indica* green synthesized silver nanoparticles: a sustainable control tool against the malaria vector *Anopheles stephensi*. Res Vet Sci 102:127–135

Nagajyothi PC, Muthuraman P, Sreeknath TVM, Kim DH, Shim J (2017) Anticancer activity of copper oxide nanoparticles against human cervical carcinoma cells. Arab J Chem 10:215–225

Naseem T, Farrukh MA (2015) Antibacterial activity of green synthesis of iron nanoparticles using *Lawsonia inermis* and *Gardenia jasminoides* leaves extract. J Chem 2015:912342

Nasrollahzadeh M, Sajadi SM, Maham M (2015) Green synthesis of palladium nanoparticles using *Hippophae rhamnoides* Linn leaf extract and their catalytic activity for the Suzuki-Miyaura coupling in water. J Mol Catal A Chem 396:297–303

Nava OJ, Soto-Robles CA, Góomez-Gutiérreza CM, Vilchis-Nestor AR, Castro-Beltrán A, Olivas A, Luque PA (2017) Fruit peel extract mediated green synthesis of zinc oxide nanoparticles. J Mol Struc 1147:1–6

Nel AE, Mädler L, Velegol D, Xia T, Hoek EM, Somasundaran P, Klaessig F, Castranova V, Thompson M (2009) Understanding biophysicochemical interactions at the nano-bio interface. Nat Mater 8:543–557

Ngo QB, Dao TH, Nguyen HC, Tran XT, Nguten TV, Khuu TD, Huynh TH (2014) Effects of nanocrystalline powders (Fe, Co and Cu) on the germination, growth, crop yield and product quality of soybean (Vietnamese species DT-51). Adv Nat Sci Nanosci Nanotechnol 5:1–7

Nhan LV, Ma C, Rui Y, Liu S, Li X, Xing B, Liu L (2015) Phytotoxic mechanism of nanoparticles: destruction of chloroplasts and vascular bundles and alteration of nutrient absorption. Sci Rep 5:11618

Nhan LV, Ma C, Rui Y, Cao W, Deng Y, Liu L, Xing B (2016) The effects of Fe_2O_3 nanoparticles on physiology and insecticide activity in non-transgenic and Bt-transgenic cotton. Front Plant Sci 6:1263

O'Brien SC, Heath JR, Curl RF, Smalley RE (1988) Photophysics of buckminsterfullerene and other carbon cluster ions. J Chem Phys 88:220

Ocsoy I, Paret ML, Ocsoy MA, Kunwar S, Chen T, You M, Tan W (2013) Nanotechnology in plant disease management: DNA-directed silver nanoparticles on graphene oxide as an antibacterial against *Xanthomonas perforans*. ACS Nano 7:8972–8980

Ojha NK, Zyryanov GV, Majee A, Charushin VN, Chupakhin ON, Santra S (2017) Copper nanoparticles as inexpensive and efficient catalyst: a valuable contribution in organic synthesis. Coordin Chem Rev 353:1–57

Ouda SM (2014) Antifungal activity of silver and copper nanoparticles on two plant pathogens, *Alternaria alternata* and *Botrytis cinerea*. Res J Microbiol 9:34–42

Pandey A, Kulkarni A, Roy B, Goldman A, Sarangi S, Sengupta P, Phipps C, Kopparam J, Oh M, Basu S, Kohandel M, Sengupta S (2014) Sequential application of a cytotoxic nanoparticle and a PI3 K inhibitor enhances antitumor efficacy. Cancer Res 74:675–685

Park HJ, Kim SH, Kim HJ, Choi SH (2006) A new composition of nanosized silica-silver for control of various plant diseases. Plant Pathol J 22:295–302

Pereira AES, Grillo R, Mello NFS, Rosa AH, Fraceto LF (2014) Application of poly(epsilon-caprolactone) nanoparticles containing atrazine herbicide as an alternative technique to control weeds and reduce damage to the environment. J Hazard Mater 268:207–215

Periasamy VS, Alshatwi AA (2013) Tea polyphenols modulate antioxidant redox system on cisplatin-induced reactive oxygen species generation in a human breast cancer cell. Basic Clin Pharmacol Toxicol 112:374–384

Poinern GEJ (2014) A laboratory course in nanoscience and nanotechnology, 1st edn. CRC Press Taylor & Francis, Boca Raton, FL, USA

Puri A, Loomis K, Smith B, Lee JH, Yavlovich A, Heldman E, Blumenthal R (2009) Lipid-based nanoparticles as pharmaceutical drug carriers: from concepts to clinic. Crit Rev Ther Drug Carrier Syst 26:523–580

Qi M, Zhang K, Li S, Wu J, Pham-Huy C, Diao X, Xiao D, He H (2016) Superparamagnetic Fe_3O_4 nanoparticles: synthesis by a solvothermal process and functionalization for a magnetic targeted curcumin delivery system. New J Chem 4480:4480–4491

Qian H, Peng X, Han X, Ren J, Sun L, Fu Z (2013) Comparison of the toxicity of silver nanoparticles and silver ions on the growth of terrestrial plant model *Arabidopsis thaliana*. J Environ Sci 25:1947–1955

Rao JP, Geckeler KE (2011) Polymer nanoparticles: preparation techniques and size-control parameters. Prog Polym Sci 36:887–913

Rastogi A, Zivcak M, Sytar O, Kalaji HM, He X, Mbarki S, Brestic M (2017) Impact of metal and metal oxide nanoparticles on plant: a critical review. Front Chem 5:78

Rawat MK, Jain A, Singh S, Mehnert W, Thunemann AF, Souto EB, Mehta A, Vyas SP (2011) Studies on binary lipid matrix based solid lipid nanoparticles of repaglinide: in vitro and in vivo evaluation. J Pharm Sci 100:2366–2378

Risom L, Møller P, Loft S (2005) Oxidative stress-induced DNA damage by particulate air pollution. Mutation Res 592:119–137

Roldán MV, Pellegri N, de Sanctis O (2013) Electrochemical method for Ag-PEG nanoparticles synthesis. J Nanopart 2013:524150

Rui M, Ma C, Hao Y, Guo J, Rui Y, Tang X, Zhao Q, Fan X, Zhang Z, Hou T, Zhu S (2016) Iron oxide nanoparticles as a potential iron fertilizer for peanut (*Arachis hypogaea*). Fron Plant Sci 7:815

Rui M, Ma C, Tang X, Yang J, Jiang F, Pan Y, Xiang Z, Hao Y, Rui Y, Cao W, Xing B (2018) Phytotoxicity of silver nanoparticles to peanut (*Arachis hypogaea* L.): physiological responses and food safety. ACS Sustain Chem Eng 5:6557–6567

Sadiq IM, Pakrashi S, Chandrasekaran N, Mukherjee A (2011) Studies on toxicity of aluminum oxide (Al2O3) nanoparticles to microalgae species: *Scenedesmus* sp. and *Chlorella* sp. J Nanopar Res 13:3287–3299

Salata OV (2004) Applications of nanoparticles in biology and medicine. J Nanobiotechnol 2:3

Shaalan M, Saleh M, El-Mahdy M, El-Matbouli M (2016) Recent progress in applications of nanoparticles in fish medicine: a review. Nanomed Nanotechnol Biol Med 12:701–710

Shankramma K, Yallappa S, Shivanna MB, Manjanna J (2016) Fe_2O3 magnetic nanoparticles to enhance *Solanum lycopersicum* (tomato) plant growth and their biomineralization. Appl Nanosci 6:983–990

Shao W, Nabb D, Renevier N, Sherrington I, Luo JK (2012) Mechanical and corrosion resistance properties of TiO_2 nanoparticles reinforced Ni coating by electrodeposition. IOP Conf Ser Mater Sci Eng 40:12043

Siddiqi KS, Husen A (2016a) Fabrication of metal nanoparticles from fungi and metal salts: scope and application. Nano Res Lett 11:98

Siddiqi KS, Husen A (2016b) Fabrication of metal and metal oxide nanoparticles by algae and their toxic effects. Nano Res Lett 11:363

Siddiqi KS, Husen A (2016c) Engineered gold nanoparticles and plant adaptation potential. Nano Res Lett 11:400
Siddiqi KS, Husen A (2016d) Green synthesis, characterization and uses of palladium/platinum nanoparticles. Nano Res Lett 11:482
Siddiqi KS, Rahman A, Tajuddin HA (2016e) Biogenic fabrication of iron/iron oxide nanoparticles and their application. Nano Res Lett 11:498
Siddiqi KS, Husen A (2017a) Recent advances in plant-mediated engineered gold nanoparticles and their application in biological system. J Trace Elements Med Biol 40:10–23
Siddiqi KS, Husen A (2017b) Plant response to engineered metal oxide nanoparticles. Nano Res Lett 12:92
Siddiqi KS, Husen A, Rao RAK (2018a) A review on biosynthesis of silver nanoparticles and their biocidal properties. J Nanobiotechnol 16:14
Siddiqi KS, Rahman A, Tajuddin HA (2018b) Properties of zinc oxide nanoparticles and their activity against microbes. Nano Res Lett 13:141
Siddiqi KS, Husen A, Sohrab SS, Osman M (2018c) Recent status of nanomaterials fabrication and their potential applications in neurological disease management. Nano Res Lett 13:231
Sigmund W, Yuh J, Park H, Maneeratana V, Pyrgiotakis G, Daga A, Taylor J, Nino JC (2006) Processing and structure relationships in electrospinning of ceramic fiber systems. J Am Ceram Soc 89:395–407
Soares C, Branco-Neves S, de-Sousa A, Pereira R, Fidalgo F (2016) Ecotoxicological relevance of nano-NiO and acetaminophen to *Hordeum vulgare* L.: combining standardized procedures and physiological endpoints. Chemosphere 165:442–452
Sondi I, Salopek-Sondi B (2004) Silver nanoparticles as antimicrobial agent: a case study on *E. coli* as a model for Gram-negative bacteria. J Colloid Interface Sci 275:177–182
Song G, Hou W, Gao Y, Wang Y, Lin L, Zhang Z, Niu Q, Ma R, Mu L, Wang H (2016) Effects of CuO nanoparticles on *Lemna minor*. Bot Stud 57:3
Sotiriou GA, Pratsinis SE (2010) Antibacterial activity of nanosilver ions and particles. Environ Sci Technol 44:5649–5654
Sotiriou GA, Teleki A, Camenzind A, Krumeich F, Meyer A, Panke S, Pratsinis SE (2011) Nanosilver on nanostructured silica: antibacterial activity and Ag surface area. Chem Eng J 170:547–554
Soundarrajan C, Sankari A, Dhandapani P, Maruthamuthu S, Ravichandran S, Sozhan G, Palaniswamy N (2012) Rapid biological synthesis of platinum nanoparticles using *Ocimum sanctum* for water electrolysis applications. Bioproc Biosys Eng 35:827–833
Stoimenov PK, Klinger RL, Marchin RL, Klabunde KJ (2002) Metal oxide nanoparticles as bactericidal agents. Langmuir 18:6679–6686
Sukirtha R, Priyanka KM, Antony JJ, Kamalakkannan S, Thangam R, Gunasekaran P, Krishnan M, Achiraman S (2012) Cytotoxic effect of green synthesized silver nanoparticles using *Melia azedarach* against in vitro HeLa cell lines and lymphoma mice model. Process Biochem 47:273–279
Syu Y, Hung J, Chen JC, Chuang H (2014) Impacts of size and shape of silver nanoparticles on *Arabidopsis* plant growth and gene expression. Plant Physiol Biochem 83:57–64
Thatoi P, Kerry RG, Gouda S, Das G, Pramanik K, Thatoi H, Patra JK (2016) Photo-mediated green synthesis of silver and zinc oxide nanoparticles using aqueous extracts of two mangrove plant species, *Heritiera fomes* and *Sonneratia apetala* and investigation of their biomedical applications. J Photochem Photobiol B Biol 163:311–318
Thomas S, Harshita BSP, Mishra P, Talegaonkar S (2015) Ceramic nanoparticles: fabrication methods and applications in drug delivery. Curr Pharm Des 21:6165–6188
Tien DC, Tseng KH, Liao CY, Huang JC, Tsung TT (2008) Discovery of ionic silver in silver nanoparticle suspension fabricated by arc discharge method. J Alloys Compd 463:408–411
Tippayawat P, Phromviyo N, Boueroy P, Chompoosor A (2016) Green synthesis of silver nanoparticles in *Aloe vera* plant extract prepared by a hydrothermal method and their synergistic antibacterial activity. Peer J 4:e2589

Tran PA, Webster TJ (2011) Selenium nanoparticles inhibit *Staphylococcus aureus* growth. Int J Nanome 6:1553–1558

Uddin I, Poddar P, Kumar U, Phogat N (2013) A novel microbial bio-milling technique for the size reduction of micron sized Gd_2O_3 particles into nanosized particles. J Green Sci Tech 1:48–53

Umar S, Anjum NA, Ahmad P, Iqbal M (2018) Drought-induced changes in growth, photosynthesis, and yield traits in mungbean: role of potassium and sulfur nutrition. In: Ozturk M, Hakeem KR, Ashraf M (eds) Crop production technologies for sustainable use and conservation: physiological and molecular advances. Apple Academic Press, Waretown, NJ, USA, pp 79–89

Venkatachalam P, Priyanka N, Manikandan K, Ganeshbabu I, Indiraarulselvi P, Geetha N, Muralikrishna K, Bhattacharya RC, Tiwari M, Sharma N, Sahi SV (2016) Enhanced plant growth promoting role of phycomolecules coated zinc oxide nanoparticles with P supplementation in cotton (*Gossypium hirsutum* L.). Plant Physiol Biochem 110:118–127

Vesenka J, Manne S, Giberson R, Marsh T, Henderson E (1993) Colloidal gold particles as an incompressible atomic force microscope imaging standard for assessing the compressibility of biomolecules. Biophys J 65:992–997

Vimala K, Sundarraj S, Paulpandi M, Vengatesan S, Kannan S (2014) Green synthesized doxorubicin loaded zinc oxide nanoparticles regulates the Bax and Bcl-2 expression in breast and colon carcinoma. Process Biochem 49:160–172

Vinković T, Novák O, Strnad M, Goessler W, Jurašin DD, Paradiković N, Vrček IV (2017) Cytokinin response in pepper plants (*Capsicum annuum* L.) exposed to silver nanoparticles. Environ Res 156:10–18

Wang J, Wang X, Song Y, Zhu C, Wang K, Guo Z (2013) Detecting and delivering platinum anticancer drugs using fluorescent maghemite nanoparticles. Chem Commun (camb) 49:2786–2788

Wang JJ, Sanderson BJ, Wang H (2007) Cyto-and genotoxicity of ultrafine TiO_2 particles in cultured human lymphoblastoid cells. Mutat Res 628:99–106

Wang Z, Xie X, Zhao J, Liu X, Feng W, White JC, Xing B (2012) Xylem- and phloem-based transport of CuO nanoparticles in maize (*Zea mays* L.). Environ Sci Technol 46:4434–4441

Weir A, Westerhoff P, Fabricius L, Hristovski K, von Goetz N (2012) Titanium dioxide nanoparticles in food and personal care products. Environ Sci Technol 46:2242–2250

Wolf R, Matz H, Orion E, Lipozencic J (2003) Sunscreens–the ultimate cosmetic. Acta Dermatovenerol Croat 11:158–162

Yousuf PY, Ahmad A, Hemant GAH, Aref IM, Iqbal M (2015) Potassium and calcium application ameliorates growth and oxidative homeostasis in salt-stressed Indian mustard (*Brassica juncea*) plants. Pak J Bot 47(5):1629–1639

Yu J, Xu D, Guan HN, Wang C, Huang LK, Chi DF (2016) Facile one-step green synthesis of gold nanoparticles using *Citrus maxima* aqueous extracts and its catalytic activity. Mat Lett 166:110–112

Yu Z, Sun X, Song H, Wang W, Ye Z, Shi L, Ding K (2015) Glutathione-responsive carboxymethyl chitosan nanoparticles for controlled release of herbicides. Mater Sci Appl 6:591–604

Zhang H, Chen G (2009) Potent antibacterial activities of Ag/TiO_2 nanocomposite powders synthesized by a one-pot sol-gel method. Environ Sci Technol 43:2905–2910

Zhang Q, Li N, Goebl J, Lu Z, Yin Y (2011) A systematic study of the synthesis of silver nanoplates: is citrate a "magic" reagent? J Am Chem Soc 133:18931–18939

Zhao G, Song S, Wang C, Wu Q, Wang Z (2011) Determination of triazine herbicides in environmental water samples by high-performance liquid chromatography using graphene-coated magnetic nanoparticles as adsorbent. Anal Chim Acta 708:155–159

Zheng Y, Sache L (2009) Gold nanoparticles enhance DNA damage induced by anti-cancer drugs and radiation. Radiat Res 172:114–119

Zur Mühlen A, Zur Mühlen E, Niehus H, Mehnert W (1996) Atomic force microscopy studies of solid lipid nanoparticles. Pharma Res 13:1411–1416

Biomass Utilization and Biorefinery By-Product from Palm Oil and Marine Resources for Animal Feed and Feed Additive

Ahmad Sofyan, Hendra Herdian, and Agung Irawan

Abstract Indonesia is the largest producer of palm oil. Biorefinery by-products from palm oil can be classified into lignocellulosic and fiber-rich biomass. Palm kernel meal (PKM), a by-product from palm kernel oil extraction contains crude protein (13–16%) and there have been ongoing efforts to improve its utilization as animal feed. The restriction of PKM used in animal feed is linked with the imbalance of amino acids, high fiber content, shell, and other physical characteristics. On the other hand, Indonesia is among the leading countries in marine industry, by-products of fish, shrimp, crustaceans, and other marine processing industries are of high potency for animal feedstuff. Chitin, the dominant by-product of shrimp production, has ahead popularity in the last decade due to its large spectrum functions, especially as antimicrobial agent, non-toxic, biodegradable, and biocompatible. This chapter discusses the recent advances in PKM and marine industry by-products availability status and utilization, and novel technologies to improve their quality for animal feed and feed additive. A practical and conceptual development of the bioproducts for implementation, especially in the context of Indonesia and other countries with similar characteristics of nature. Biological processes including solid-state fermentation, mechanical processing, and valorization techniques can be integrated to process the biomass from palm oil industry. Chemical treatments including green chemistry techniques could improve chitosan functionality. Implementation of biorefinery techniques of biomass and by-products of palm oil and marine resources promise supporting raw material stock and sustainability for the feed and feed additive of animals.

Keywords Agroindustrial by-product · Biorefinery · Feed technology · Livestock · Palm oil

A. Sofyan (✉) · H. Herdian
Research Center for Animal Husbandry, National Research and Innovation Agency, Cibinong, Bogor, West Java, Indonesia
e-mail: ahma053@brin.go.id

A. Irawan
Vocational School of Animal Production, Universitas Sebelas Maret (UNS), Surakarta, Central Java, Indonesia

M. A. R. Lubis et al. (eds.), *Biomass Conversion and Sustainable Biorefinery*, Green Energy and Technology, https://doi.org/10.1007/978-981-99-7769-7_5

1 Introduction

Availability of feed ingredients in terms of quality, quantity, and continuity (sustainability) is the key in supporting livestock productivity. Therefore, efforts to continuously increase the availability of feed ingredients are paramount to support national meat self-sufficiency. This is in line with the national program "Efforts for the Special Acceleration of Increasing the Population of Pregnant Cattle and Buffalo (Upsus Siwab)" in increasing the national livestock population. However, the problem arises when conversion of land-producing forages or feedstock is continuously occurring. On the other hand, increasing effluent residues is identified to be a major problem in the biomass-generating industry (biorefinery). Therefore, integrating biorefinery industries into the livestock industry coupled with implementing novel technologies can help to aid the problem for both.

There are many underutilized biomass or waste generated from agricultural processing and biorefinery industry that are plenty and economical. Based on the biomass characteristics, they can be classified into lignocellulosic and non-lignocellulosic biomass. It is well known that Indonesia is the leading producer of palm oil (Safi et al. 2022). Many by-products of palm oil biorefinery are left-over including oil palm trunks, empty fruit bunches, oil palm fronds, palm kernel shells, and palm kernel meal, that can be classified into lignocellulosic and fiber-rich biomass (Kahar et al. 2022). Those biomasses are produced up to 70 tons per hectare of palm oil. Among those biomass, palm kernel meal (PKM), a by-product generates from the extraction of palm kernel oil, receives growing interest due to its potential quality and quantity, particularly in Indonesia and Malaysia that contributes to over 80% of the world's PKM supply (Safi et al. 2022). PKM contains moderate crude protein (13–16%) and there have been ongoing efforts to improve its utilization as animal feed. The problem associated with palm kernel meal is related to the imbalance of amino acids, high fiber content, and physical characteristics (Sundu et al. 2008). Lack of access to technology is a key problem to optimally utilize the PKM. At present, most of them are exported to European countries and used as an economical feed ingredient for livestock (Azizi et al. 2021). For other biomasses of palm oil biorefinery as mentioned above, they are generally burned and used as fertilizer in the plantation which can increase pollution to the air (Kahar et al. 2022).

Valorization of PKM is a critical point to enhance the quality of valuable animal feed. Several strategies that can be used to improve the nutritive value of PKM include mechanical, biological, and chemical treatments. Using multiple alternative processing technologies, PKM can be processed as a whole biomass or subjected to protein extraction to obtain various functional protein concentrate. In addition, integrating livestock into the palm oil crop system has also been introduced in Indonesia as this approach is potentially more economically and environmentally beneficial (Agus and Widi 2018). However, much more effort is needed to valorize the low-quality biomass using novel technologies that can be implemented in the plantation area, together with the livestock.

On the other hand, reflecting that Indonesia is among the leading countries for its aquaculture industry, by-products of the fish processing industry are of high potency for feed of livestock. In particular, Indonesia is the leading country for shrimp production and is becoming the top five exporter countries for fresh shrimp in the world due to the large production volume supported by nature and available resources (Aneesh et al. 2020). Chitin is a major by-product of shrimp production and a primary source of chitosan, biopolymer obtained from diacylation of chitin (Shan et al. 2012). Chitosan is gaining popularity in the last decade due to its large spectrum of functions, especially as antimicrobial agent. It is also considered non-toxic, biodegradable, and biocompatible (Anggraeni et al. 2022). In addition, chitosan has been widely used in various industries including pharmaceuticals, foods, textiles, and agriculture. Recently, the use of chitosan as an additive for livestock has been extensively investigated for its potential as rumen modulator and methane suppressor (Harahap et al. 2022).

Optimization of extraction process and efforts for improving the functionality of chitosan have been introduced in the last decade, in addition to undergoing efforts to optimize the bio-functionality. To produce chitosan, chitin is commonly subjected to different steps including demineralization, deproteinization, and deacetylation. Recent advances in processing chitin into chitosan using novel technology can provide insight into the progress of technological development and future priorities of product development, especially in the context of feed additives for livestock.

In this chapter, we discuss the recent advances of PKM and fish industry by-products availability status and utilization, their potency, and novel technologies to improve the quality of by-products as an animal feed. In addition, we provide practical and conceptual development of the bioproducts for implementation, especially in the context of Indonesia and other countries with similar characteristics of nature. Collaborative framework among stakeholders should be established to take full advantage of such bioresources as a feedstock and to connect the farmers with industry.

2 Biomass and By-Product Characteristic and Potency

As a country with comparison of the ocean and land area of 6.3 versus 1.9 M km^2 makes Indonesia known as the largest archipelagic country in the world (Indonesian Geospatial Information Agency 2013). So it is not surprising that Indonesia has great potential for marine resources. Crustacean as one of the marine products has an important meaning for Indonesia's economic growth. No less than US $ 1.55 billion in year 2021 the export values of crustacean products from Indonesia that make Indonesia the sixth largest country as crustacean producer in the world, while the total sales of crustacean products in the world reach US $ 33.2 billion (OEC 2023). As much as 40–50% of these crustacean products can end up as waste (Muthu et al. 2021), so, the potential to disturb the environment is massive if there is no further utilization process. Likely, waste from crustaceans still has the potential to be reused

Table 1 Chemical composition of shrimp shell (Rødde et al. 2008)

Compound	Concentration (dry weight)
Protein	33 and 40%
Chitin	17 and 20%
Mineral ($CaCO_3$)	34 ± 2%
Lipid	0.3–0.5%
Astaxanthin	14–39 mg kg^{-1}

either as food or feed (Senel and McClure 2004; Özogul et al. 2019). Crustacean shell waste (lobster, shrimp, crab) generally contains protein parts, chitin minerals, lipids, and a small portion of astaxanthin (Table 1). This part of waste has a number of potentials that can lead to certain functions either as human food or animal feed (Özogul et al. 2019; Suryawanshi et al. 2018).

Protein portions from crustacean skin waste can be obtained through the fermentation of lactic acid bacteria to the cephalothorax and exoskeleton sections, resulting in hydrolysates ranging in proteins from 8.43 ± 0.22 to 46.73 ± 1.29 (Bueno-Solano et al. 2009), using protease (alcalase) enzyme following pH-stat method (Dey and Dora 2014). Crustacean skin extraction using the alcalase enzyme (Mizani et al. 2005) is described in Fig. 1. Protein sources from hydrolysate crustacean skin waste have the potential in addition to provide a source of nitrogen from crude protein also has the potential to contain amino acids that are quite complete for animal feeds Penaeus shrimp shells have a fairly complete amino acid balance and this is almost equivalent to soybean meal (Yan and Chen 2015). The use of hydrolysate protein in addition to supplying protein and amino acids also improves other functions because there is a compensatory bioactive peptide (Alvarez et al. 2015). Chitin is a polysaccharide found in the exoskeleton of an insect, crustaceans, mushroom structures, fungi, and yeast (Pighinelli et al. 2019), this condition causes chitin to be the most common polysaccharide found after Cellulosa (Senel and McClure 2004).

Chitin is a polymer of d-glucosamine while its deacetylated form is known as chitosan polymer N-acetyl-d-glucosamine monomers (Amiri 2022). Chitosan has a higher binding activity with several molecules compared to chitin. As a natural biopolymer, chitin is a material that has biocompatibility, biodegradability, and non-toxicity, has antibacterial properties, and anti-coagulant, absorbent molecules in nature, the form of chitin consists of αchitin, β chitin and ϒ chitin (Pighinelli et al. 2019). Conventionally to recover chitin and chitosan from the crustacean shell the extraction process by the chemical compound is described in Fig. 2. In ruminants, chitosan has extensive biological functions including antimicrobial, and protein-protecting agents to improve silage quality with various mechanisms of action (Fig. 3).

Although not made de novo by crustaceans, the pigment substance astaxanthin is obtained by crustaceans from the supply of feed material that contains a lot of carotenoids and astaxanthin, in crustacean astaxanthin obtained in free form or binds to proteins in the form of carotenoproteins (Özogul et al. 2019). The astaxanthin extraction method using acetone solvents yields the best results compared to

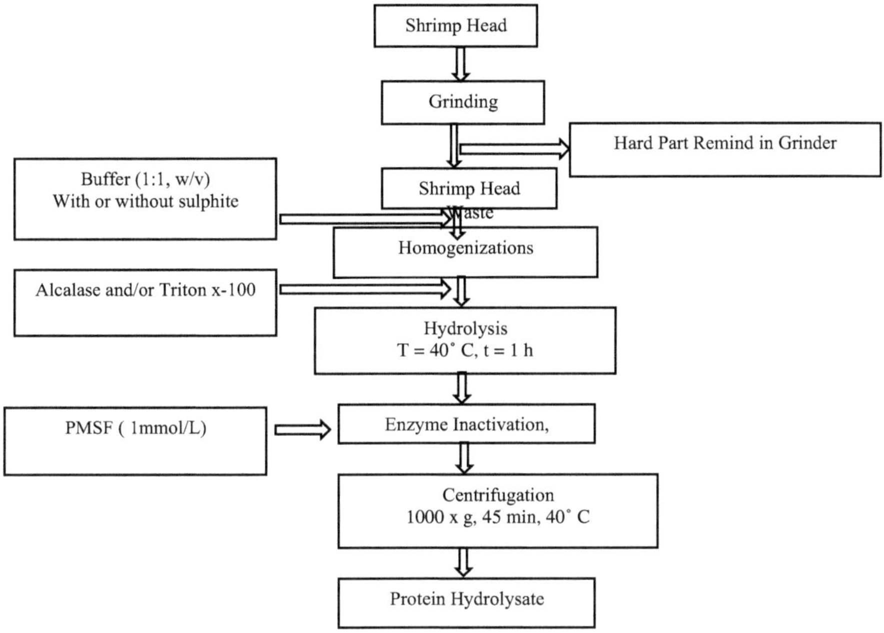

Fig. 1 Scheme of extraction crustacean shell protein procedure (Mizani et al. 2005)

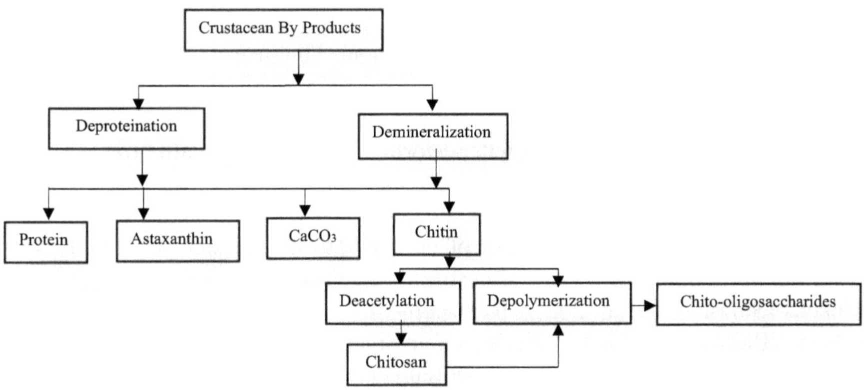

Fig. 2 Biorefinery of crustacean by-product to valuable compound (Özogul et al. 2019)

other conventional polar solvents (Dalei and Debasish 2015). Giving astaxanthin to Holstein Friesian dairy cows that are lactating can improve immune function and livestock reproduction (Da Costa et al. 2021).

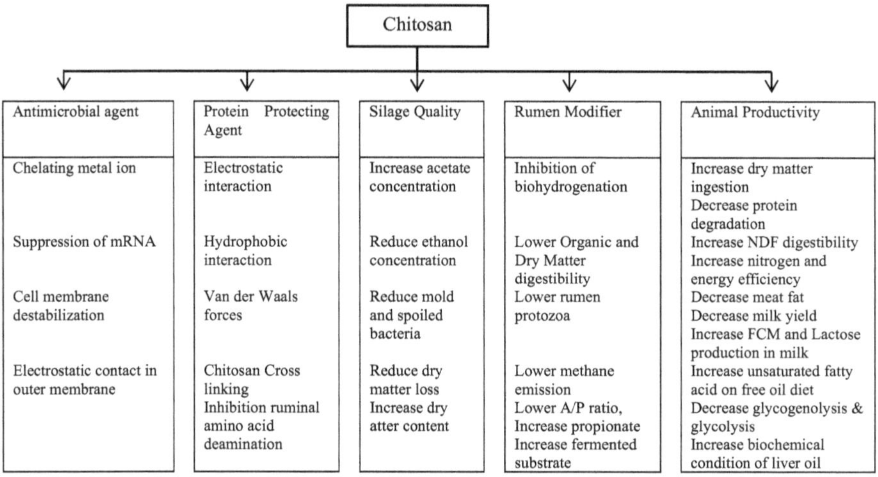

Fig. 3 Applications and effect of chitosan as feed additive in ruminants (Anggraeni et al. 2022)

3 Method for Improving Quality

3.1 By-Product of Palm Oil Processing Industry

Major limitation of PKM as a feedstuff for both poultry and ruminants is related to the imbalance of amino acids and high fiber content (Sundu et al. 2008). For ruminant animals, such high fiber content is not an issue. Several studies have demonstrated that inclusion of PKM or palm kernel cake in ruminant diets did not affect nutrient utilization, ingestive behavior, growth performance, and meat quality (da Silva et al. 2020, 2021). However, pretreatment processing is needed when it is used as a feed for poultry animals. The high content of cellulase is co-limiting the availability of amino acids by blocking the access of enzymes to utilize them especially when fed to poultry. Amino acid (AA) digestibility is considered as moderate in broiler chickens (Suprayogi et al. 2022). Replacement of a soybean meal-corn-based diet with 25% PKM resulted in low crude protein digestibility (41.6%) and moderate AA digestibility (averaged 75%) (Reza Abdollahi et al. 2015). The average essential and non-essential AA digestibility commonly falls into the value between 80 and 85% (Kim et al. 2022). The low CP digestibility is particularly a problem that can impair the growth of broiler chickens. To this end, methods to aid the low nutrient profile of PKM are of prime importance. Multiple strategies can be used to improve the nutritive value of PKM, such as fermentation, thermomechanical, thermochemical treatments, and their combinations.

3.1.1 Enzymatic Treatment

Enzymatic hydrolysis is more practical when compared to protein isolation in the context of its utilization as an animal feed. The main purpose of enzymatic treatment is to break down resistant starch of the PKM mainly lignin, cellulose, and β-mannan. PKM contains considerable amounts of β-mannan (up to 30%) (Navidshad et al. 2016) that can cause hyperplasia due to the high viscosity characteristics. A growing number of studies have been performed for this purpose. For instance, treatment with a mixture of *Aspergillus niger* and β-mannanase was reported to increase the true metabolizable energy (TME) by 32.1% and crude protein by 9% and concomitantly decrease the crude fiber content by 48.3% (Navidshad et al. 2016). Other carbohydrase enzymes such as xylanase, α-galactosidase, and β-mannosidase were also demonstrated to effectively enhance the molecular structure of protein of PKM (Moreira and Filho 2008; Álvarez-Cervantes et al. 2013; Chen et al. 2013).

A comprehensive review is available for the enzymatic treatment effect of PKM (Alshelmani et al. 2021). Overall, enzymatic treatment either using crude enzyme or microbial fermentation resulted in significant improvement in the quality of PKM, especially most of AA components. Compared to direct enzyme supplementation to the feed, enzymatic treatment to the PKM is more effective to increase the value of PKM which subsequently can improve productive performance of poultry. When these above enzymes were supplemented to broiler feed, no improvement was observed (Aftab and Bedford 2018; Chen et al. 2019). On the other hand, weight gain of broiler chickens was improved when fed with PKM treated enzyme compared to untreated PKM (Navidshad et al. 2016).

3.1.2 Solid-Sate Fermentation

Solid-state fermentation has gained more attention in the last few years due to its ability to improve the value of biomass, including feed ingredients for poultry. The main advantages of SSF are its effect to enhance protein quality, produce functional peptides, decrease antinutritional properties, and thus increase functionality (Suprayogi et al. 2021, 2022). SSF is known to enhance antioxidant activity of the substrate. There are not many available reports regarding the application of SSF to improve the value of PKM. There is a huge opportunity to implement this technology for PKM. The process of SFF is generally consisted of several steps including (1) preparation of microbial culture; (2) substrate incubation; and (3) harvesting. The schematic process of the SSF is provided in Fig. 4.

To date, several reports related to the effect of SFF were available on the quality of palm kernel cake (PKC) while the study on PKM is scarce. Alshelmani et al. (2017) reported that SSF of PKC using cellulolytic bacteria *Paenibacillus polymyxa* ATCC 842 and *P. curdlanolyticus* DSMZ 10248 resulted in significant decrease of NDF (−12.9%) and hemicellulose (−20.7%) and also increased several AA contents such as isoleucine, histidine, phenylalanine, threonine, methionine, arginine, and glycine. SFF of PKC using four fibrolitic bacteria *Bacillus amyloliquefaciens* DSMZ 1067,

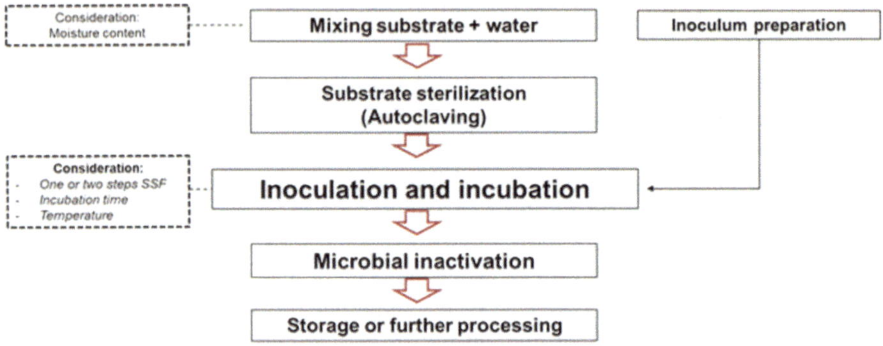

Fig. 4 Schematic representation of solid-state fermentation to enhance quality of PKM

Paenibacillus curdlanolyticus DSMZ 10248, *P. polymyxa* ATCC 842, and *B. megaterium* ATCC 9885 showed similar trends of nutritional profile enhancement (Alshelmani et al. 2014). In addition, PKC was also suggested as an economical substrate for enzyme production through SFF process, as reported in Ong et al. (2004). Another study reported that *Aspergillus niger*, *Aspergillus oryzae*, and *Aspergillus awamori* were able to grow in PKC, and xylanase was successfully produced. The obtained enzyme was able to improve the nutritive value of PKC as shown by the increase of CP and NFC and the decrease in the non-starch polysaccharides (Mohamad Asri et al. 2020). Studies in this area using PKM and also various by-products generated from palm oil biorefinery are worth investigating.

Different from PKM which is generated from solvent extraction process in palm oil processing, PKC is obtained from expeller press to extract the oil. Nevertheless, the characteristics of PKM and PKC are relatively similar. They contain considerable amounts of antinutritional factors (ANFs) such as β-mannan, cellulose, arabinoxylan, and glucuronoxylan (Alshelmani et al. 2021), while the first was considered as a prebiotic and is beneficial when added to the diet in low amount (Salami et al. 2022). Results of SFF are different depending on the type of substrate, inoculants, water-to-substrate ratio, temperature, and incubation time. To obtain the best conditions of SFF for PKM, a response surface model is the best starting point.

3.1.3 Valorization of PKM into Concentrated Protein

Although the percentage is considerably low, the protein quality of PKM is highly attractive. Various methods can be used for protein extractions from PKM including fractionation or using alkaline solution to extract and separate soluble and non-soluble protein contents. A series process of enzymatic treatment is effective to extract and purify soluble protein from PKM. The hydrolysis can be conducted in a reactor by utilizing 5–10% dry weight of PKM at 60 °C using a certain level of proteolytic enzymes for up to 4 h of stirring. After enzyme inactivation, centrifugation is needed to obtain the supernatant-containing protein. Next, the supernatant is purified using ultrafiltration to eliminate non-protein components using a specific membrane size with the aid of pressure and continuous stirring. Using this technique, soluble protein fraction is obtained (Safi et al. 2022). Known for different hydrolysis capabilities, selecting highly effective enzyme is an important decision. In their work, Safi et al. (2022) reported that alcalase is the most superior enzyme producing the highest purity and yield of protein from PKM, up to >50% and 70% of the supernatant, respectively. They also reported that using this technique, protein with high solubility was obtained. In addition, characterization of protein isolates and protein hydrolysate of PKM processed using the Osborne-type method have demonstrated a significant AA profile improvement than unprocessed PKM. In their study, protein fraction of the PKM was obtained by extracting the PKM using cold distilled water at room temperature with continuous stirring for 1 h.

After a series of extraction processes, supernatant and soluble protein from residue were dialyzed where the hydrolysate (protein concentrate) was obtained. Protein isolate, on the other hand, is obtained by further processing the protein concentrate. The protein concentrate is obtained using a dispersion method in a NaOH solution for 2 h stirring to obtain supernatant. The supernatant is then filtered, and the filtrate is subjected to protein precipitation in pH 4.6 and the precipitate is centrifuged to obtain protein concentrate. Protein hydrolysate can also be produced using microbial hydrolysis, as described in Aluko and Monu (2003) and Chang et al. (2014). Protein isolate from PKM can contain up to 75.6% protein with 15–50 kDa of polypeptide mass (Chang et al. 2014). Such protein isolate can be used as a highly valuable protein for poultry feed. Additionally, SSF can also be employed to produce low molecular weight functional peptides by using PKC as the main substrate. The peptides exhibit antifungal activity as shown in the inhibitory activity against *Aspergillus flavus*, *Aspergillus niger*, *Fusarium* sp., and *Penicillium* sp. (Mohamad Asri et al. 2020).

3.2 By-Product from Marine Industry

The increase in the human population increases the need for food consumption, one of which is the increase in crustacean consumption. This problem certainly has a linear value with the increasing waste by-product of this product. The challenge is how to increase reuse without increasing disruption to the environment. The conventional Chitin chitosan manufacturing process always involves the use of chemical reagents which besides being expensive can also interfere with the environment. The approach is carried out using a number of enzymatic, physical, and biological processes.

3.2.1 Biological Process

Lactobacillus acidophilus SW01 from shrimp waste for deproteinated fermentation of shrimp wastewas used. As a result, the minerals and protein were quickly removed with their contents decreasing to 0.73% and 7.8%, respectively, after 48 h fermentation (Duan et al. 2012). A protease-producing strain, *Exiguobacterium profundum*, and a lactic acid-producing strain, *Lactobacillus acidophilus,* were used to extract the chitin. The yields for the chitin were 47.82 and 16.32%, respectively (Xie et al. 2021). Two strains of bacterial BAO-01 and BAO-02 were isolated and observed for the chitinase activity. These *Salinivibrio* spp. did not show bioamine production, hemolytic activity, and mucin degradation. Therefore, the in vitro screening results suggested that these bacteria could be widely used as new candidates for chitin hydrolyzation and seafood fermentation (Le et al. 2018).

3.2.2 Physical Treatment

The extraction process could use ball milling processes combined with steam explosion that worked for woody biomass or biorefinery action, or other uses converting chitin into small nitrogen-containing chemicals—such as derivatives of ETA and of the widely used organic solvent furan (Yan and Chen 2015).

4 Feed and Feeding Containing Biomass

As a feed additive in ruminants, chitosan administration has two functions, namely directly affecting the microbial fermentation process of the rumen, digestive system, and livestock metabolism and indirectly, namely improving the quality of feed to be consumed by livestock.

4.1 Direct Effect

4.1.1 Anti-Inflammatory and Immunomodulator

The intrauterine infusion of a chitosan solution could speed up the healing process of uterine after parturition in dairy cows (Okawa et al. 2021). It happened because of the antimicrobial effect of chitosan. Diet supplementation of chitosan in dairy cows diet reveals the response of suppressed the activity and expression level of gene and protein of inducible nitric oxide synthase, and linearly decreased interleukin-1 content and gene expression. The supplementation of CHI linearly increased the proportion of CD4+ and CD3+ (T lymphocytes). The supplementation of CHI linearly down-regulated the expression of nuclear factor-κBp65 gene and phosphorylation level in peripheral blood mononuclear cells and linearly inhibited the activity of inducible nitric oxide synthase and the production of nitric oxide (Zheng et al. 2021). The infusion of Chito-oligosaccharides on peripheral blood mononuclear cells (PBMCs) in dairy cows at a concentration of 160 μg/mL exhibited the strongest effect decreased the content and gene expressions of interleukin-1β and tumor necrosis factor-α, the mRNA and protein expression and activity of the inducible nitric oxide synthase, and NO production compared with the values observed in the control group (Zheng et al. 2020).

4.1.2 Protecting Agent

Rumen-protected amino acid coating chitosan on the particles of hydrogenated fat entrapping amino acid could protect arginine particles coated with chitosan by the coacervation using NaOH could retain about 80% of arginine, with the addition of ethanol employed, the chitosan-coated particles could retain even 97% of arginine, and the chitosan-coated methionine particles prepared by addition of ethanol led to 85% (Chiang et al. 2009). Minimizing the ruminal biohydrogenation process of unsaturated fatty acids by the use of encapsulation of flaxseed oil with chitosan (14 and 7%) to encapsulate fats and their effect on in vitro fermentation and fatty acid biohydrogenation with the result increased oleic unsaturated fatty acid significantly compare to control (Besharati et al. 2022).

4.1.3 Methane Reducer and Rumen Modulator

Related to the metabolism of energy supply from crude fiber sources, normally CH_4 gas is releaseed from the rumen, losses in the form of loss energy and emission of greenhouse gas causing this process to be attempted to be reduced. chitosan addition was able to reduce enteric methane emissions such decrease was accompanied by a decline in the protozoa population and a tendency of methanogen reduction also (Harahap et al. 2020). The methanogens could react with the chitosan as the result of different cell charges leading to leakage of the cell (Zanferari et al. 2018). The proportion of VFA was affected by the addition of chitosan. While dry matter intake was not affected the addition of chitosan changed the carbohydrate fermentation resulting in the increase of propionate and decrease of acetate (Araujo et al. 2015) this condition was the same result as monensin was used (Goiri et al. 2010; Zanferari et al. 2018).

4.2 Indirect Effect

Chitosan has a function as an antimicrobial, based on this concept, the addition of chitosan is given to the silage-making process to control fermentation processes that are not desired by microbes. Chitosan addition at 4.47–7.47 g/kg of dry matter (DM) increased the efficiency process by reducing the fermentation losses and increasing the nutritional content of sugarcane silage (Del Valle et al. 2022). The infusion of chitosan and a microbial inoculum to soybean whole plant silage enhances nutritional and fermentative quality, increases all bacteria, and decreases yeast and mold on the silage product (Gandra et al. 2018). The addition of 2% chitosan on the alfalfa silage Neutral detergent fiber (NDF), neutral detergent insoluble crude protein (NDICP) values increase a lactic acid and butyric acid contents decreased, in conclusion, chitosan negatively influenced fermentation quality of alfalfa silage, but reduced mold and clostridial development (Sırakaya and Büyükkılıç-Beyzi 2022).

5 Future Strategy for Biomass Utilization

Biomass transformation via novel bioconversion technology is new frontiers for biorefinery concept, with the main objective to produce value-added underutilized by-products. At present, large volume of palm oil and fish processing by-products need a serious attention for further processing. Solid evidence from the capability of various technological approaches to increase the quality of by-products provides a realistic target for future implementation. From an industrial and economic point of view, there is a need to design a business model for palm kernel cake and palm kernel meal, prior to the implementation of technological processing. At this step, identification of stakeholders that are in need of the final products and potential market volume is fundamental. Next, re-routing the biorefinery process can be determined by considering the industrial-consumer chain characteristics.

For the palm oil processing industry, integrated applications of novel technology to process PKC, PKC, and other potential biomass is the first critical step to produce competitive products. Solid-state fermentation, coupled with mechanical processing, and various valorization techniques can be integrated to process the biomass from palm oil industry. Meanwhile, chemical treatments to improve chitosan functionality are now promising. At the same time, on-farm trials are needed to generate data and validate the quality of the products, that may also help with future improvement strategy. In addition, life cycle assessment may also help to assess the economical, environmental, and social impacts of the implementation of biorefinery technologies for palm oil and fish processing industries. The successful implementation of integrated biorefinery technology to produce valuable by-products from these industries is promising to establish a better livestock industry that is environmentally friendly while decreasing the demand for imported raw materials.

6 Conclusion

Utilization of biomass and biorefinery by-products as feedstock is emerging as a potentially viable alternative to support livestock development and sustainability. Feed ingredients in terms of quality, quantity, and continuity (sustainability) is the key in supporting livestock productivity. Implementation of technology for improving biomass and by-product quality could be carried out by several techniques. Biological processes including solid-state fermentation, mechanical processing, and valorization techniques can be integrated to process the biomass from palm oil industry. Chemical treatments including green chemistry technique could improve chitosan functionality. Consequently, utilization of biomass and by-products of palm oil and marine resources promises supporting raw material stock and sustainability for the feed and feed additive of animals. Further, life cycle assessment should be carried out to evaluate the economic, environmental, and social impacts of the implementation of biorefinery technologies for palm oil and marine processing industries and their relationship with the livestock industry.

Acknowledgements The authors thank the Research Collaboration Center of Biomass and Bioproduct Collaborated Between UNPAD and BRIN. Research Center for Animal Husbandry for facilitating and supporting during manuscript preparation.

References

Aftab U, Bedford MR (2018) The use of NSP enzymes in poultry nutrition: myths and realities. Worlds Poult Sci J 74:277–286. https://doi.org/10.1017/S0043933918000272

Agus A, Widi TSM (2018) Current situation and future prospects for beef cattle production in Indonesia-A review. Asian-Austral J Anim Sci 31:976–983. https://doi.org/10.5713/ajas.18.0233

Alshelmani MI, Abdalla EA, Kaka U, Abdul Basit M (2014) Nontraditional feedstuffs as an alternative in poultry feed. Adv Poult Nutr Res 2014. https://doi.org/10.5772/intechopen.95946

Alshelmani MI, Loh TC, Foo HL et al (2017) Effect of solid state fermentation on nutrient content and ileal amino acids digestibility of palm kernel cake in broiler chickens. Indian J Anim Sci 87:1135–1140

Alshelmani MI, Kaka U, Abdalla EA et al (2021) Effect of feeding fermented and non-fermented palm kernel cake on the performance of broiler chickens: a review. Worlds Poult Sci J 77:377–388. https://doi.org/10.1080/00439339.2021.1910472

Aluko RE, Monu E (2003) Functional and bioactive properties of quinoa and amaranth. Food Chem Toxicol 68:1254–1258. https://doi.org/10.3136/fstr.16.163

Álvarez-Cervantes J, Hernández-Domínguez EM, Arana-Cuenca A et al (2013) Purification and characterization of xylanase SRXL1 from *Sporisorium reilianum* grown in submerged and solid-state fermentation. BioResources 8:5309–5318. https://doi.org/10.15376/biores.8.4.5309-5318

Alvarez OM, Susana C, Agustin B (2015) Protein hydrolysates from animal processing by-products as a source of bioactive molecules with interest in animal feeding: a review. Food Res Int 73:204–212. https://doi.org/10.1016/j.foodres.2015.04.005

Amiri H, Aghbashlo M, Sharma M et al (2022) Chitin and chitosan derived from crustacean waste valorization streams can support food systems and the UN sustainable development goals. Nature Food 3:822–828. https://doi.org/10.1038/s43016-022-00591-y

Aneesh PA, Anandan R, Kumar LRG et al (2020) A step to shell biorefinery-extraction of astaxanthin-rich oil, protein, chitin, and chitosan from shrimp processing waste. Biomass Convers Biorefin. https://doi.org/10.1007/s13399-020-01074-5

Anggraeni AS, Jayanegara A, Laconi EB et al (2022) Marine by-products and insects as a potential chitosan source for ruminant feed additives. Czech J Anim Sci 67:295–317. https://doi.org/10.17221/42/2022-cjas

Araujo APC, Venturelli BC, Santos MCB et al (2015) Chitosan affects total nutrient digestion and ruminal fermentation in Nellore steers. Anim Feed Sci Technol 206:114–118

Azizi MN, Loh TC, Foo HL, Chung ELT (2021) Is palm kernel cake a suitable alternative feed ingredient for poultry? Animals 11:1–15. https://doi.org/10.3390/ani11020338

Besharati M, Giannenas, I, Palangi V et al (2022) Chitosan/calcium–alginate encapsulated flaxseed oil on dairy cattle diet: in vitro fermentation and fatty acid biohydrogenation. Animals 12:1400. https://doi.org/10.3390/ani12111400

Bueno-Solano CB, López-Cervantes J, Campas-Baypoli ON, et al (2009) Chemical and biological characteristics of protein hydrolysates from fermented shrimp by-products. Food Chem 112:671–675. https://doi.org/10.1016/j.foodchem.2008.06.029

Chiang Y-W, Wang T-H, Lee W-C (2009) Chitosan coating for the protection of amino acids that were entrapped within hydrogenated fat. Food Hydrocolloids 23(3):1057–1061. ISSN 0268-005X. https://doi.org/10.1016/j.foodhyd.2008.04.007

Chang SK, Ismail A, Yanagita T et al (2014) Biochemical characterisation of the soluble proteins, protein isolates and hydrolysates from oil palm (*Elaeis guineensis*) kernel. Food Biosci 7:1–10. https://doi.org/10.1016/j.fbio.2014.05.003

Chen WL, Jahromi MF, Candyrine SCL et al (2019) Enzymatic hydrolysis drastically reduces fibre content of palm-kernel expeller, but without enhancing performance in broiler chickens. Anim Prod Sci 59:2131–2137. https://doi.org/10.1071/AN17860

Chen WL, Liang JB, Jahromi MF et al (2013) Optimization of multi-enzyme production by fungi isolated from palm kernel expeller using response surface methodology. BioResources 8:3844–3857. https://doi.org/10.15376/biores.8.3.3844-3857

Da Costa RA, dos Santos MLP, Castro IC, et al (2021) Effect of astaxanthin supplementation on in vitro embryo development and immune response of dairy cows. Reprod Fertil Dev 33:171–171. https://doi.org/10.1071/RDv33n2Ab126

da Silva LO, de Carvalho GGP, Tosto MSL et al (2020) Digestibility, nitrogen metabolism, ingestive behavior and performance of feedlot goats fed high-concentrate diets with palm kernel cake. Livest Sci 241:104226. https://doi.org/10.1016/j.livsci.2020.104226

da Silva LO, de Carvalho GGP, Tosto MSL et al (2021) Effects of palm kernel cake in high-concentrate diets on carcass traits and meat quality of feedlot goats. Livest Sci 246. https://doi.org/10.1016/j.livsci.2021.104456

Dalei J, Debasish S (2015) Extraction and characterization of astaxanthin from the crustacean shell waste from shrimp processing industries. Int J Pharm Sci Res 6:2532–2537

Del Valle TA, Campana M, Pereira NR, et al (2022) Lactobacillus buchneri inoculation compared to chitosan and facultative heterofermentative lactic acid bacteria improves sugarcane silage conservation. J Agric Sci 160(5):317–324. https://doi.org/10.1017/S0021859622000417

Dey SS, Dora KC (2014) Antioxidative activity of protein hydrolysate produced by alcalase hydrolysis from shrimp waste *(Penaeus monodon and Penaeus indicus)*. J Food Sci Technol 51:449–457. https://doi.org/10.1007/s13197-011-0512-z

Duan S, Li L, Zhuang Z, et al (2012) Improved production of chitin from shrimp waste by fermentation with epiphytic lactic acid bacteria. Carbohydr Polym 89(4):1283–1288. https://doi.org/10.1016/j.carbpol.2012.04.051

Gandra JR, Takiya CS, Del Valle TA, et al (2018) Soybean whole-plant ensiled with chitosan and lactic acid bacteria: microorganism counts, fermentative profile, and total losses. J Dairy Sci 101(9):7871–7880. https://doi.org/10.3168/jds.2017-14268

Goiri I, Oregui LM, Garcia-Rodriguez A (2010) Use of chitosans to modulate ruminal fermentation of a 50:50 forage-toconcentrate diet in sheep. J Anim Sci. Feb 1;88(2):749–755

Harahap RP, Setiawan D, Nahrowib S et al (2020) Enteric methane emissions and rumen fermentation profile treated by dietary chitosan: a metaanalysis of in vitro experiments. Trop Anim Sci J 43(3):233–239

Harahap RP, Suharti S, Ridla M et al (2022) Meta-analysis of dietary chitosan effects on performance, nutrient utilization, and product characteristics of ruminants. Anim Sci J 93:e13676. https://doi.org/10.1111/asj.13676

Indonesian Geospatial Information Agency / BIG (2013) Rencana strategis badan informasi geospasial tahun 2013–2014. Badan Informasi Geospasial (BIG) Jakarta

Kahar P, Rachmadona N, Pangestu R et al (2022) An integrated biorefinery strategy for the utilization of palm-oil wastes. Bioresour Technol 344:126266. https://doi.org/10.1016/j.biortech.2021.126266

Kim JH, Choi HS, Choi WJ et al (2022) Determination of metabolizable energy and amino acid digestibility in various hatchery byproducts for broiler chickens. Poult Sci 101:101544. https://doi.org/10.1016/j.psj.2021.101544

Le B, Gyuhwa C, Yang, SH (2018). Chitinase producing *Salinivibrio* bacteria isolated from salt-fermented shrimp with antimicrobial and safety assessments J Appl Biol Chem 61(3):233–238. https://doi.org/10.3839/jabc.2018.034

Mizani M, Aminlari M, Khodabandeh M (2005) An effective method for producing a nutritive protein extract powder from shrimp-head waste. Food Sci Tech Int 11:49–54 https://doi.org/10.1177/1082013205051271

Mohamad Asri N, Muhialdin BJ, Zarei M, Saari N (2020) Low molecular weight peptides generated from palm kernel cake via solid state lacto-fermentation extend the shelf life of bread. Lwt 134:110206. https://doi.org/10.1016/j.lwt.2020.110206

Moreira LRS, Filho EXF (2008) An overview of mannan structure and mannan-degrading enzyme systems. Appl Microbiol Biotechnol 79:165–178. https://doi.org/10.1007/s00253-008-1423-4

Muthu M, Gopal J, Chun S, et al (2021) Crustacean waste-derived chitosan: antioxidant properties and future perspective. Antioxidants 10:228. https://doi.org/10.3390/antiox10020228

Navidshad B, Liang JB, Jahromi MF et al (2016) Effects of enzymatic treatment and shell content of palm kernel expeller meal on performance, nutrient digestibility, and ileal bacterial population in broiler chickens. J Appl Poult Res 25:474–482. https://doi.org/10.3382/japr/pfw029

Observatory of Economic Complexity (OEC) (2023) Crustaceans in Indonesia. https://oec.world/en/profile/bilateral-product/crustaceans/reporter/idn. Download 24th Oct 2023

Ong LGA, Abd-Aziz S, Noraini S et al (2004) Enzyme production and profile by Aspergillus niger during solid substrate fermentation using palm kernel cake as substrate. Appl Biochem Biotechnol—Part A Enzym Eng Biotechnol 118:73–79. https://doi.org/10.1385/abab:118:1-3:073

Özogul F, Hamed I, Özogul Y, Regenstein JM (2019) Crustacean by-products. In: Melton L, Shahidi F, Varelis P (eds) Encyclopedia of food chemistry. Elsevier Inc., pp 33–38

Okawa H, Wijayagunawardane MMP, Vos PLAM et al (2021) Effects of intrauterine infusion of a chitosan solution on recovery and subsequent reproductive performance of early postpartum dairy cows with endometritis: a pilot field trial. Animals 11(1):9

Pighinelli LBJ, Zanin BG, Flach AM, et al (2019) Methods of chitin production a short review. Am J Biomed Sci Res 3:308–314. https://doi.org/10.34297/AJBSR.2019.03.000682

Reza Abdollahi M, Hosking B, Ravindran V (2015) Nutrient analysis, metabolisable energy and ileal amino acid digestibility of palm kernel meal for broilers. Anim Feed Sci Technol 206:119–125. https://doi.org/10.1016/j.anifeedsci.2015.05.019

Rødde RH, Aslak E, Kjell MV (2008) A seasonal study of the chemical composition and chitin quality of shrimp shells obtained from northern shrimp *(Pandalus borealis)* Carbohydr Polyms 71:388–393. https://doi.org/10.1016/j.carbpol.2007.06.006

Safi C, Humblet NP, Geerdink P et al (2022) Valorisation of proteins from palm kernel meal. Bioresour Technol Reports 18:16–22. https://doi.org/10.1016/j.biteb.2022.101050

Salami SA, Ross SA, Patsiogiannis A et al (2022) Performance and environmental impact of egg production in response to dietary supplementation of mannan oligosaccharide in laying hens: a meta-analysis. Poult Sci 101:101745. https://doi.org/10.1016/j.psj.2022.101745

Şenel S, McClure SJ (2004) Potential applications of chitosan in veterinary medicine. Adv Drug Deliv Rev 56:1467–1480. https://doi.org/10.1016/j.addr.2004.02.007

Shan D, Lei Li, Zhuang Z et al (2012) Improved production of chitin from shrimp waste by fermentation with epiphytic lactic acid bacteria. Carbohydr Polym 89(4):1283–1288. https://doi.org/10.1016/j.carbpol.2012.04.051

Sırakaya S, Büyükkılıç-Beyzi S (2022) Treatment of alfalfa silage with chitosan at different levels to determine chemical, nutritional, fermentation, and microbial parameters. J Anim Feed Sci 31(1):73–80. https://doi.org/10.22358/jafs/147014/2022

Sundu B, Kumar A, Dingle J (2008) Amino acid digestibilities of palm kernel meal in poultry. J Indones Trop Anim Agric 33:139–144

Suprayogi WPS, Ratriyanto A, Akhirini N et al (2021) Evaluation of thermo-mechanical and microbial-facilitated processing on the chemical composition of soybean meal. IOP Conf Ser Earth Environ Sci 902:012026. https://doi.org/10.1088/1755-1315/902/1/012026

Suprayogi WPS, Ratriyanto A, Akhirini N et al (2022) Changes in nutritional and antinutritional aspects of soybean meals by mechanical and solid-state fermentation treatments with *Bacillus subtilis* and *Aspergillus oryzae*. Bioresour Technol Reports 17:100925. https://doi.org/10.1016/j.biteb.2021.100925

Xie J, Wancui X, Yu J et al (2021) Extraction of chitin from shrimp shell by successive two-step fermentation of *Exiguobacterium profundum* and *Lactobacillus acidophilus*. Front Microbiol 12:677126. https://doi.org/10.3389/fmicb.2021.677126

Yan N, Chen X (2015) Sustainability: don't waste seafood waste. Nat News 524:M155-156. https://doi.org/10.1038/524155a

Zanferari F, Vendramini THA, Rentas MF et al (2018) Effects of chitosan and whole raw soybeans on ruminal fermentation and bacterial populations, and milk fatty acid profile in dairy cows. J Dairy Sci 101(12):10939–10952

Zheng Y, Sumei Y, Jingyu Q, et al (2020) Protective effect of chitosan oligosaccharide against oxidative damage of peripheral blood mononuclear cells in dairy cows induced by diethylenetriamine/nitric oxide via NF-κB signalling Italian. J Anim Sci (19). https://doi.org/10.1080/1828051X.2020.1772131

Zheng YG, Zhang BQ, Qi JY et al (2021) Dietary supplementation of chitosan affects milk performance, markers of inflammatory response and antioxidant status in dairy cows. Anim Feed Sci Technol 277. https://doi.org/10.1016/j.anifeedsci.2021.114952

Aquatic Biomass Conversion and Biorefinery for Value-Added Products

Novia Amalia Sholeha, Nova Rachmadona, Fajriana Shafira Nurrusyda, Nanang Masruchin, and Khatarina Meldawati Pasaribu

Abstract The biorefinery concepts that merge technology and methods to transform aquatic biomass require the efficient utilization of most of the components. The presence of lipids, protein, and carbohydrates in aquatic biomass makes it a suitable feedstock for biofuel generation. Aquatic biomass's sugar and lignin components might be used to produce gas, heat, and bio-oil using thermochemical processes. The sugar component might be fermented to generate bio-butanol, bio-methanol, and bioethanol. The aquatic biomass lipid component could be used to manufacture biodiesel. Aquatic biomass might also be converted through biological processes into bio-methane and bio-hydrogen. Thermochemical processing (hydrothermal, pyrolysis, torrefaction) is a potential clean method for converting aquatic biomass and lignocellulosic materials to high-added value chemicals and bioenergy.

Keywords Aquatic biomass · Torrefaction · Pyrolysis · Biochar · Biogas · Bio-oil

---

N. A. Sholeha (✉)
College of Vocational Studies, Bogor Agricultural University (IPB University), Jalan Kumbang No. 14, Bogor 16151, Indonesia
e-mail: noviaamaliasholeha@apps.ipb.ac.id

N. A. Sholeha · N. Masruchin · K. M. Pasaribu
Research Center for Biomass and Bioproducts, National Research and Innovation Agency of Indonesia (BRIN), Cibinong 16911, Indonesia

N. A. Sholeha · N. Rachmadona · N. Masruchin · K. M. Pasaribu
Research Collaboration Center for Biomass and Biorefinery Between BRIN and Universitas Padjadjaran, Jatinangor 45363, Indonesia

N. Rachmadona · F. S. Nurrusyda
Department of Chemistry, Faculty of Mathematics and Natural Sciences, Universitas Padjadjaran, Jatinangor 45363, Indonesia

M. A. R. Lubis et al. (eds.), *Biomass Conversion and Sustainable Biorefinery*, Green Energy and Technology, https://doi.org/10.1007/978-981-99-7769-7_6

1 Introduction

Bioenergy is an environmentally conscious approach for minimizing reliance on oil, coal, and natural gas fuel. Energy is a commodity; hence a massive amount of feedstock is necessary. The search for innovative feedstocks for bioenergy production has been prompted by the rising energy demand and challenges associated with traditional feedstocks (Malode et al. 2022). In the recent past, researchers, policymakers, and the energy business have shown considerable interest in identifying and creating novel feedstocks. Current research focuses mainly on identifying acceptable feedstocks, creating effective conversion procedures, and reducing production costs overall. As biofuel feedstocks, lignocellulosic waste, municipal trash, microalgae, fungus, and other biomass have recently received considerable attention (Fakayode et al. 2023). These feedstocks have demonstrated biofuel production potential (Patel et al. 2021).

As a biofuel feedstock, aquatic biomass, including macro-, micro-algae, aquatic plants, and cyanobacteria, have the capacity to produce far more biomass per hectare than terrestrial crops; certain species produce fuel directly (Biller 2018). Advantages of such aquatic biomass include cultivating on the non-arable ground or even offshore and employing industrial carbon dioxide as a carbon source or wastewater as a fertilizer input (nitrogen and phosphorus). Aquatic biomass refers to energy crops that do not thrive with food crops for land or other resources. Numerous factors influence the productivity and composition of aquatic biomass, including nutrients, salinity, dark/light cycles, pH, irradiance levels, temperature, CO_2, and O_2 concentration. The composition of aquatic biomass includes proteins, lipids, carbohydrates, vitamins, and pigments, with lipids being the most interesting portion for biofuel production (Azwar et al. 2022). The practical implementation of regulating aquatic biomass as the raw material for diverse value-added products, as well as its biorefinery process (Fig. 1), has increasingly attracted the attention of researchers worldwide.

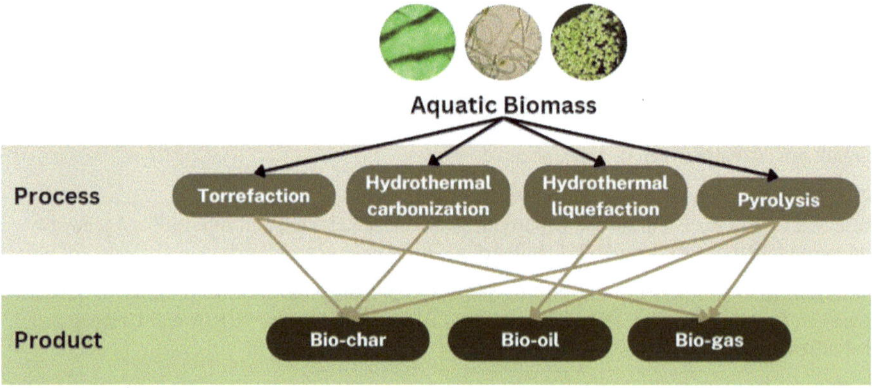

Fig. 1 The aquatic biomass biorefinery and its products

2 Aquatic Biomass as a Bioenergy Feedstock

Aquatic biomass has been identified as a viable renewable biomass feedstock in the production of bio-ethanol due to its high area-specific yields and photosynthetic efficiency. Microalgae have historically been explored for their potential high levels of lipids for biodiesel production (Ruiz et al. 2013). Among the significant benefits of microalgae versus terrestrial biomass is their elevated efficiency of photosynthetic to enhance CO_2 abatement and contributes to larger growth rates. Phototrophic and heterotrophic are the two functional classes of algae. Photoautotrophic algae absorb light and carbon dioxide for photosynthesis, while heterotrophic algae require oxygen and organic carbon sources.

Seaweed refers to a category of eukaryotic, photosynthetic marine organisms known as macroalgae (Alam et al. 2021). Both in terms of physical and biochemical properties, they differ significantly from microalgae. There are numerous species of them throughout the oceans and coastal waterways of the world. Aquatic weeds contain considerable levels of biofuel-convertible lignin, hemicellulose, and cellulose. The bulk of aquatic plant biomass is composed of lignin and carbohydrates. Lignin might be utilized to create combustible gases, bio-oil, and heat energy through thermochemical processes. The sugar component is immediately fermentable to produce bioethanol (Khammee et al. 2021). There is also a substantial opportunity to make bio-methanol and bio-butanol, among other alcohol molecules.

Additionally, aquatic weeds have lipids and a waxy covering composed primarily of modified fatty acids. Through the process of transesterification, these lipids may be converted into biodiesel. Through anaerobic digestion and biological processes, aquatic weeds can be used to produce biomethane and biohydrogen, respectively. Since aquatic weeds are fast-growing plants, they can provide greater biomass yields than most terrestrial energy crops. Macroalgae generally consist of a stipe, lamina or blade, and an anchoring and sustaining holdfast in marine conditions. Due to their size, the bulk of macroalgae must be ground before pumping. For instance, freshwater macrophytes are a diverse group of aquatic plants consisting of mono- or multi-cellular forms that frequently contain chlorophyll but lack genuine roots and stem in some cases. (Anyanwu et al. 2022). Algae and floating macrophytes (submerged, floating, and developing) are tiny, fast-growing plants typically found in watery habitat, and would not require agricultural land for agriculture, and several varieties are capable of living in freshwater, so preventing competing for water and land essential to produce food. Recent studies have revealed that the aquatic macrophyte *Ledermanniella schlechteri* (LS) and the green macroalgae *Ulva lactuca* (UL), prevalent throughout the Democratic Republic of the Congo, may be utilized to produce sustainable bioenergy (Mayala et al. 2022). Using biochemical methane potential (BMP) assays to evaluate how their anaerobic digestion functioned, it was revealed that the typical CH_4 levels for LS and UL are 262 and 162 mL gVS^{-1}, respectively. Moreover, LS's average HHV is 14.1 MJ kg^{-1} and UL's average HHV is 10.5 MJ kg^{-1}. Due to their negative ash behavior and high ash content, both biomasses would be challenging to convert thermally. The biochemical analyses revealed a high percentage of anaerobically digestible proteins and carbohydrates and a low quantity of lipids and lignin. The average biodegradability (BI) for LS was 76.5%, compared to 43.5% for UL.

Other kinds of aquatic biomass that are not algae but instead aquatic plants include water lettuce (Pistia stratiotes), water hyacinth (Eichhornia crassipes), cattail (Typha latifolia), salvinia (Salvinia molassis), and duckweed (Lemnaceae). Typically, all of those are invasive species that could colonize large bodies of water and prevent sunlight from reaching the bottom. They must be physically or mechanically removed from different rivers where they significantly damage the ecosystem to retain an intact biological system in many parts of the globe. These aquatic plants may also be grown expressly for biofuel generation; they use nutrients from wastewater and can generate approximately 20 tons (dry)/ha/year in only 24 h. As a result of its high moisture content, wet biomass cannot be burnt, making its disposal problematic and potentially expensive.

One kind of aquatic biomass, cyanobacteria, uses photosynthetic processes to convert carbon dioxide and solar energy into chemical compounds efficiently. Gram-negative photosynthetic bacteria called cyanobacteria are essential for global processes, including nitrogen fixation, carbon sequestration, and oxygen evolution. The natural cyanobacterial host system must often be better understood to boost goal output. In recent years, the accumulation of invaluable insights into the biochemistry and metabolism of cyanobacteria has propelled the development of cyanobacteria as cell factories for biochemical synthesis, including the synthesis of biofuels (Liu et al. 2022). Among the cyanobacteria that have been extensively examined for their ability to make biofuels are a marine species (*Synechococcus sp.* PCC 7002) and two freshwater species (*Synechocystis sp.* PCC 6803 and *Synechococcus elongatus sp.* PCC 7492) (Kumar et al. 2022).

That aquatic biomass (micro-, macro-algae, aquatic plants, and cyanobacteria) was processed through liquefaction, hydrothermal carbonization, hydrothermal torrefaction, and pyrolysis. The products from the process are summarized in Table 1.

2.1 Bio-Oil

A combination of several organic compounds known as "bio-oil" is often utilized as a raw material to manufacture pure chemicals like phenol, alcohol, organic acids, and aldehydes. Gasoline might be made from bio-oil via a processing step. Additionally, it contains chemicals that may be utilized for various applications at economically recoverable levels. Bio-oil has numerous manufacturing and selling benefits in the areas of combustion, preservation, transportation, adaption, and refurbishment. The literature is severely lacking in information on the creation of bio-oil from aquatic biomass. Several researchers have tried using thermochemical methods to extract bio-oil and other byproducts from aquatic biomass. Water hyacinth by pyrolysis method at 450 °C produced carboxylic acids, aldehydes, ketones, alkenes, quinines, alcohols, phenols, and aromatics, significant bio-oil components (Wauton and Ogbeide 2019).

Table 1 Value-added products from biomass refinery (Boro et al. 2022; Chacon et al. 2022)

Products	Properties	Limitations
Biogas (bio-methane and bio-hydrogen)	–Approximately consists of CH_4 (70%), CO_2 (25%), and other gases such as H_2S, NH_3 (5%) –Ignition temperature around 700 °C in anaerobic tanks –There is no smoke or residue produced during combustion (carbon neutral) –It can be utilized as cleaner fuel to generate electricity in the form of compressed natural gas (CNG) –Zero carbon dioxide and greenhouse gas emissions –Highly flammable and effective in producing energy –The only byproducts produced are water and heat	–Large bioreaction tanks increase the land area needed –Contain contaminants that, when used as fuel, might damage the engine systems of automobiles –Maintenance energy, optimal temperature, and a considerable amount of organic biomass are required –Foul odor –The production procedure is costly –Need compression due to its extremely low density
Bio-oil (bioethanol and bio-butanol)	–Utilized as an alternative fuel for automobiles by blending with gasoline –Must increase the combustion rate while cleaning the emissions –A transparent, colorless liquid –As a result of the low vapor pressure in comparison to gasoline, the rate of evaporation is low –Can utilize any substrate containing sugar; thus, agro-based lignocellulosic waste biomass usage is highly regarded for reducing challenges associated with the disposal of such waste –Nature-friendly and readily dilutable	–Low efficiency compared to gasoline –Implementation in vehicles necessitates engine modifications for older vehicles –Due to bioethanol's low vapor pressure, it is difficult to use it as a fuel at low temperatures, resulting in cold-start issues for vehicles –It has a high capacity to absorb moisture, which increases the risk of fuel pump corrosion
Biochar	–Enhance soil permeability –Increasing the water-holding capacity makes it simpler for plants to absorb water, nutrients, and oxygen –Boost soil pH levels	–Land loss due to erosion –Compaction of the soil during application –Elimination of crop residues

Moreover, duckweed at 350–700 °C of pyrolysis yields bio-oil around 35.5–45%, char around 30–50%, and gas around 11–20% (Djandja et al. 2021). Various aquatic plant biofuels' calorific value, such as Azolla (38.2 MJ kg^{-1}) and *Salvinia molesta* (39.73 MJ kg^{-1}), is more than that of biogas (30 MJ kg^{-1}), shown in Table 2 (Arefin et al. 2021).

In recent years, bioethanol has surpassed bio-oil as the primary alternative to fossil fuels, contributing up to 75% of global biofuel demand with an approximate extensive distribution rate of 86,000 kt/year. Aquatic biomass is handled using conventional hydrolysis techniques, much like any other bioethanol feedstock, and the resultant sugars are subsequently fermented to produce bioethanol. Aquatic vegetation is a favorable feedstock for bioethanol synthesis due to its richness of both cellulose and hemicellulose and the absence of lignin. *Limnocharis flava* was converted to bioethanol using several alkaline treatments (0% alkaline, 2% NaOH, and 1–2% CaO) to determine the most effective pretreatment for degrading cellulose, hemicellulose, and lignin to sugars fermentation. Significantly, 1% CaO resulted in a satisfactory total sugar, ethanol yield, and reducing sugar of 50.81, 0.72, and 28.88 g/L, respectively (Mejica et al. 2022). Prior research indicates the significance of NaOH for bioethanol in terms of Brachiaria mutica (Para grass) and Alternanthera philoxeroides (Alligator weed) (Aarti et al. 2022). In 12–96 h, the biomass from pre-treatment process showed that saccharification degree increased by 44.46 0.7%, 55.53 0.8%, 73.26 0.7%, 94.41 0.8%, and 73.3 0.7%. Bioethanol production from

Table 2 Aquatic biofuel properties comparisons with conventional fuel (Arefin et al. 2021)

Fuel properties	Aquatic plant biofuels					Conventional fuel		
	Azolla	Water hyacinth	Salvinia molesta	Water lettuce	Duckweed	Diesel	Gasoline	Biogas
Calorific value (MJ/kg)	38.2	35.8	39.73	24.93	21.7	45.5	45.8	30
Density (kg/m^3)	~880	834	792.23	952	800	850	715–780	1.15–1.25
Fire point (°C)	120	600–1370	300	–	–	210	280	650–750
Flashpoint (°C)	108	246	139	120	169	60	−43	−188
Pour point (°C)	3	−5	1.4	17	6	−2 to −12	−4 to −20	–
Cloud point (°C)	8	−1	1.5	–	–	−12	−22	–
Viscosity (cP)	4.3	9.85	3.657	26.4	~4.9	2.40	0.48	0.01–0.06
pH	3.5–10	2.93	6–7.7	6.6	7.8	5.5–8	5.9–6.8	6.8–7.2
Water	40	1.8	5	94.6	63.46	2	10	1–5

pre-treated aquatic weeds was evaluated utilizing yeast cells immobilized in sodium alginate for simultaneous saccharification and fermentation.

2.2 Biogas

Biogas in the aquatic biomass biorefinery comprises bio-hydrogen and bio-methane. Bio-hydrogen is seen as a feasible renewable energy source and an alternative to fossil fuels due to its higher energy content (122–142 kJ g^{-1}) in contrast to biomethane (56 kJ g^{-1}) and biodiesel (37 kJ g^{-1}). Biohydrogen can be produced at ordinary pressures and temperatures with low energy input, and its combustion simply creates water. Biohydrogen is already used in fuel cells, gasoline, and automobile engines (Yu et al. 2020). Currently, fossil fuels, which are expensive and inefficient, account for 96% of H_2 production. Single-celled algal species, including blue- and green algae such as *Chlorella sp*, *Platymonas subcordiformis*, and *Chlamydomonas reinhardti*, are typically used in H_2 production systems (enzymes such as the family Enterobacteriaceae). In anaerobic processes, the proton reduction by electronic hydrogenase of ferredoxin is necessary for biohydrogen production. The release of electrons from the breakdown of glucose to pyruvate leads to acetyl-CoA oxidation and carbon dioxide (Debowski et al. 2021).

One of the most flexible and clean-burning biofuels is biomethane, which is created through the anaerobic digestion of various feedstock materials (Zhang et al. 2021). Biomethane has advantages in easily transported and distributed by the same pipes as natural gas due to its easy storage after liquefaction. The byproduct of the manufacturing process may also be used on agricultural land as fertilizer. All types of biomass can be used to make biomethane, which offers advantages over other feedstocks, not just in terms of renewability but also in terms of waste management. Therefore, aquatic biomass has significant potential as a feedstock because it may be used immediately for biomethane production. AcD, also known as anaerobic co-digestion, is a promising strategy for boosting the biomethane manufacturing process's efficiency and overcoming the constraints of single digestion using catalysts. One of the AcD investigations found that adding Co_3O_4-NPs (3 mg/L) to water hyacinth (WH) increased biogas production by 27.2%. In addition, the production of methane (CH_4) was raised by 89.96% for the CD method and by 43.4% for the co-digestion method. The techno-economic analysis reveals that this method would generate 428.05 kWh of revenue based on the maximum net energy content of biogas, with such a sales revenue of 67.66 USD per m^3 of substrate (Ali et al. 2023).

2.3 Biochar

Biochar is black carbon or carbon-rich charcoal derived from organic matter through pyrolysis process; however, it can also be formed from a feedstock a feedstock via flash carbonization, torrefaction, or gasification (Janiszewska et al. 2021). Biochar has the capacity to hold carbon for millennia through enhancing water and nutrient retention and reducing greenhouse gas emissions from fertilized soils, hence enhancing the condition of the soils to which it is applied. As a feedstock for biochar, lignocellulosic ("woody") biomass has been the subject of most of the study. This feedstock produces biochar with low mineral concentration and high fixed-C content. Marine and freshwater macroalgae are alternate feedstocks for biochar manufacturing. Algal biochar contains considerable amounts of macronutrients and essential trace elements, including nitrogen, phosphorus, calcium, magnesium, potassium, and molybdenum, while having less carbon than lignocellulosic biochar. Due to the nutrient retention effects of micronutrients (Mo) and macronutrients (Ca, N, Mg, P, and K) on the soil, algal biochar has the potential to produce more significant increases in the quantity of certain types of soil than lignocellulosic biochar. According to previous findings, the biochar of the freshwater macroalgae Oedogonium formed at 750 °C has the most resistant carbon and leaches the least amount of metal (Roberts et al. 2015). The retention of fertilizer nutrients (Mo, Ca, N, Mg, P, and K) and the growth of radishes are both boosted by 35–40% when this biochar is applied to poor-quality soil. Radishes grown in biochar-modified soil exhibited comparable or lower metal concentrations than radishes grown in unmodified soil but had significantly greater concentrations of essential macronutrients (Mg, K, and P) and trace elements (Mo).

3 Thermochemical Process of Aquatic Weeds

3.1 *Torrefaction*

Torrefaction, one of thermochemical processes with slow heating, has been utilized extensively to volatilize biomass and can be classified as dry and wet, with biocoal (biochar and hydrochar) as the main products (Yek et al. 2022). Without the use of solvents, dry torrefaction (DT) takes place in oxidizing (flue gas or air) or non-oxidizing (CO_2 or N_2) atmospheres between 200 and 300 °C. Compared to non-oxidative torrefaction, oxidative torrefaction has a quicker reaction rate and shorter torrefaction duration due to exothermic reactions in the biomass thermal breakdown (Viegas et al. 2021). Additionally, the ultimate separation of nitrogen and air is unnecessary for oxidative torrefaction. A large part of the ash content remained in the torrefied aquatic biomass following dry torrefaction pretreatment, leading to undesirable agglomeration, fouling, and slagging despite the good potential for biofuel production. Aquatic biomass has been pre-treated to lower its ash content before torrefaction. At a reactor temperature of 440 °C, the pyrolysis process was carried out after pretreatment of the water hyacinth biomass at 200, 250, and 300 °C. Torrefaction severity significantly impacted the yields of char classified as brown coal (high quality) or peat. ST-Raw non-torrefied sample had a char yield of

27.4%, whereas the ST-300, ST-250, and ST-200 torrefied samples had char yields of 59.4%, 51.2%, and 42.3%, respectively. However, when the torrefaction temperature increased, syngas and bio-oil yield declined. GC-MS and FTIR analyses both showed that the bio-oil acidity had significantly decreased and that the torrefaction temperature had increased. Torrefied bio-oils are therefore assured to be less corrosive than un-torrefied bio-oils (Parvej et al. 2022).

Water causes wet torrefaction (WT) when it is present at temperatures between 180 and 260 °C for 10–24 min (Das et al. 2021). When later wet torrefaction happens in a wet situation, the conventional pre-drying stage for thermal conversion processes may be avoided, particularly for highly moist biomass such as manure, sewage, and aquatic biomass. When water is heated to 180 °C, its properties (density, viscosity, ion products, and dielectric constant) change in a manner that is favorable for thermochemical conversion in the aqueous phase (Nazos et al. 2022). The dissolution of the ash's minerals in the liquid reduces the quantity of ash in the solid result. In addition, steam torrefaction can operate at greater temperatures (200–260 °C) with the assistance of a high-pressure steam explosion that expands the lignocellulosic components and separates individual fibers. Carbon content and calorific value of the biomass increase as low molecular weight volatiles are eliminated during the steam explosion, although the product's bulk density, equilibrium moisture content, and mean particle size decline. The lowest production costs (without carbon credits) were associated with grape pomace's dry and wet torrefaction, at 2.29 and 4.14 $/GJ, respectively. It is more difficult to create pellets from biochar than from raw biomass because biochar is more brittle, dry, and volatile. Because hydrochar has a higher concentration of oxygen functional groups than biochar, it has a higher water affinity (hydrophilicity) on the surface, which enhances the soil's ability to retain water when immersed (Akbari et al. 2020).

3.2 Hydrothermal Carbonization (HTC)

Hydrothermal carbonization (HTC) or wet thermal process takes place at pressures higher than 1 MPa and temperatures between 180 and 300 °C (Akbari et al. 2020). Although HTC has a shorter residence time and a lower temperature than HTL, both processes are carried out in subcritical water conditions. Furthermore, HTC produces hydrochar with the same yield and energy content as a torrefied solid product at far lower temperatures. Biomass/water ratio, temperature, and duration of 42 wt%, 232 °C, and 99 min, were determined to be optimal for producing high HHV (22 MJ/kg) and low char generation (47 wt%), respectively (Lynam et al. 2015). The carbonization processes quicken as the temperature increases, leading to quicker kinetics and less hydrochar generation. If the length of the stay is increased, the temperature may yet have a distinct impact. The HTC research with fresh aquatic plants such as cattail and water hyacinth use an autoclave reactor at 180–220 °C. Following HTC treatment at 220 °C, the carbon content of aquatic biomass (cattail and water hyacinth) increased by 30.2–41.7%. Greater H/C and O/C ratios in the feedstocks relative to the comparable hydrochars may have resulted from the dissociation of the dehydration and decarboxylation processes that occur throughout the HTC process. As the temperature rose, the H/C and O/C atomic ratios fell, and the 220 °C hydrochar sample exhibited peat-like characteristics (Poomsawat and Poomsawat 2021).

3.3 Hydrothermal Liquefaction (HTL)

Hydrothermal liquefaction (HTL) is among the most major advancement promising processes for aquatic biomass upgrading, which directly converts biomass into bio-oil (Guo et al. 2017). HTL has several advantages, including obtained bio-oil having lower oxygen content and not requiring drying as the required microalgae concentration is around 20 wt.% (Biller 2018; Biswas et al. 2021). The drying process was known as the main economic and energetic obstacle before further processing of aquatic biomass conversion into biofuel. However, in HTL, the cost of the drying process can be reduced because water functions as a solvent in the system (Biswas et al., 2021).

Various products, such as aqueous-phase product, bio-oil, volatiles, gas, and solid residual, are the primary constituents of the HTL process' hydrothermally decomposed biomass conversion (Guo et al. 2017). Species of feedstock and processing parameters, including temperature, residence time, and solvent, determine HTL product. Numerous studies have been investigating these various parameter effects on different aquatic biomass. Due to its potential lipid content and enhanced photosynthetic efficiency, aquatic biomass has been recognized as a possible renewable biofuel source (Biller 2018). Furthermore, aquatic biomass has a higher growth rate than terrestrial plant biomass with less demanding cultivation and land use (Biswas et al. 2022).

Some studies show that microalgae, as a species of aquatic biomass, had been utilized in the production of biofuel using HTL, such as *Chlorella, Nannochloropsis*, and *Sargassum* sp. (He et al. 2020; Moazezi et al. 2022). Microalgae with a high lipid content will be completely converted to bio-oil; therefore, algae species with a high lipid content will be more valued (Biller 2018). Table 3 demonstrates that Sargassum sp. is rich in lipid and protein; hence, it is more susceptible to being transformed into bio-oil. While, in *Nannochloropsis* sp., ash contents are much higher, mainly contributing to solid residue production (He et al. 2020). As shown in Table 4, the bio-oil yields of *Sargassum* sp. (16.3% wt.) were significantly less than those of *Nannochloropsis* sp. (39.0% wt.).

Table 4 also presents bio-oil yields for different biomass species and operational parameters. Temperature has an important influence on the production of bio-residues, gas, and bio-oil. At lower temperatures, the degradation of biomass will be incomplete and unreacted. Thus, the bio-oil formation will be suppressed while increasing the solid products. An increase in temperature should be beneficial for bio-oil formation due to the acceleration of biomass decomposition. The yield of

Table 3 Composition data of algal biomass used for hydrothermal liquefaction

Biomass	Component (%wt)									References
	Protein	Lipid	Cellulose	Hemicellulose	Lignin	Poly-saccharide	Carbo-hydrate	Moisture	Ash	
Sargassum sp	9.9	0.80	9.04	38.6	13.0	–	–	–	–	He et al. (2020)
Nannochloropsis sp.	45.6	6.20	0.30	0.91	0.52	–	–	–	–	
Sargassum sp	7.5	1.33	–	–	–	–	50.7		27	Ardiansyah et al. (2018)
Gracilaria corticata	22.8	7.07	–	–	–	49.6	8.30	8.40	8.10	Rosemary et al. (2019)
Azolla filiculoides	4.6	0.72	–	–	–	–	0.82	91.8	2	Bhaskaran and Kannapan (2015)
Azolla filiculoides	19.7	4.2	–	–	–	10.3	–	–	18.5	Datta (2011)
Chlorella vulgaris	58.0	11.5	–	–	–	–	19	5	6.5	Moazezi et al. (2022)
Chlorella vulgaris	45.0	20.0	–	–	–	5	20	–	10	El-Naggar et al. (2020)

Table 4 Bio-crude yields and properties of distinct aquatic feedstocks under various HTL conditions

Biomass	Description	Solvent	Operational parameter		Product (%wt)			References
			T (°C)	Reaction Time (min)	Solid	Liq	Gas	
Sargassum tenerrimum	Brown macroalgae	H_2O	260	15	61.2	11.5	11.6	Biswas et al. (2018)
			280	15	32.3	16.3	12.1	
			300	15	24.2	14.7	9.0	
Gracilaria corticata	Red macroalgae	H_2O	260	15	21.7	3.9	11.8	Fernandes et al. (2021)
			280	15	23.0	2.8	4.9	
			300	15	26.0	5.2	11.2	
		Methanol	300	15	–	8.2	–	
		Ethanol	300	15	–	14	–	
		Acetone	300	15	44.0	16.2	–	
		Ethanol–water	300	15	–	13.3	–	
Azolla filiculoides	Aquatic plants	H_2O	280	15	38.0	21.5	5.0	Biswas et al. (2021)
			260	30	47.0	13.6	12.4	
			300	60	39.0	15.2	6.9	
		Methanol	260	15	41.2	28.7	12.9	
			260	30	37.5	26.3	7.6	
			280	60	36.2	24.3	11.9	
		Ethanol	300	15	29.5	26.5	2.1	
			300	30	33.8	26.3	14.8	
			280	60	36.7	28.8	15.5	

(continued)

Table 4 (continued)

Biomass	Description	Solvent	Operational parameter		Product (%wt)			References
			T (°C)	Reaction Time (min)	Solid	Liq	Gas	
Chlorella vulgaris	High lipid, microalgae	H_2O	287	40	–	56.2		Moazezi et al. (2022)
Nannochloropsis sp.	Low lipid, microalgae	H_2O	260	30	11.0	39.1	15.9	He et al. (2020)
			280	30	6.7	43.5	18.8	
			300	30	4.8	45.3	23.9	
			320	30	3.1	54.1	20.8	
			340	30	4.39	41.73	36.97	
Sargassum sp.	Brown macroalgae	H_2O	260	30	40.04	3.11	22.22	He et al. (2020)
			280	30	36.36	5.99	22.39	
			300	30	36.12	6.93	23.96	
			320	30	32.08	8.43	25.38	
			340	30	32.04	9.49	26.20	

bio-residue and bio-oil decreases at the temperature above 280 °C, as shown in Table 4 (Biswas et al. 2018). The relationship between re-polymerization and hydrolysis has a substantial impact on the HTL process's temperature. Extensive depolymerization will occur at high temperatures to activate the bond-breaking activation energy. The bond-breaking increases free radicals and repolymerizes the pieces that have been degraded (Moazezi et al. 2022).

Based on bio-oil yield using *Gracilaria corticata* as biomass, the relative efficiency of solvent used during the HTL process could be reported as follows: water < methanol < ethanol-water < ethanol < acetone. Table 4 demonstrates that the type of solvent affects conversion yield. Organic solvents promote the solubility and stability of chemical intermediates due to their lower dielectric constants, resulting in a greater yield. Additionally, it will facilitate esterification and alkylation between intermediate molecules and solvents (Fernandes et al. 2021).

In addition to being affected by the type of biomass employed, the duration of reaction times can determine the products derived from HTL as well as the feedstock conversion rate. As indicated in Table 4, a relatively short reaction time is suited for efficient biomass breakdown since the HTL process rapidly hydrolyzes biomass. In a longer reaction period, liquid products will undergo greater decomposition and repolymerization, hence contributing to the creation of gaseous products and biochar (Moazezi et al. 2022).

3.4 *Pyrolysis*

The thermochemical conversion processes can be split into four categories based On the basis of operating features such as temperature, pressure, heating rate, and reaction environment: gasification, combustion/incineration, liquefaction, and pyrolysis (Vuppaladadiyam et al. 2022). Pyrolysis, often known as thermal decomposition in an inert atmosphere, has been widely used to transform biomass into products with added value (Gao et al. 2020). Pyrolysis is a type of thermolysis or carbonization that employs intense heat in a low or oxygen-free (O_2) atmosphere to thermally decompose biomass into a number of pyrolytic chemicals (Tripathi et al. 2016; Lee et al. 2020). This thermochemical conversion yields biochar, bio-oil, and bio-syngas as its principal by-products (Azizi et al. 2018). The features of the aquaculture biomass, the operational parameters, and the kind of pyrolysis reaction influence the number of products and the HHV (Chen et al. 2015).

Table 5 summarizes the experimental parameters for pyrolysis techniques. The pyrolysis process has been classified into 2 categories; conventional and advanced approaches (Lee et al. 2020), presented in Fig. 2. Conventional pyrolysis can be divided into three distinct types: slow pyrolysis, fast pyrolysis, and flash pyrolysis, depending on the operational parameters employed during the process. Slow pyrolysis is a crucial synthesis technique that is mostly used to produce biochar with byproducts such as syngas and bio-oil (Lee et al. 2017). Slow pyrolysis settings emphasize slow heating rates (30 °C/min), moderate temperatures (550–950 °C), and slow reaction time. According to Table XZ, the yields of biochar, bio-oil, and bio-syngas produced by pyrolysis at 600 °C in which bio-syngas is the dominant product obtained in this technique (Maddi et al. 2011). Fast pyrolysis, the counterpart of slow pyrolysis, is frequently utilized for biomass under the following pyrolysis conditions: rapid heating rate (>60 °C/min), high temperature (850–1200 °C), and brief pyrolysis period (0.5–10 s) (Campanella and Harold 2012; Ly et al. 2015). Fast pyrolysis aims to optimize bio-oil synthesis, readily stored, or transported, and contains less nitrogen and sulfur (Roddy and Manson-Whitton 2012).

To improve the pyrolysis process, advanced pyrolysis techniques are often modified to create new methods, e.g. co-pyrolysis, catalytic pyrolysis, and microwave-assisted pyrolysis, that make the pyrolysis process superior to conventional techniques and enhanced the yield, quality, and characteristics the pyrolysis products. Under a catalyst, catalytic pyrolysis is a directed control method for obtaining high-quality liquid fuel and high-value-added chemicals with a high yield (Qiu et al. 2022). In a fixed-bed reactor, *Pavlova* microalgae were pyrolyzed at various temperatures in the presence of titania-based catalysts. When Ni/TiO_2 (22.55 wt%) was present at 500 °C, the bio-oil output increased by 20% (Aysu et al. 2017).

In parallel to catalytic pyrolysis, co-pyrolysis (Duan et al. 2015; Uzoejinwa et al. 2018, 2019) and microwave-assisted pyrolysis (Beneroso et al. 2013; Hong et al. 2017) have been identified as a promising strategy for enhancing the performance of biomass pyrolysis processes through synergistic interactions. Co-pyrolysis is the process of heating together two or more organic materials in the absence of oxygen to produce the bio-oil, and it is also the synergistic effect in terms of gas, liquid, and solid product distribution and product composition modifications (Ma et al. 2022). (Duan et al. 2015) reported a good synergistic impact between the waste rubber tire (WRT) and microalgae. The largest synergistic impact value (37.8%) was recorded at a mass ratio of 1:1 R:M. During co-pyrolysis, the interaction between microalgae and WRT promoted denitrogenation and deoxygenation, hence enhancing the quality of the bio-oil. The heating values of bio-oils derived from the co-pyrolysis of microalgae and WRT were between 35.80 and 42.03 MJ/kg.

On the other hand, microwave-assisted pyrolysis is regarded as a straightforward processes with direct control (Zhang et al. 2016). Hong et al. found that porphyra was a more ideal raw material for syngas-rich gas production (85.6–87.1 wt%) by using microwave-assisted pyrolysis because of its high carbohydrate content (47.7 wt%), but spirulina and chlorella were more advantageous for oil production due to their higher protein levels. *Scenedesmus almeriensis* was also found to be an appropriate feedstock for microwave-assisted pyrolysis to create gas products (Beneroso et al. 2013). By reducing CO_2 and light hydrocarbons, it has been claimed that the maximum output of syngas at 800 °C with the highest H_2/CO ratio can approach 94% by volume.

The pyrolysis of algal biomass generates and disperses a variety of organic and inorganic chemicals. As pyrolysis fuel, the chemical components of aquatic biomass such as cyanobacteria, duckweed, micro- and macroalgae are acceptable. As measured by pyrolysis, they may affect the HHV values, viscosities, pH, densities, and product composition (Bharathiraja et al. 2015). By a significant margin,

Table 5 Biofuel production from different aquatic feedstocks under various conditions of pyrolysis

Algal biomass	Type of Pyrolysis	Operational reaction parameters				Product (wt.%)			HHV (MJ/kg)	References
		T (°C)	t (min)	Carrier gas flowrate (mL/min)	Heating rate (°C/min)	Solid	Liquid	Gas		
Lyngbya sp.	Slow	600	20	He: 200	30	17	12	44	Bio-char: 25.6	Maddi et al. (2011)
Cladophora sp.	Slow	600	20	He: 200	30	26	20	38	Bio-char: 22.7	
Saccharina japonica	Fast	350	2 s	N_2: 4500	–	34.2	45	20.8	Bio-oil: 24.8	Ly et al. (2015)
Green algae	Fast	500	1.5 s	N_2: 250	–	26	58.6	17.8	Bio-oil: 26.7	Campanella and Harold (2012)
Green-blue algae	Fast	500	1.5 s	N_2: 250	–	28.4	54	19.9	Bio-oil: 26.8	
Chlorella	Fast	500	1.5 s	N_2: 250	–	29	53.9	17.3	Bio-oil: 25.5	
Pavlova sp.	Catalytic (Titania)	450–550	60	N_2: 545	100	35.9–49.0	14.1–22.5	36.5–46.3	Bio-char: 4.8–6.9 Bio-oil: 33.3–37.1	Aysu et al. (2017)
Saccharina japonica	Fast catalytic (HZSM-5)	500	–	–	–	22.3	39.1	39.3	Bio-oil: 27.2	Ly et al. (2019)

(continued)

Table 5 (continued)

Algal biomass	Type of Pyrolysis	Operational reaction parameters				Product (wt.%)			HHV (MJ/kg)	References
		T (°C)	t (min)	Carrier gas flowrate (mL/min)	Heating rate (°C/min)	Solid	Liquid	Gas		
Enteromorpha prolifera	Co-pyrolysis (Rice Husk)	400–600	–	N_2: 100	5–25	22.8–31.4	39.2–47.2	28.7–31.4	Bio-char: 26.9–31.6 Bio-oil: 25.5–30.6	Uzoejinwa et al. (2018)
Chlorella pyrenoidosa	Co-pyrolysis (waste rubber tyre)	290–370	10–120	–	–	19–49.7	37.5–65.4	4.6–14	Bio-oil: 33.7–42.9	Duan et al. (2015)
Scenedesmus almeriensis	Microwave	400	30	He: 100	–	–	–	87.7 wt.%	Bio-syngas: 3.36 Wh/g	Beneroso et al. (2013)
Chlorella	Microwave	400	30	N_2: 100	–	8	8	84	Bio-syngas: 5.6	Hong et al. (2017)
Spirulina	Microwave	700	30	N_2: 100	–	10	6.3	83.7	Bio-syngas: 2.9	
Porphyra	Microwave	700	30	N_2: 100	–	10.4	2.5	87.1	Bio-syngas: 3.1	

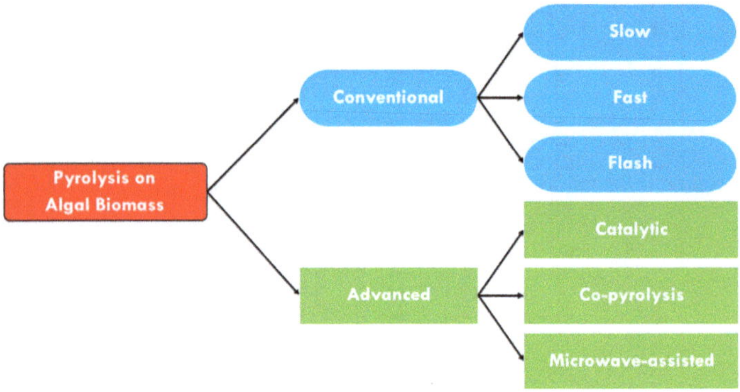

Fig. 2 The classification of the pyrolysis process in algal biomass

aquatic biomass confirms its suitability as pyrolysis feedstock for the eventual commercialization of energy-dense goods.

4 Conclusion

Aquatic biomass is emerging as a resource to produce biofuels and other goods with added value. Biomass derived from aquatic organisms offers significant potential for biomethane, bio-oil, and bioethanol production. However, the scientific community must address the following concerns.

- Research is necessary to develop an effective pre-treatment and conversion process.
- Collecting biomass, high processing costs for scaling up, poor hydrolysis, and conversion are challenges that must be overcome.
- Biological and other hybrid pretreatment approaches, as well as the intensification of the process, can be utilized to increase biofuel output.
- To achieve economic viability, the whole potential of aquatic weed biomass must be utilized.

References

Aarti C, Khusro A, Agastian P, Kuppusamy P, Al FDA (2022) Synthesis of gold nanoparticles using bacterial cellulase and its role in saccharification and bioethanol production from aquatic weeds. J King Saud Univ Sci 34:101974. https://doi.org/10.1016/j.jksus.2022.101974

Akbari M, Oyedun AO, Kumar A (2020) Techno-economic assessment of wet and dry torrefaction of biomass feedstock. Energy 207:118287. https://doi.org/10.1016/j.energy.2020.118287

Alam SN, Singh B, Guldhe A (2021) Aquatic weed as a biorefinery resource for biofuels and value-added products: challenges and recent advancements. Clean Eng Technol 4:100235. https://doi.org/10.1016/j.clet.2021.100235

Ali SS, Zagklis D, Kornaros M, Sun J (2023) Cobalt oxide nanoparticles as a new strategy for enhancing methane production from anaerobic digestion of noxious aquatic weeds. Bioresour Technol 368:128308. https://doi.org/10.1016/j.biortech.2022.128308

Anyanwu IN, Okeke CS, Nwankwo SC, Nwachukwu MO, Michael MO, Opara VC, Anorue CO, Azuama OC, Oti PO, Ekechukwu LE, Ezenwa CM, Chamba EB (2022) Aquatic macrophytes (Spirogyra porticalis and Nymphaea L.) as substrates for biofuel production: potentials and challenges. Sci African 18:e01412. https://doi.org/10.1016/j.sciaf.2022.e01412

Arefin MA, Rashid F, Islam A (2021) A review of biofuel production from floating aquatic plants: an emerging source of bio-renewable energy. Biofuels Bioprod Biorefining 5:574–591. https://doi.org/10.1002/bbb.2180

Ardiansyah D, Hartinah I et al (2018) Improvement of the nutritive quality of Sargassum powder through Aspergillus niger, Saccharomyces cerevisiae, and Lactobacillus spp. fermentations. AACL Bioflux 11:753–764

Aysu T, Ola O, Maroto-Valer MM, Sanna A (2017) Effects of titania based catalysts on in-situ pyrolysis of Pavlova microalgae. Fuel Process Technol 166:291–298. https://doi.org/10.1016/j.fuproc.2017.05.001

Azizi K, Keshavarz Moraveji M, Abedini Najafabadi H (2018) A review on bio-fuel production from microalgal biomass by using pyrolysis method. Renew Sustain Energy Rev 82:3046–3059. https://doi.org/10.1016/j.rser.2017.10.033

Azwar E, Adibah W, Mahari W, Rastegari H, Tabatabaei M (2022) Progress in thermochemical conversion of aquatic weeds in shellfish aquaculture for biofuel generation: Technical and economic perspectives. Bioresour Technol 344:126202. https://doi.org/10.1016/j.biortech.2021.126202

Beneroso D, Bermúdez JM, Arenillas A, Menéndez JA (2013) Microwave pyrolysis of microalgae for high syngas production. Bioresour Technol 144:240–246. https://doi.org/10.1016/j.biortech.2013.06.102

Bharathiraja B, Chakravarthy M, Ranjith Kumar R et al (2015) Aquatic biomass (algae) as a future feed stock for bio-refineries: a review on cultivation, processing and products. Renew Sustain Energy Rev 47:634–653. https://doi.org/10.1016/j.rser.2015.03.047

Bhaskaran SK, Kannapan P (2015) Nutritional composition of four different species of Azolla. Pelagia Res Library Euro J Exp Biol 5:6–12

Biller P (2018) Hydrothermal liquefaction for aquatic feedstock. In: Zanol R (ed) Direct Thermochemical liquefaction for energy applications. Joe Hayton, India, pp 101–125

Biswas B, Arun Kumar A, Bisht Y et al (2021) Role of temperatures and solvents on hydrothermal liquefaction of Azolla filiculoides. Energy 217:119330. https://doi.org/10.1016/j.energy.2020.119330

Biswas B, Fernandes AC, Kumar J et al (2018) Valorization of Sargassum tenerrimum: value addition using hydrothermal liquefaction. Fuel 222:394–401. https://doi.org/10.1016/j.fuel.2018.02.153

Biswas B, Sahoo D, Sukumaran RK et al (2022) Co-hydrothermal liquefaction of phumdi and paragrass an aquatic biomass: characterization of bio-oil, aqueous fraction and solid residue. J Energy Inst 102:247–255. https://doi.org/10.1016/j.joei.2022.03.013

Boro M, Verma AK, Chettri D, Yata VK, Verma AK (2022) Strategies involved in biofuel production from agro-based lignocellulose biomass. Environ Technol Innov 28:102679. https://doi.org/10.1016/j.eti.2022.102679

Campanella A, Harold MP (2012) Fast pyrolysis of microalgae in a falling solids reactor: effsects of process variables and zeolite catalysts. Biomass Bioenergy 46:218–232. https://doi.org/10.1016/j.biombioe.2012.08.023

Chacon FJ, Cayuela ML, Cederlund H, Sanchez-Monedero MA (2022) Overcoming biochar limitations to remediate pentachlorophenol in soil by modifying its electrochemical properties. J Hazard Mater 426:127805. https://doi.org/10.1016/j.jhazmat.2021.127805

Chen W-H, Lin B-J, Huang M-Y, Chang J-S (2015) Thermochemical conversion of microalgal biomass into biofuels: a review. Bioresour Technol 184:314–327. https://doi.org/10.1016/j.biortech.2014.11.050
Das P, Chandramohan VP, Mathimani T, Pugazhendhi A (2021) Recent advances in thermochemical methods for the conversion of algal biomass to energy. Sci Total Environ 766:144608. https://doi.org/10.1016/j.scitotenv.2020.144608
Datta SN (2011) Culture of Azolla and its efficacy in diet of Labeo rohita. Aquaculture 310:376–379. https://doi.org/10.1016/j.aquaculture.2010.11.008
Debowski M, Dudek M, Zielinski M, Nowicka A, Kazimierowicz J (2021) Microalgal hydrogen production in relation to other biomass-based technologies—A review. Energies 14:6025. https://doi.org/10.3390/en14196025
Djandja OS, Yin L, Wang Z, Guo Y, Zhang X, Duan P (2021) Progress in thermochemical conversion of duckweed and upgrading of the bio-oil: a critical review. Sci Total Environ 769:144660. https://doi.org/10.1016/j.scitotenv.2020.144660
Duan P, Jin B, Xu Y, Wang F (2015) Co-pyrolysis of microalgae and waste rubber tire in supercritical ethanol. Chem Eng J 269:262–271. https://doi.org/10.1016/j.cej.2015.01.108
El-Naggar NE-A, Hussein MH, Shaaban-Dessuuki SA, Dalal SR (2020) Production, extraction and characterization of Chlorella vulgaris soluble polysaccharides and their applications in AgNPs biosynthesis and biostimulation of plant growth. Sci Rep 10:3011. https://doi.org/10.1038/s41598-020-59945-w
Fakayode OA, Wahia H, Zhang L, Zhou C, Ma H (2023) State-of-the-art co-pyrolysis of lignocellulosic and macroalgae biomass feedstocks for improved bio-oil production-A review. Fuel 332:126071. https://doi.org/10.1016/j.fuel.2022.126071
Fernandes AC, Biswas B, Kumar J et al (2021) Valorization of the red macroalga Gracilaria corticata by hydrothermal liquefaction: product yield improvement by optimization of process parameters. Bioresour Technol Rep 15:100796. https://doi.org/10.1016/j.biteb.2021.100796
Gao Z, Li N, Wang Y et al (2020) Pyrolysis behavior of xylan-based hemicellulose in a fixed bed reactor. J Anal Appl Pyrolysis 146:104772. https://doi.org/10.1016/j.jaap.2020.104772
Guo S, Dong X, Zhu C et al (2017) A simple modeling approach for characteristics analysis of hydrothermal liquefaction products from low-lipid aquatic plants. Appl Therm Eng 125:394–400. https://doi.org/10.1016/j.applthermaleng.2017.07.042
He S, Zhao M, Wang J, et al (2020) Hydrothermal liquefaction of low-lipid algae nannochloropsis sp. and Sargassum sp.: effect of feedstock composition and temperature. Sci Total Environ 712:135677. https://doi.org/10.1016/j.scitotenv.2019.135677
Hong Y, Chen W, Luo X et al (2017) Microwave-enhanced pyrolysis of macroalgae and microalgae for syngas production. Bioresour Technol 237:47–56. https://doi.org/10.1016/j.biortech.2017.02.006
Janiszewska D, Olchowski R, Nowicka A, Zborowska M, Marszałkiewicz K, Shams M, Giannakoudakis DA, Anastopoulos I, Barczak M (2021) Activated biochars derived from wood biomass liquefaction residues for effective removal of hazardous hexavalent chromium from aquatic environments. GCB Bioenergy 13:1247–1259. https://doi.org/10.1111/gcbb.12839
Khammee P, Ramaraj R, Whangchai N, Bhuyar P, Unpaprom Y (2021) The immobilization of yeast for fermentation of macroalgae Rhizoclonium sp. for efficient conversion into bioethanol. Biomass Convers Biorefinery 11:827–835. https://doi.org/10.1007/s13399-020-00786-y
Kumar N, Liu Y, Chen S (2022) Advances in metabolic engineering of cyanobacteria for production of biofuels. Fuel 322(July 2021):124117. https://doi.org/10.1016/j.fuel.2022.124117
Lee XJ, Lee LY, Gan S et al (2017) Biochar potential evaluation of palm oil wastes through slow pyrolysis: thermochemical characterization and pyrolytic kinetic studies. Bioresour Technol 236:155–163. https://doi.org/10.1016/j.biortech.2017.03.105
Lee XJ, Ong HC, Gan YY et al (2020) State of art review on conventional and advanced pyrolysis of macroalgae and microalgae for biochar, bio-oil and bio-syngas production. Energy Convers Manag 210:112707. https://doi.org/10.1016/j.enconman.2020.112707

Liu X, Xie H, Roussou S, Lindblad P (2022) Current advances in engineering cyanobacteria and their applications for photosynthetic butanol production. Curr Opin Biotechnol 73:143–150. https://doi.org/10.1016/j.copbio.2021.07.014

Ly HV, Kim S-S, Woo HC et al (2015) Fast pyrolysis of macroalga Saccharina japonica in a bubbling fluidized-bed reactor for bio-oil production. Energy 93:1436–1446. https://doi.org/10.1016/j.energy.2015.10.011

Ly HV, Choi JH, Woo HC et al (2019) Upgrading bio-oil by catalytic fast pyrolysis of acid-washed Saccharina japonica alga in a fluidized-bed reactor. Renew Energy 133:11–22. https://doi.org/10.1016/j.renene.2018.09.103

Lynam JG, Reza MT, Yan W, Vásquez VR, Coronella CJ (2015) Hydrothermal carbonization of various lignocellulosic biomass. Biomass Convers Biorefinery 5:173–181. https://doi.org/10.1007/s13399-014-0137-3

Ma M, Xu D, Zhi Y et al (2022) Co-pyrolysis re-use of sludge and biomass waste: development, kinetics, synergistic mechanism and industrialization. J Anal Appl Pyrolysis 168:105746. https://doi.org/10.1016/j.jaap.2022.105746

Maddi B, Viamajala S, Varanasi S (2011) Comparative study of pyrolysis of algal biomass from natural lake blooms with lignocellulosic biomass. Bioresour Technol 102:11018–11026. https://doi.org/10.1016/j.biortech.2011.09.055

Malode SJ, Aaqueeb S, Gaddi M, Kamble PJ, Nalwad AA, Muddapur UM, Shetti NP (2022) Recent evolutionary trends in the production of biofuels. Mater Sci Energy Technol 5:262–277. https://doi.org/10.1016/j.mset.2022.04.001

Mayala TS, Ngavouka MDN, Douma DH, Hammerton JM, Ross AB, Brown AE, Passi-mabiala BM, Lovett JC (2022) Characterisation of congolese aquatic biomass and their potential as a source of bioenergy. Biomass 2:1–13. https://doi.org/10.3390/biomass2010001

Mejica GFC, Unpaprom Y, Whangchai K, Ramaraj R (2022) Cellulosic-derived bioethanol from Limnocharis flava utilizing alkaline pretreatment. Biomass Convers Biorefinery 12:1737–1743. https://doi.org/10.1007/s13399-020-01218-7

Moazezi MR, Bayat H, Tavakoli O, Hallajisani A (2022) Hydrothermal liquefaction of Chlorella vulgaris and catalytic upgrading of product: effect of process parameter on bio-oil yield and thermodynamics modeling. Fuel 318:123595. https://doi.org/10.1016/j.fuel.2022.123595

Nazos A, Politi D, Giakoumakis G, Sidiras D (2022) Simulation and optimization of lignocellulosic biomass wet- and dry-torrefaction process for energy fuels and materials production. Energies 15:9083. https://doi.org/10.3390/en15239083

Parvej AM, Rahman MA, Reza KMA (2022) The combined effect of solar assisted torrefaction and pyrolysis on the production of valuable chemicals obtained from water hyacinth biomass. Clean Waste Syst 3:100027. https://doi.org/10.1016/j.clwas.2022.100027

Patel SKS, Das D, Chang S, Cho B, Chandra V, Lee J (2021) Integrating strategies for sustainable conversion of waste biomass into dark-fermentative hydrogen and value-added products. Renew Sustain Energy Rev 150(May):111491. https://doi.org/10.1016/j.rser.2021.111491

Poomsawat S, Poomsawat W (2021) Analysis of hydrochar fuel characterization and combustion behavior derived from aquatic biomass via hydrothermal carbonization process. Case Stud Therm Eng 27:101255. https://doi.org/10.1016/j.csite.2021.101255

Qiu B, Tao X, Wang J et al (2022) Research progress in the preparation of high-quality liquid fuels and chemicals by catalytic pyrolysis of biomass: a review. Energy Convers Manag 261:115647. https://doi.org/10.1016/j.enconman.2022.115647

Roberts DA, Paul NA, Cole AJ, De NR (2015) From wastewater treatment to land management: conversion of aquatic biomass to biochar for soil amelioration and the fortification of crops with essential trace elements. J Environ Manag 157:60–68. https://doi.org/10.1016/j.jenvman.2015.04.016

Roddy DJ, Manson-Whitton C (2012) Biomass gasification and pyrolysis. In: Comprehensive renewable energy. Elsevier, pp 133–153

Rosemary T, Arulkumar A, Paramasivam S et al (2019) Biochemical, micronutrient and physicochemical properties of the dried red seaweeds gracilaria edulis and gracilaria corticata. Molecules 24:2225. https://doi.org/10.3390/molecules24122225
Ruiz HA, Rodrıguez-Jasso RM, Fernandes BD, Vicente AA, Teixeira JA (2013) Hydrothermal processing, as an alternative for upgrading agriculture residues and marine biomass according to the biorefinery concept: a review. Renew Sustain Energy Rev 21:35–51. https://doi.org/10.1016/j.rser.2012.11.069
Tripathi M, Sahu JN, Ganesan P (2016) Effect of process parameters on production of biochar from biomass waste through pyrolysis: a review. Renew Sustain Energy Rev 55:467–481. https://doi.org/10.1016/j.rser.2015.10.122
Uzoejinwa BB, He X, Wang S et al (2019) Co-pyrolysis of macroalgae and lignocellulosic biomass. J Therm Anal Calorim 136:2001–2016. https://doi.org/10.1007/s10973-018-7834-2
Uzoejinwa BB, He X, Wang S et al (2018) Co-pyrolysis of biomass and waste plastics as a thermochemical conversion technology for high-grade biofuel production: recent progress and future directions elsewhere worldwide. Energy Convers Manag 163:468–492. https://doi.org/10.1016/j.enconman.2018.02.004
Viegas C, Nobre C, Correia R, Gouveia L, Goncalves M (2021) Optimization of biochar production by co-torrefaction of microalgae and lignocellulosic biomass using response surface methodology. Energies 14:7330. https://doi.org/10.3390/en14217330
Vuppaladadiyam AK, Vuppaladadiyam SSV, Awasthi A et al (2022) Biomass pyrolysis: a review on recent advancements and green hydrogen production. Bioresour Technol 364:128087. https://doi.org/10.1016/j.biortech.2022.128087
Wauton I, Ogbeide SE (2019) Characterization of pyrolytic bio-oil from water hyacinth (Eichhornia crassipes) pyrolysis in a fixed bed reactor. Biofuels 899–904. https://doi.org/10.1080/17597269.2018.1558838
Yek PNY, Mahari WAW, Kong SH, Foong SY, Peng W, Ting H, Liew RK, Xia C, Sonne C, Tabatabaei M, Almomani F, Aghbashlo M, Lam SS (2022) Pilot-scale co-processing of lignocellulosic biomass, algae, shellfish waste via thermochemical approach: recent progress and future directions. Bioresour Technol 347:126687. https://doi.org/10.1016/j.biortech.2022.126687
Yu M, Wang K, Vredenburg H (2020) Insights into low-carbon hydrogen production methods: green, blue and aqua hydrogen. Int J Hydrogen Energy 46:21261–21273. https://doi.org/10.1016/j.ijhydene.2021.04.016
Zhang R, Li L, Tong D, Hu C (2016) Microwave-enhanced pyrolysis of natural algae from water blooms. Bioresour Technol 212:311–317. https://doi.org/10.1016/j.biortech.2016.04.053
Zhang B, Wang L, Ghimire S, Li X, Scott M, Abolghasem T (2021) Enhanced biomethane production via thermophilic anaerobic digestion of cattail amended with potassium phosphate- and magnesium-modified biochar. Clean Technol Environ Policy 23:2399–2412. https://doi.org/10.1007/s10098-021-02154-4

The Production of Microalgae and Cyanobacteria Biomass and Their Valuable Bioproducts

Awalina Satya, Azalea Dyah Maysarah Satya, Tjandra Chrismadha, Nofdianto Nofdianto, Gunawan Gunawan, Ika Atman Satya, Souvia Rahimah, Efri Mardawati, Sara Kazemi Yazdi, and Pau-Loke Show

Abstract Microalgae and cyanobacteria produced abundant high-valued bioproducts in small arable land in a short time. The bioproducts range from their biomass for food, feed, and biofuels to extractable fine bioproducts. The growing market and techno-economical aspect support the viability of this biomass production. Microalgae and Cyanobacteria are also highly diverse thus progression of its current usage in biomass production served as a challenge of its own but also an opportunity. In this book chapter, progression in cultivating and screening of technologies on bioprocess engineering of microalgae and cyanobacteria will be discussed with their high-demands on food, feed, and energy industry. This chapter further discusses the advance and manufacturer of different valuable bioproducts through technologies and production platforms for Microalgae and Cyanobacteria.

A. Satya (✉) · T. Chrismadha · N. Nofdianto · G. Gunawan
Research Centre for Limnology and Water Resources, National Research and Innovation Agency, KST Soekarno, Cibinong, Indonesia
e-mail: awalina@brin.go.id

A. D. M. Satya · S. K. Yazdi · P.-L. Show
Department of Chemical and Environmental Engineering, Faculty of Science and Engineering, University of Nottingham Malaysia, Jalan Broga, 43500 Semenyih, Selangor Darul Ehsan, Malaysia

I. A. Satya
Research Centre for Computation, National Research and Innovation Agency, KST Soekarno, Cibinong, Indonesia

S. Rahimah · E. Mardawati
Department of Agro-Industrial Technology, Faculty of Agro-Industrial Technology, Padjadjaran University, Jatinangor, Indonesia

P.-L. Show
Department of Sustainable Engineering, Saveetha School of Engineering, SIMATS, Chennai 602105, India

Zhejiang Provincial Key Laboratory for Subtropical Water Environment and Marine Biological Resources Protection, Wenzhou University, Wenzhou 325035, China

M. A. R. Lubis et al. (eds.), *Biomass Conversion and Sustainable Biorefinery*, Green Energy and Technology, https://doi.org/10.1007/978-981-99-7769-7_7

Keywords Microalgae · Cyanobacteria · Cultivation technology · Fine biochemicals

1 Introduction

Microalgae and cyanobacteria consist of a broad spectrum of photoautotrophic microorganisms which grow through photosynthesis. The conversion of chemical energy from solar as a unicellular form made them known as the oldest life form/thallophytes (primitive plants) known (Abreu et al. 2022). As primitive plants, they exhibit an absence of roots, stems, and leaves and possess chlorophyll-a as the main photosynthetic pigment for energy conversion. This characteristic enables them to adapt toward the predominant environmental circumstances, hence allowing said lifeforms to flourish for a long period of time (Kumar et al. 2020). As for the types of the cells, cyanobacteria—as the oldest of the two—have prokaryotic cells which are characterized by the lack of membrane-bound organelles usually present around plastids, mitochondria, nuclei, Golgi bodies, and flagella. Meanwhile, microalgae are eukaryotic cells which have these organelles that control the function of cells. Microalgae are mainly eukaryotes grouped into many classes defined accordingly to their pigmentation, life cycle, and basic cellular structure; with green algae (Chlorophyta), red algae (Rhodophyta), and diatoms (Bacillariophyta) posing as the three most important classes in microalgae. Both microalgae and cyanobacteria can be either autotrophic (inorganic compounds such as CO_2, salts, and light as prerequisites) or heterotrophic (with an external source of organic compounds and nutrients due to their non-photosynthetic nature). Several photosynthetic algae or cyanobacteria are mixotrophic (capable to perform both photosynthesis and acquire exogenous organic nutrients). Autotrophs rely on photosynthesis for their survival, as they convert solar irradiance and absorbed CO_2 by chloroplast into adenosine triphosphate (ATP) and O_2. Both ATP and O_2 are usable energy packages at the cellular level, which are then used in respiration to produce energy to support their growth (Farhan et al. 2017).

Recently there are numerous studies related to techniques and large-scale production of microalgal and cyanobacterial biomass (Ugoala et al. 2012). In general, there are two types of cultivation techniques, namely open pond system and closed photobioreactor system. Both systems have advantages and disadvantages, therefore preference for the being used system depends on the characteristics of targeted products (Milledge 2011). Difficulties in controlling the contamination and predation in open pond systems frequently occur. On the contrary, photobioreactor systems enable to control of nutrients for growth and operating parameter such as pH, temperature, dissolved $CO_{2,}$ and contamination/predation. Unfortunately, photobioreactor systems need a high investment cost and are quite specific to strains of microalgal and cyanobacterial physiologies which being cultivated (Acién Fernández et al. 2013). Therefore, a consideration of the production to facilitate an optimum production for specific microalga or cyanobacterium is quite important. The harvesting process is conducted for dewatering the algal and cyanobacterial biomass. Methods for this purpose are flocculation, centrifugation, and filtration. A favorable dewatering process for harvesting must apply to a wide range of microalgal and cyanobacterial strains. This aims to concentrate biomass recovery and cost-effective production. The important matter to consider for mass biomass production is combining cultivation and dewatering processes at the possibly lowest cost while maximizing the microalgal and or cyanobacterial biomass production. Bioprocess engineering in microalgal and cyanobacterial biomass production provides various downstream products for commercial purposes. Microalgal and cyanobacterial biomass contains an abundant bioactive compound that useful being used in many important industries such as pharmaceuticals (for the manufacture of antioxidants, antibiotics, immunomodulators, etc.) (Mobin and Alam 2017; Kholssi et al. 2021). Meanwhile, for human consumption, microalgae biomass can be extracted to obtain its high protein contents, vitamins, and polysaccharides (Catone et al. 2021; Hernández et al. 2015). Some microalgae and cyanobacteria are known to contain high lipids which are then extracted (oil press, solvent extraction, supercritical fluid extraction and ultrasound) and converted (trans-esterified) into biofuels (Castro et al. 2021; Felix et al. 2019). Furthermore, the residue of lipid-extracted microalgal and cyanobacterial biomass can be converted into other forms of biofuels, such as biomethane, bioethanol, and biohydrogen (Felix et al. 2019; Nitsos et al. 2020).

Microalgae and cyanobacteria also display a capability to overcome emerging environmental issues, for example, the greenhouse effect and water (industrial, domestic, and agricultural) pollution (Gil-izquierdo et al. 2021; Liu et al. 2021). These microorganisms can sequester CO_2 from flue gas for their photosynthetic activity and reduce aquatic nutrients efficiently from wastewater at minimal cost (Song et al. 2019). Some species of microalgae and cyanobacteria show the capability to fix nitrogen and absorb phosphorus as well as heavy metals from wastewater (Gonçalves et al. 2017; Singh and Ahluwalia 2013; Satya et al. 2017, 2021a, b; Vijayaraghavan and Balasubramanian 2015). Those facts demonstrated that microalgae and cyanobacteria can provide a promising solution to address emerging environmental problems and concomitantly generate many valuable consumer products. Conceptually, the production of microalgal and cyanobacterial biomass can be generated from the carbon recycling process (Fig. 1).

This chapter book discusses the different cultivation, harvesting, and processing methods of microalgae and cyanobacteria for producing several bioproducts. These materials involve biofuels, fine biochemicals, and food/functional food. The prospect of microalgae and cyanobacteria overcoming emerging environmental problems is also delivered in this chapter book.

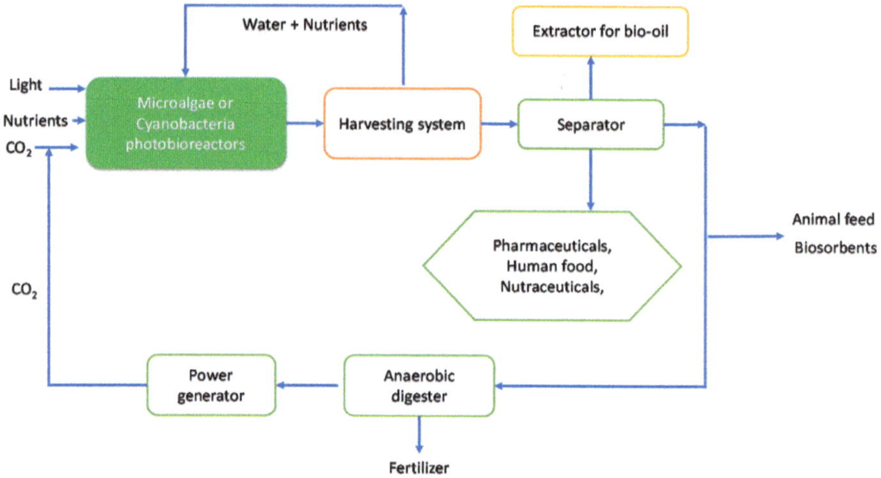

Fig. 1 Conceptual diagram for producing microalgae and cyanobacteria biomass adapted from Chisti (2007)

2 Microalgae Cultivation

2.1 Open System

An open system for cultivating microalgae and cyanobacteria is generally found as an open pond in a variety of shapes and sizes. There are some advantages and disadvantages related to the implementation of this cultivation system. Examples of open pond types are raceway pond with paddle wheels (Fig. 2a), open tanks (Fig. 2b), shallow big ponds (Fig. 2c), circular ponds, etc. The place where the pond is located becomes a determining factor for choosing the proper type of pond the selected strain of microalgal or cyanobacterial being used and the availability of light for photosynthesis. The open pond is a function of the local climate; therefore, the chosen location will affect the achievement of cultivation. Several key growth parameters (solar irradiance, temperature, pH, and concentration of dissolved oxygen) are limiting the performance of the open pond (Ashokkumar et al. 2014). Another critical problem for an open pond system is the occurrence of predation due to the higher risk of contamination. Only several microalgal or cyanobacterial can be grown successfully in an open pond even in severe environmental conditions (such as *Dunaliella* in high salinity, *Arthrospira* in high alkalinity and *Chlorella* in high aquatic nutrients). *Dunaliella salina* was cultivated for producing carotenoids, this compound in nature protects this microalga against the high intense sunlight during grown in open pond. Other reports mentioned that *Chlorella sp* achieved a decent photosynthetic activity in raceway system, while *Muriellopsis sp.* was cultured for producing lutein in an open tank equipped with a paddle wheel (Blanco et al. 2007). These biomasses can

Fig. 2 Open ponds system for microalgal and cyanobacterial cultivation: raceways with paddle wheel (**a**); open tanks (**b**); shallow big pond (**c**). Both (**a**) and (**b**) locates in RC for Limnology and water resources-NRIA, Cibinong while (**c**) locates in PT. Albitech, Semarang Regency-Indonesia

be used as food colorant, feed additives in aquaculture, and poultry. Cost for cultivating microalga and cyanobacterium is an important factor for considering a choice between using an open system and photobioreactor. The investment and maintenance costs in open pond construction is less than photobioreactor, even so the biomass productivity in open pond is lower than photobioreactor and biomass quality in open pond is more variable compared to photobioreactor (Ugoala et al. 2012; Debeni Devi et al. 2022). Consequently, for providing the needs of bulk requirement of biomass (e.g., for producing biofuel), an open pond form is preferable.

2.2 Photobioreactor System

Photobioreactor provides better control on most operational parameter compared to open pond system, therefore extensive research on designing photobioreactors for cultivating microalgae and cyanobacteria is a must (Huang et al. 2017; Duan and Shi 2014). Higher biomass productivity can be achieved in a controlled environment which is the advantage of using this system. Careful considerations, however, are needed since the performance of a bioreactor is assessed from its productivity. These considerations stem from difficulties in comparing the productivity among photobioreactors because the difference in microalgae strains and scales of the photobioreactors used. In the principle, photobioreactor types are differentiated into tubular and plate shapes. The tubular reactors are considered to be more suitable for outdoor cultivation. It is because of the use of transparent material for configuring the tubes, which exhibits large illuminated surfaces. Various configurations can be made depending on the specification of the system and its purposes. Generally, those tubing configurations may be found in straight-line forms and or coiled forms (Johnson et al. 2018; Nwoba et al. 2019). Geometry of the photobioreactor also determines its performance, as tubular reactors can be configured in inclined, horizontal, or vertical planes. The vertical design enables better mass transfer and requires lower energy supply. In the case of the horizontal photobioreactor design, a larger area is required than vertical design. In terms of scalability, horizontal design is more preferred (Sirohi et al. 2022; Xia et al. 2013).

Tubular photobioreactors for culturing microalgae and cyanobacteria can be configured as vertical, horizontal, and helical forms. Another configuration is flat-plate photobioreactors which are characterized by narrow lightpath and therefore able to maintain a higher cell concentration up to an order magnitude than another configuration. Moreover, this configuration type of photobioreactor is favorable since it allows a lower power energy consumption and high mass transfer capacity. It also allows a reduction in oxygenic accumulation, providing no dark volume and high photosynthetic efficiency. A proper photobioreactor design is needed to obtain the maximum biomass cell production. The flat-plate photobioreactor for culturing microalgae and cyanobacteria may be constructed in the form of glass, thick transparent polyvinyl chloride materials, V-shaped, and inclined. The translucent material gives a maximum light penetration meanwhile other materials for designing are inexpensive and easy to construct. Figure 3 describes the basic diagram of a tubular photobioreactor (3a) and a real view of a tubular photobioreactor system on a pilot scale.

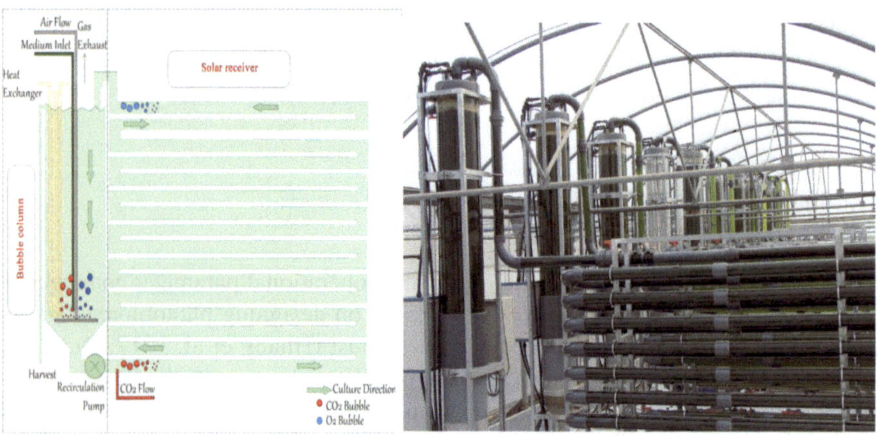

a b

Fig. 3 Schematic basic diagram of a tubular photobioreactor (3a) and a real scene of tubular photobioreactor at a pilot scale. After Fernandez et al. (2014)

The two existing cultivation systems present their own advantages and disadvantages with few similarities in operation as seen in Table 1. Optimization of the two systems can be done by either change of several operative conditions or even combining the two systems. The goal of combining the two systems is to gain higher biomass yield with few design parameters. Important design parameters for tubular photobioreactors are mixing, gas hold-up, bubble diameter, and intensity of light and dark cycles. Meanwhile, important design parameters for flat-plate photobioreactors are mixing, gas hold-up, bubble diameter, light-to-dark cycle efficiency, and illuminated surface-to-volume ratio (Cui et al. 2021). The factors affecting missing in flat-plate photobioreactors are much more than that in tubular reactors. In flat-plate photobioreactors, aside from resident time distribution (RTD) and circulation time which are present in a tubular reactor's mixing factor, shear rate played a bigger role as sheer stress affects cells, nutrients, temperature, and eventually toxic levels of dissolved oxygens and carbon demands with novel system dealing with the particular problem showed significant improvement of biomass production up to 61% (Yaqoubnejad et al. 2021). There are also significantly more hardware designs needed in developing a tubular reactor. These harder designs in addition to sparger designs are methods of mixing and pumping. However, the tubular reactor offered much more sparger designs (orifice, ring, foam types) to allow some enhancement methods. These enhancement methods are the incorporation of static mixers, static kinetics mixers, helical mixers, or swirl flow for a tangential inlet (Sirohi et al. 2022; Sung et al. 2022). Flat-plate photobioreactors provide easier control in massive microalgae biomass production but its maintenance and large light areas are their major bottleneck. For outdoor cultivation, solar-irradiation filtration technologies should be considered for garnering multiple advantages in addition to combining the existing systems (Wu et al. 2023; Huang et al. 2023).

3 Harvesting of Microalgae

3.1 Flocculation

Flocculation is the first stage in the bulk harvesting process. This stage aims to aggregate the microalgae and cyanobacteria cells to increase the effective particle size for harvest. This method is usually applied as the initial step in harvesting (dewatering) which significantly facilitates the next processing steps. Their effectiveness depends on their ionic charge. Microalgal and cyanobacterial cells pose a negative charge, therefore, repulsed themselves from aggregating into suspension. The surface charge of microalgal and cyanobacterial cells can be neutralized by adding chemicals such as flocculating agents (flocculants). These cationic compounds coagulate the suspended microalgal and cyanobacterial cells without affecting the composition and harming the cells and not poisoning the product as can be seen in Fig. 4. The typical flocculants are multivalent salts such as $FeCl_3$, $Al_2(SO_4)_3$, and $Fe_2(SO_4)_3$.

Table 1 Advantages and Disadvantages of Tubular and Flat-panel photobioreactors

Types of PBRs	Tubular reactors	Flat-plate reactors
Advantages	• Lower land needs to produce a considerable amount of biomass • The photostage loop and mixing ensure higher concentrations of biomass cultures • Reducing power consumption and prevention of cell injury is possible by controlling liquid velocity from 0.1 to 0.8 ms^{-1}	• Higher photosynthetic efficiency due to its more surface-per-volume ratio • Possible and easier optimization with a reduction in energy consumption by factoring in environmental conditions (inclination, panel spacing, and light pathways) • Comparatively minimal deposition of dissolved O_2
Disadvantages	• Transmission of mass is lower with the increase in size • Prone to oxygen build-up • Optimizing scale-up between width and height	• Lower biomass per areal yields • Scale-up process requires a safely laminated surface • Requires a larger area and one more design parameter
Similarities	• Use of transparent material is preferred to ensure illuminance • Sparger design is applicable to enhance their mixing effectivity	• Use of transparent material is preferred to ensure illuminance • Sparger design is applicable to enhance their mixing effectivity

Combined information obtained from Sirohi et al. (2022), Yaashikaa et al. (2022)

Cationic polymers (those are polyelectrolytes) also can be used as flocculants by physically linking cells together, thus advancing little to no disruptions on the cells. Key polymer characteristics involved are charge, molecular weight, and concentration. The preference for polymer types depends on the properties of microalgae or cyanobacteria cultures such as charge in suspension, pH, and biomass concentration (Wu et al. 2012). Harvesting using Fe [III] flocs induced with pH improved efficiency up to 80% (Knuckey et al. 2006). Flocculation using $FeCl_3$ can be suppressed by exopolymers released by *Aphanotece halophytica* therefore extra addition of this flocculant is needed (Chen et al. 2009). *Chlorella* was found better to be flocculated using cationic polyelectrolytes, while anionic polyelectrolytes gave no flocculation.

The use of organic flocculants is advantageous due to their stability by being less sensitive to pH, allowing them for a wide range of applications and in most cases needing a lower dose of flocculant depending on the presence of ions in mediums. Brackish and saline waters, for example, needed more chemical flocculants due to the presence of competing cations (Abbaslou et al. 2020). Aside from cationic ions, natural flocculants can also be used such as in the case of *Oscillatoria, Spirulina, Chlorella,* and *Synechocystis* flocculated using chitosan. The dose of given flocculants depends on algal or cyanobacterial species. For example, dose of 40 mg/L chitosan is effective to flocculate *Tetraselmis chui, Thalassiosira pesudonana*, and *Isochrysis sp*, while *Chaetoceros muellaris* required 150 mg/L of chitosan (Divakaran and Sivasankara Pillai 2002). Auto-flocculation can also naturally occur by interrupting the CO_2 supply in culture microalgal or cyanobacterial broth. This fact attributes to elevated pH through photosynthetic CO_2 consumption related to

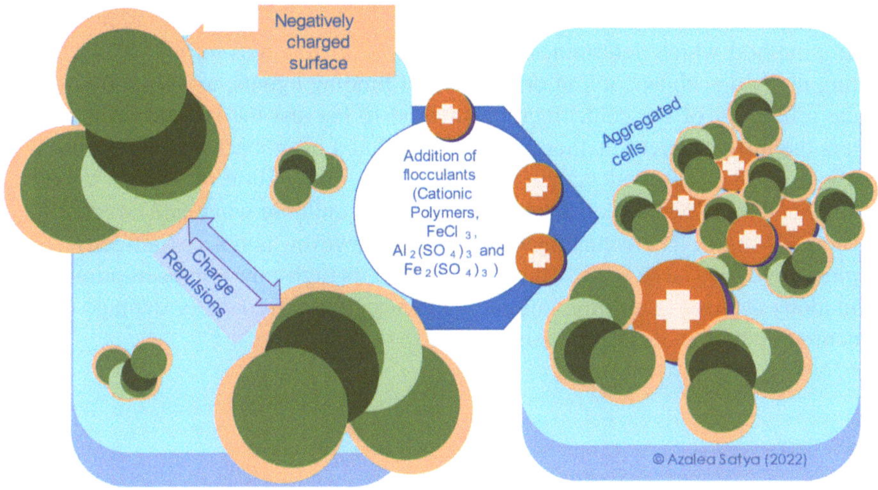

Fig. 4 Flocculants in aggregating microalgal and cyanobacterial cells. The neutralization provided by flocculants allowed for microalgae to aggregate

salt precipitations containing calcium, magnesium, phosphate, and carbonate. The positive charge of calcium phosphate is prone to react with the negative charge of microalgal or cyanobacterial cells then leads to flocculation (Sukenik and Shelef 1984).

3.2 Centrifugation

The most preferred method for harvesting microalgal and cyanobacterial cells is centrifugation. Centrifugation is conducted by using centripetal acceleration to separate the microalgal growth medium into sections depending on their densities which correlate with their growth. Mature cells will be obtained in the lower part of the centrifugation vessel. The separated mature cells of microalgae or cyanobacteria biomass (supernatant) can later be obtained by simply draining the medium solution from the saturated supernatant at the bottom of the centrifuge cells. This method is reasonable for microalgae harvesting with one drawback: the shear forces during the spinning can disrupt cells. That prevents the faster centrifugation speed which would generate higher separation capability. There are several key parameters to look for in applying centrifugation for large-scale harvesting of microalgae. These factors are concentration which determines the rate of rotation needed to separate; energy consumption which correlates with the amount of energy needed to power the rotation; the cost which is how much of a percentage the whole harvesting process would require; operation mode which determines the size of the operation; concentrating method which determines whether or not there needs to be a prior concentrating method such as cooling or addition of binding agents; and last but not least is the reliability of the centrifugation method to harvest the biomass. Centrifugation is the most efficient method for microalgal and cyanobacterial harvesting (95–100% with 88–100% cell viability) compared to drum filtration and dissolved air floatation. Centrifugation on a laboratory scale is suitable when cell concentration is about 30 mg/L. The centrifugation method, however, is not cost-effective when implemented on a large scale. This is because of its high-power consumption. Other limitations are its higher gravitational and shear stress frequently damage the cell structure (Ashraf et al. 2020; Shao et al. 2015).

3.3 Filtration

The filtration method is preferable in harvesting microalgae and cyanobacteria compared to other harvesting methods. Filtration methods consisted of several types such as dead-end filtration, microfiltration, ultra-filtration, pressure filtration, vacuum filtration, and tangential filtration. In the principle, filtration is implemented by passing through the culture broth through filters then biomass will accumulate on the filter membrane while the medium passes through the filter continuously until the filter contains a cake of algal or cyanobacterial biomass (Mohan and Sivasubramanian 2010). To further suitably concentrate the biomass, filter presses under pressure or vacuum are added into the system for microalgae or cyanobacterium with larger sizes for example *Arthrospira (Spirulina) platensis*, but not for smaller sizes like *Chlorella* and *Dunaliella.* Tangential flow filtration and pressure filtration are considered energy-efficient methods for separating microalgal or cyanobacterial biomass from its medium culture. This fact is suggested by the amount of output and input of the feedstock (Shao et al. 2015). Back mixing becomes a drawback to the use of dead-end filtration; however, this simple filtration method can be combined with centrifugation for improving the separation process. In general, the filtration method is frequently associated with wide-ranging running costs and concealed pre-concentration requirements (Senatore et al. 2022; Morais et al. 2020).

4 Bioproducts from Microalgae

4.1 Biodiesel

Generation of biodiesel is a process of breaking down vegetable oils or animal fats that contain triglycerides which comprise of three fatty acid chains linked with a glycerol molecule. In the process of generating biodiesel, glycerol substitutes with methanol which forms fatty acid methyl ester more commonly known as biodiesel. A phase separation method was then implemented to separate glycerol (as a by-product) from biodiesel. The process is denoted as transesterification which is a process of replacing methanol for glycerol in a chemical reaction with an acid or alkali catalyst. The encounters of replacing conventional diesel with biodiesel are: (1) biodiesel feedstock must be sufficient at a commercial scale, (2) must have a lower price than conventional fossil fuel, and (3) meet standard specifications of fuel quality. Those reasons are met by the microalgal and cyanobacterial biomasses for biofuel since it can provide raw material at a cheaper yet faster biomass productivity rate reaching up to 50 times magnitudes than ordinary terrestrial plants with adequate lipid contents fraction for biodiesel. Their lipids are mostly neutral lipids with a lower level of unsaturated grade. According to Chisti (2007) *Botryococcos braunii, Nannochloropsis sp,* and *Schizochytrium sp.* respectively contain lipids ranging between 25–75%, 31–68%, and 50–77%, respectively.

Many methods for algal/cyanobacterial lipid extraction are known, but the most commonly practiced methods of algal/cyanobacterial lipid extraction are by the use of an expeller, supercritical fluid extraction (SFE), solvent extraction (liquid–liquid extraction), and various ultrasound techniques. In the method of expeller/oil press, the biomass of microalgae must be dried for effective extraction, cell breaking was then conducted by pressure which squeezed the oil out. In addition to that, this method was only capable to extract 75% of oil with a longer extraction time. Solvent extraction is excellent to extract from microalgae and cyanobacteria, faster and simpler than the SFE method (Jacob-Lopes and Franco 2013).

In the solvent extraction method, organic solvents (benzene, cyclohexane, hexane, acetone, or chloroform) are added to algal/cyanobacterial paste. The solvent's function is to destroy algal/cyanobacterial cell walls which allows the extraction process to ensue. The extraction will result in the formation of a lipid layer on top of the aqueous medium since their higher solubility in organic solvents than in water (medium). Then the solvent extract can be distilled to separate the oil from the solvent. The solvent can later be reused. Among the reusable solvent in this process, Hexane is the most efficient solvent due to its lower cost and higher extraction capacity. Lipid extraction also can be conducted in two steps using ethanol then followed secondly with hexane. This procedure is aimed to purify extracted lipids with a yield recovery of 80%. Temperatures can also improve this extraction method. Some studies reported that fatty acids were always nearly extractable at 100 °C compared to ambient temperature mainly saturated acids (16:0, 18:0), but polyunsaturated fatty acids (18:2;18:3) or PUFA resulted in lower yield with hot propanol-water (3:1 v/v). However, fatty acids content varied with microalgal strains, and the solvents for extraction (such as chloroform, and methanol) were also hazardous and destructive to the environment and human health (Slade and Bauen 2013).

Supercritical extraction (SFE) ruptures the microalgal cells through high pressures and temperatures. This method is extremely time efficient and is commonly used. The implemented high pressures and temperatures did not give any effect on the yield of the extracted compounds but affected the rate of extraction. It was observed in lipid extraction from *Nannochloropsis sp* for obtaining PUFA using SFE at 45 and

55 °C, 400–700 bar. A higher yield result was found when SFE was used in *Spirulina platensis* for obtaining PUFA compared to extraction using solvent (Amorim et al. 2020; Bleakly and Hayes 2017). Ultrasound method for lipid extraction from microalgae and cyanobacteria is also promising. This method treats microalgae and cyanobacteria with high-intensity of ultrasonic waves which form minute cavitation bubbles around cells. The shockwaves resulting from collapsing bubbles will shatter and rupture the cells which then released desired compounds into solution. Fatty acids and pigment extraction from *Scenedesmus obliquus* using ultrasound showed over 90% without any changes or breakdown in the product (related to time storage), while almost complete extraction of lipid was achieved in *Chaetoceros gracilis.* Ultrasonic can increase the rate of extraction of oil content in microalgae at a laboratory scale, further study on its feasibility for commercial production in this method is needed (Vandamme et al. 2013; Kumar et al. 2017).

4.2 Bioethanol

Bioethanol is commonly produced through a biochemical process (fermentation) and thermochemical process (gasification) of biomass sources. The conventional biomass sources (sugar cane, corn, and bit) have a general acute problem that is high value for food utilization and the requirement on land to be cultivated. Therefore, this problem is a constraint for expanding biofuel production. Alternatively, microalgal and cyanobacterial biomass as feedstock for the fermentation process in bioethanol production eludes those problems. These biomasses contain carbohydrates and proteins (Table 2) which enable being used as carbon sources in the fermentation process.

Other microorganisms such as fungi, bacteria, and yeast (*Saccharomyces saravesei*) are ordinarily used for fermenting the microalgal and cyanobacterial carbohydrates under anaerobic conditions for the production of bioethanol. Theoretically, the maximum yield is 0.51 kg ethanol and 0.49 kg CO_2 per kg of glucose. The produced bioethanol then can be purified for producing biofuel while produced CO_2 can be recycled for cultivating microalgae as a growth nutrient source. In the second stage, the remaining biomass after fermentation can be used as feedstock for the anaerobic digestion process resulting in methane (CH_4) gas which can later be converted into electrical power (Demirbas 2010, 2011).

Hon-Nami (2006) mentioned that *Chlamydomonas periglanulata* can be fermented for producing ethanol, butanediol, acetic acid, and CO_2. They also reported that the recovery of H_2 and carbon was attained by 139 and 105%. Some advantages of microalgal and cyanobacterial fermentation for producing bioethanol are the less requirement for energy consumption and more simplicity of the process compared to the conventional biodiesel production system. Even so, the production of bioethanol from microalgal and cyanobacterial biomass still needs further research for commercialization.

Table 2 Carbohydrates and Proteins (%dry weight) in several Microalgal and Cyanobacterial biomasses

Cyanobacteria	Carbohydrates (% dry weight)	Proteins (% dry weight)
Spirulina platensis	8–14	46–63
Spirulina maxima	13–16	60–71
Synechoccus sp.	15	63
Anabaena cylindrica	25–30	43–56
Microalgae		
Scenedesmus obliquus	10–17	50–56
Scenedesmus quadricauda	–	47
Scenedesmus dimorphus	21–52	8–18
Chlamydomonas reinhardtii	17	48
Chlorella vulgaris	12–17	51–58
Chlorella pyrenoidosa	26	57
Spirogyra sp.	23–64	6–20
Dunaliella bioculata	4	49
Dunaliella salina	32	57
Euglena gracilis	14–18	39–61
Prymnesium parvum	25–33	28–45
Tetraselmis maculata	15	52
Porphyridium cruentum	40–57	28–39

Modified from Becker (1994)

4.3 Biomethane

The methane fermentation technology on microalgae *Chroococcus sp.* and *Tetraselmis sp. Chlorella sp.* (Koutra et al. 2018) and cyanobacteria (*Aphanizomenon ovalisporum* and *Anabaena planktonica*) biomass are perspective since they can produce economical by-products. One of these by-products is biogas (Catone et al. 2021). Biogas consists of a mixture of 55–75% of CH_4 and 25–45% of CO_2 during microbial anaerobic digestion. The CH_4 can later be converted into electricity and fuel gas. As for the residual biomass that is left after the process, they can be processed into biofertilizer. Therefore, these processes yield support renewable and sustainable agricultural production systems by improving efficient practices and lowering microalgal/cyanobacterial production costs. Microalgae and cyanobacteria are considered to have no lignin and lower cellulose ingredient, thus processes of turning their biomass would be considered faster than conventional biomass sources. This character gives excellent conversion efficiency and stability for the anaerobic digestion process. In the anaerobic digestion method, the biogas production from microalgal/cyanobacterial biomass is determined by its organic loadings, temperature, pH, and retention time in the bioreactor. In the principle, long solid retention and high organic loading can significantly affect CH_4 yield. This method can be performed in either mesophilic or thermophilic conditions. Integration between microalgae and cyanobacteria cultivation in a wastewater treatment pond and harvested its biomass anaerobically digested for producing biogas can offer good potentials in overcoming environmental problems (water pollution) and commercializing biogas production from microalgal and cyanobacterial biomass (Ramos-Suárez et al. 2014; Tijani et al. 2015).

4.4 *Fine Biochemicals*

Carbohydrate content in microalgae and cyanobacteria biomass has the potential in Acetone-Butanol fermentation (fermented using bacteria such as *Clostridium sp*) for producing biobutanol with bio acetone as a by-product. These fine biochemicals are valuable organic solvents. Biobutanol belongs to renewable transportation fuel, while bio acetone is utilized as a multi-purpose solvent such as a cleaning agent, an extraction solvent, and other laboratory works. Applying biobutanol as fuel for vehicles is reported to not require any engine modification as it can be directly blended in higher concentrations with gasoline compared to other biofuels. Blending biobutanol with gasoline is aimed to lower vapor pressure. Production of butanol with *Neochloris aquatica* CL-M1 was done by using wastewater medium to yield 0.89 g/(L.h) of butanol with 96.2% efficient removal of NH_3-N. This process' success in yielding butanol meant circular usage for extraction is possible by using the produced butanol. Meanwhile, on the genus of *Dunaliella (D.tertiolectra, D.primoelectra, D.parva, D.bardawil, and D.salina*) fermented by *Clostridium pasteurianum* was found to yield four different kinds of organic solvents. These four organic substances namely propanediol, acetic acid, ethanol, and n-butanol were present as a mixture with a concentration of 14–16 g/L. However further study still needs to be conducted related to the mechanism of the process (Veza et al. 2021; Nakas et al. 1983).

4.5 *Food and Functional Foods*

4.5.1 Omega 3 Oil

Naturally, microalgae and cyanobacteria contain omega-3 fatty acids that can be purified into high-value added bioproducts such as food supplement. The sources of omega-3 fatty acid in microalgae and cyanobacteria biomass are eicosapentanoic acid (EPA) and docosahexaenoic acid (DHA). These compounds are widely found in fish oil, but due to its low supply, unpalatable taste, and inadequate oxidative stability of fish oil made it not a convenient source for omega-3 fatty acids. Microalgae and cyanobacteria are self-producing omega-3 fatty acids, therefore processing on microalga biomass is simpler than fish biomass. The use of EPA is widely known in human health purposes for curing heart and inflammatory diseases (asthma, arthritis, migraine, and psoriasis). According to Hu et al. (2008) outdoor cultivation of *Pavlova viridis* gave lower total fatty acid but higher EPA than of indoor cultivation system, hence concluded that outdoor cultivation is more favorable in producing EPA. *Nannochloropsis sp.* also show similar result according to Cheng-Wu et al. (2001). Production yields of EPA are also determined by the season. It was found that the production of EPA is higher as 35% in summer than in winter. Temperature and irradiance were not significantly impacting on yield of EPA (4% of dry biomass) produced by *Nannochloropsis* sp. in an outdoor tubular reactor, and this report showed the potential of this eustigmatophyte as an alternative source of EPA (Chini Zittelli et al. 1999). Vazhappily and Chen (1998) reported that the highest EPA proportion (% of total fatty acids) was produced by *Monodus subterraneus* UTEX 151 (34.2%), followed by *Chlorella minutissima* UTEX 2341 (31.3%) and *Phaeodactylum tricornutum* UTEX 642 (21.4%). But further studies are still needed to ensure the feasibility of this EPA production system. In terms of DHA, this compound is also useful to fight against cancer, AIDS, and heart diseases by reducing cholesterol, boosting immune system, and detoxification of the body. Production of DHA depends on the cultivated species of both microalgae and cyanobacteria. Marine species of both microalgae and cyanobacteria have significantly higher DHA (mainly consist of saturated or monosaturated fatty acids) content than freshwater species (Patil et al. 2007). Marine microalga, *Schizochytrium mangrove*, contains DHA in the range of 33–39% of total fatty acids (Jiang et al. 2004), whileVazhappily and Chen (1998) reported that the highest DHA proportion (% of total fatty acids) was obtained in *Crypthecodinium cohnii* UTEX L1649 (19.9%), followed by *Amphidinium carterae* UTEX LB 1002 (17.0%) and *Thraustochytrium aureum* ATCC 28211 (16.1%). Another report by Patil et al. (Patil et al. 2007) mentioned that *Isochrysis galbana* contained a significant amount of DHA with a specific productivity of around 0.16 g/(L.d). The amount of CO_2, light intensity, and operation modes (batch and continuous) significantly affect the productivity of DHA (which was found of 1.29 mg/(L.d) under optimized conditions in the cultivation of microalga *Pavlova lutheri* (Carvalho and Malcata 2005).

4.5.2 Chlorophyll-a

In microalgae and cyanobacteria, there are mainly two main types of chlorophyll being produced. They consist of chlorophyll a and chlorophyll b. The presence of chlorophyll as a photosynthetic pigment is found in all photoautotrophic organisms. Chlorophyll has been widely used as a medicinal drug due to its ability to stimulate liver function recovery and increase bile secretion. It also possesses the ability to repair damaged cells, increase the amount of hemoglobin in blood and encourage rapid cell growth. Chlorophyll has also been reported to have various properties (antimutagenic, anticarcinogenic, and antioxidant). Traditionally, chlorophyll has been used in the food industry as a natural pigment due to the increasing consumer demands for natural foods. Its use as a natural pigment possesses the same properties as listed previously. Chlorophyll is best extracted by the use of the supercritical fluid extraction (SFE) method. High-performance liquid chromatography was found to be the most accurate and sensitive technique to fractionate and quantify chlorophyll along with its derivatives (Silva and Sant'Anna 2016).

Microalgae (*Chlorella sp, Scenedesmus sp*) and cyanobacteria (*Spirulina sp.*) seem to be promising alternative sources for chlorophyll. The growth stage of a particular algal species was found to be highly linked with the amount of chlorophyll extracted. Microalgae extracted during the stationary growth phase were shown to have a substantially higher amount of chlorophyll as compared to the same species extracted during the logarithmic phase. Chlorophyll a has been recently revealed to be a key compound in the treatment of ulcers which makes it vital in the postoperative treatment of rectal surgery on humans. During the removal of large areas of tissue, recovery can be difficult and the area near said removed tissue tends to be painful. With the application of chlorophyll, the stimulation of cells in the host and the consequent acceleration in tissue formation increases the rate of recovery, in many cases up to 25%. Moreover, the use of chlorophyll was also found to eliminate foul odor emanating from the wound after a few administrations. Chlorophyll's non-toxic nature, antibacterial properties, and ability to deodorize make it a prominent product in treating oral sepsis (Stirbet et al. 2018). In conclusion, a downstream process needs to be developed to purify chlorophyll a and b from microalgae and cyanobacteria.

4.5.3 Phycocyanin

Phycocyanin is a colorant present in cyanobacteria and red microalgae. When purified, phycocyanin shows a brilliant blue color in the solution. Phycocyanin is composed of two different subunits α and β combined into one. Its presence in nature exists as monomers, trimers, or hexamers; small quantities of oligomers have been found as well. In general, phycocyanin includes C-phycocyanin and allophycocyanin. Both possess different maximum absorption peaks, which is at 620 and 650 nm, respectively. Studies have suggested that both C-phycocyanin (C-PC), and allophycocyanin each have their own ratio of absorbance to indicate their purity; with C-PC possessing a ratio of 620 and 280 nm, and allophycocyanin possessing a ratio of 650 and 280 nm. Phycocyanin shows numerous special bioactivities which are gradually recognized to have potential as raw materials for healthy food products. Studies have indicated that phycocyanin also has various bioactivities such as anti-neoplasm, antioxidant, and anti-inflammatory effects. Several reports have also suggested that C-PC has anticancer bioactivity.

The high content of protein in *Arthrospira (Spirulina) sp.* (~70% in dry cell weight) poses as a valid reason to consider cyanobacteria as an alternate source of protein, which is vital due to its presence at several cell locations as enzymes, structural component, linked carbohydrate, among others under various forms. Phycobiliproteins correspond to nearly 60% of all the soluble protein in cyanobacteria among the protein pool, while about 20% corresponds to C-PC. Pigment purity within the cultivation of microalgae is of utmost importance, especially when it's used as a fluorescent marker in biomedical research where the presence of impurities can severely impair the quality of the extract. Extract of C-PC obtained from *A. platensis* which was grown in a nitrogen-reduced medium indicated higher levels of purity (0.80) followed by the extract cultivated within a control group (0.55) and lastly within a nitrogen-free medium (0.21). Several studies have demonstrated crude extraction of C-PC (with varying levels of purity between 0.19 and 1.4) extracted from *A. platensis*. What is considered to be food grade C-PC has a purity of 0.7 (with a commercial value of ~US$ 0.13 per mg), while 3.9 is considered as reactive grade C-PC (value varies ~US$ 5 per mg) and a purity level of greater than 4.0 as analytical grade C-PC (value can be as high as US$ 15 per mg. All in all, C-PC obtained from *A. plantesis* grown within a nitrogen-reduced medium presents a very promising use within the food industry (Qiang et al. 2021; Pagels et al. 2019).

4.5.4 Carotene

Carotenoids (Carotene) are known as one major class of photosynthetic pigments. In microalgae, there are generally three known pigments which are chlorophyll, phycocyanin, and carotenoids. Carotenoids, like chlorophyll, are water-soluble. They consisted of terpenoid pigments derived from a 40-carbon chained polyene. This distinctive molecular structure is associated with their chemical properties which allowed electron transfers induced by light-absorption which is essential in photosynthetic activities. This pigment may be complemented by cyclic groups and oxygen functional groups. The oxygenated derivatives are particularly known as xanthophylls as there is the presence of hydroxyl groups (e.g. lutein), oxy groups (e.g., canthaxanthin), or both combinations (e.g. astaxanthin). However, carotenoids are usually typed as two, namely primary and secondary carotenoids. Primary carotenoids come from their structural and functional components in the cellular photosynthetic apparatus (i.e., xanthophylls). Meanwhile, secondary carotenoids consisted of those produced at a large level after exposure to specific environmental stimuli. Relatively, xanthophyl are hydrophobic therefore, they are typically found linked on the membranes or noncovalently bound to specific proteins (Srivastava et al. 2022; Begum et al. 2016).

There are more than 400 variants of carotenoids within nature. Among them, β-carotene is considered to be the most prominent. Moreover, some carotenoids contain provitamin A and possess a broad range of biological functions and actions, most notably in relation to human health (Pisal and Lele 2005). Researchers have reported the benefits of β-carotene for the human body as the human body converts β-carotene to vitamin A via the body tissue. The necessity of vitamin A in the immunity of the human body is to prevent cataract, night blindness, and skin diseases. In the context of multivitamin preparations, β-carotene is often used as pro-vitamin A (retinol) and as an ingredient in the formulation of healthy foods. Alternative use of β-carotene within the context of food production, is as a food colorant to improve the appearance of margarine, cheese, fruit juices, confectionary, and other food products to increase appeal toward customers, such as the case of the β-carotene cultivated from

Dunaliella. β-carotene has also been reported to decrease the hazard of several degenerative diseases such as cancer (Nethravathy et al. 2019). Studies have found that β-carotene from *Dunaliella sp.* contains 40% 9-cis and 50% all-trans stereoisomers which play a crucial role in lowering incidence of several varities of cancer and other degenerative diseases. Furthermore, an investigation of the antioxidant properties of β-carotene was found allowing it to help mediate the harmful effects of free radicals thus preventing life-threatening diseases such as arthritis, coronary heart diseases, premature aging, and various forms of cancer. Another study has also shown that β-carotene has the ability to stimulate the immune system, potentially preventing various kinds of life-threatening diseases. In addition, it can also reduce the cognitive impairment linked with Alzheimer's which is caused by persistent oxidative stress within the brain (Nethravathy et al. 2019; Murthy et al. 2005).

4.6 Microalgae for Phytoremediation

Phytoremediation is a process of remediating environmental contaminants or excess nutrients through the use of plants. Phytoremediation usually focuses on controlling Total Phosphorus (TP), Total Nitrogen (TN), and their related constituents such as PO_4^{3-}, NO_3^-, NH_4^+, and much more which can be seen as Total Dissolved Salts (TDS). Conventionally, complex plant systems available in the environment especially in aquatic environments are used to remediate environmental contaminants. Phytoremediation often uses available plants with little value in the market such as common reeds, water lilies, and pteridophytes (Pandey 2012; Wang et al. 2022; Riggio et al. 2015). The downside of the conventional route is the process of developing enough biomasses to effectively remediate an area required years of acclimation and cultivation in the environment. In the wake of sustainable development and circular economy, phytoremediation using microalgae and cyanobacteria fits the criteria. In developing a sustainable business, the technology applied should not further factor into the depletion of arable lands and not compete with the existing Food-Energy-Water nexus (Olabi et al. 2023). The adaptive capability and structural simplicity of microalgal and cyanobacterial cells make them a perfect phytoremediation agent. Owing to their adaptive capability, microalgae can live in numerous conditions depending on the strains and their evolutionary pathways in combating extreme conditions. A class of microalgae is even equipped with an additional protective layer called frustules to accommodate their cells to live in extreme heat or sudden temperature changes (Kooistra et al. 2007; Kim et al. 2017). This adaptability of microalgae and cyanobacterial cells also resulted from their cellular simplicity which allowed for an optimum growth rate attained in a shorter period than complex plants.

In determining the process for microalgae or cyanobacteria phytoremediation, there are three main factors to consider. These factors are interconnected and serve as important determinants in the growth of algae and thus the success of phytoremediation. These considering factors are Environmental, Biological, and Operational factors (Nie et al. 2020). The connections between them go as follows: Environmental conditions provide biological activities, biological activities ensure the optimum operating conditions are chosen, and the chosen operating conditions are chosen to sustain the environmental conditions (Fig. 5). Nine different genera of Chlorophyta phylum: *Asterarcys, Chlorella, Chloroccoum, Chlorosarcinopsis, Coelastrella, Desmodesmus, Micratinium, Parachlorella,* and *Scenedemus* are often considered the native to a freshwater environment involving aquaculture (Couto et al. 2022). Their removal capacity has been studied in Galicia and continental Spain (without Coalastrella, Asterarcys or Parachlorella genera as they are not usually present in their freshwater streams) and showed removal efficiencies of 99, 92, and 49% for ammonium, nitrite, and nitrate, respectively, on aquaculture-derived effluents in a raceway pond with microalgae biomass production around 30–40 mg/L on the 7th day.

Aside from Chlorophyta, Bacillariophyta (Diatoms), Cyanophyta (blue-green algae), and Chrysophytae (golden algae) are the most abundant microalgae classes in aquatic ecosystems (Vieira et al. 2020). In a present study of treating municipal wastewater with three native microalgae species (*Navicula veneta* -Diatom-, *Chlorella vulgaris* -Chlorophyta-, and *Nostoc muscorum*-Cyanophyta-), the *Navicula veneta* treatment was found to produce reusable effluent with high-rate removal of COD, TP, and TN by 95.75%, 99.8%, and 96.96%, respectively (Sisman-Aydin 2022).

Microalgae phytoremediation does not require large areas with easier controllability as the internet of things (IoT) can be incorporated into monitoring its growth and even the environment in a working system (Peter et al. 2021; Abdul-Hadi et al. 2013). Recent research also suggested a phytoremediation system that included energy

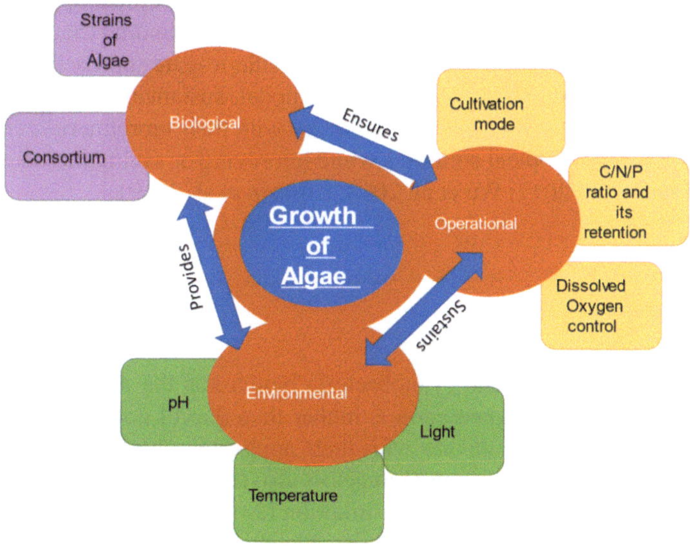

Fig. 5 Factors at play in microalgae & cyanobacteria phytoremediation

production. This was done by the construction of microbial fuel cells (MFCs) where microalgae functioned as biocathode (Mathuriya et al. 2016). The energy conversion efficiency of microalgae MCFs showed a maximum output of up to 9% while other photosynthetic plants were at 4.6–6% (Shukla and Kumar 2018). Generally, the external resistance of this system is at 1000 Ω with pollutant removal focusing on Chemical Oxygen Demand (COD). There seems to be a relationship between working volume and the type of wastewater playing a role in the maximum power density as seen in recently published data regarding wastewater treatment and maximum bioelectricity in Table 3 (Sharma et al. 2022). This type of phytoremediation is regarded as a complete recycling machine (Greenman et al. 2019) with the potential for high-yield hydrogen gas production (Logan et al. 2008). Further improvement in phytoremediation using microalgae and cyanobacteria cells was also done by the introduction of immobilized systems in wastewater treatment to improve retention time in optimizing nutrient capture (Han et al. 2022; Shen et al. 2017).

Phytoremediation of microalgae usually utilized the use of an open pond cultivation system. That, however, resulted in lower biomass yield which is why the development of tubular and flat-panel photobioreactors in bioremediation is of interest (Luo et al. 2017). The system would require flue gas submersions into the photobioreactor allowing for the capture of CO_2 from the environment-utilizing the sequestration or CO_2 fixing pathways which have been done in a consortium with *Clostridium sp., E.coli,* and *Saccharomyces cerevisiae* (Hu et al. 2019). Sequestration of CO_2 for the valorization of waste mitigation has also been analyzed for massive algal biomass production which showed positive outcomes with the integration of (Ma et al. 2022; Yadav et al. 2019). In addition to sequestration, microalgae are also reportedly able to interact with known emerging contaminants such as heavy metals, antibiotics, and microplastics which all resulted in oxidative stresses that either boost or decrease their lipid content or antioxidant levels to adapt to the surrounding (Satya et al. 2023). These positive outcomes are overall improved cradle-to-gate approach of microalgae biorefineries to renew the current economic model, sustainable recycling of water and supporting food security as well as energy with their biorefineries, and helping economic growth in tropical countries while improving or saving their environment (Hosseinizand et al. 2017; Wu et al. 2018; Hossain et al. 2019).

5 Conclusion

Production of microalgae and cyanobacteria biomass is the new key to a sustainable future. The biomass generation is higher than that of conventional plants yet requires smaller land to cultivate with their applicable closed cultivation system using photobioreactors. The photobioreactors also ensure the quality of the biomass and biorefineries are safe for even consumption levels. There are many potentials in the biomass of microalgae. As bioproducts (bioethanol, biomethane, and biodiesel), microalgae and cyanobacterial biomass have unique properties which are their higher lipid levels with a faster rate of growth than conventional biomass sources. As they are

Table 3 Microalgae-MFCs for Phytoremediation

Wastewater type	Location	Microalgae-electrode	Working volume (mL)	External resistance (Ω)	Pollutant removal	Maximum power density (mW m^{-3})
Industrial wastewater	Denmark	*Chlorella vulgaris* with Ti-electrode mesh coated with platinum/carbon	200	1000	COD (66.8%), TN (69%), TP (48.5%)	–
Synthetic wastewater	India	*Chlorella sorokiniana* with carbon felt	300	1000	COD (95%)	2320
Synthetic wastewater	India	*Chlorella vulgaris* with graphite plate	–	700	COD (96%), NH_4^+ (85%), PO_4^{3-} (69%), NO_3^- (68%)	33.14
Synthetic wastewater	Thailand	*Chlorella vulgaris* with carbon cloth	1000	1000	COD (71%), NH_4^+ (79%)	199.12
Domestic wastewater	Iran	*Chlorella vulgaris* with stainless steel	–	1000	–	126
Pharmaceutical wastewater	India	*Scenedesmus abundans* with Graphite rods	–	100	COD (97.24%), TN (97.12%), TP (93.71%)	838.68
Kitchen wastewater	India	*Synechococcus sp.* and *Chlorococcum sp.* with plain graphite plate	250	–	COD (68%)	41.5 (*Synechococcus sp.*) 30.5 (*Chlorococcum sp.*)

Adapted from Sharma et al. (2022)

simple cells, the cost of separating the required biomass from contaminants (stems, leaves, etc.) is relatively cheaper. The same can be said in applying microalgae as a living cell in phytoremediation, replacing the conventional phytoremediators which needed a longer time to acclimate to the conditions. The water conditions coming from microalgae phytoremediation have also been shown to be within the permissible limit of water reuse. The cradle-to-gate approach of using microalgae and cyanobacteria products also showed higher interest in the availability of omega-3 oils and colorants (Phycocyanins) which are in high demand for food industries and the new sustainable industries that follow.

Acknowledgements The authors would like to acknowledge PKR-Biomassa and Biorefinery UNPAD-BRIN for funding the publication of this chapter book.

Author Contributions Awalina Satya, Azalea Dyah Maysarah Satya, and Tjandra Chrismadha were the main contributors (data curation, writing original draft); others were supporting contributors: Nofdianto Nofdianto (review), Gunawan Gunawan (review), Ika Atman Satya (review), Souvia Rahimah (review), Efri Mardawati (review), Sara Kazemi Yazdi (review), and Pau-Loke Show (review).

References

Abbaslou H, Hadifard H, Ghanizadeh AR (2020) Effect of cations and anions on flocculation of dispersive clayey soils. Heliyon 6:e03462. https://doi.org/10.1016/J.HELIYON.2020.E03462

Abdul-Hadi A, Mansor S, Pradhan B, Tan CK (2013) Seasonal variability of chlorophyll-a and oceanographic conditions in Sabah waters in relation to Asian monsoon—A remote sensing study. Environ Monit Assess 185:3977–3991. https://doi.org/10.1007/s10661-012-2843-2

Abreu AP, Morais RC, Teixeira JA, Nunes J (2022) A comparison between microalgal autotrophic growth and metabolite accumulation with heterotrophic, mixotrophic and photoheterotrophic cultivation modes. Renew Sustain Energy Rev 159:112247. https://doi.org/10.1016/J.RSER.2022.112247

Acién Fernández FG, Fernández Sevilla JM, Molina Grima E (2013) Photobioreactors for the production of microalgae. Rev Environ Sci Biotechnol 12:131–151. https://doi.org/10.1007/s11157-012-9307-6

Amorim ML, Alimentos BB, Soares J, Leite MDO (2020) Microalgae proteins: production, separation, isolation, quantification, and application in food and feed microalgae proteins: production, separation, isolation, quantification and application in food and feed. https://doi.org/10.1080/10408398.2020.1768046

Ashokkumar V, Rengasamy R, Deepalakshmi S et al (2014) Mass cultivation of microalgae and extraction of total hydrocarbons: a kinetic and thermodynamic study. Fuel 119:308–312. https://doi.org/10.1016/j.fuel.2013.11.062

Ashraf O, Chang C, Yi H et al (2020) Energy conversion and management: X prospects of industry 5.0 in algae: customization of production and new advance technology for clean bioenergy generation. Energy Convers Manag X:100048. https://doi.org/10.1016/j.ecmx.2020.100048

Becker EW (1994) Microalgae: biotechnology and microbiology. Cambridge University Press

Begum H, Yusoff FMD, Banerjee S et al (2016) Availability and utilization of pigments from microalgae. Crit Rev Food Sci Nutr 56:2209–2222. https://doi.org/10.1080/10408398.2013.764841

Blanco AM, Moreno J, Del Campo JA et al (2007) Outdoor cultivation of lutein-rich cells of Muriellopsis sp. in open ponds. Appl Microbiol Biotechnol 73:1259–1266. https://doi.org/10.1007/s00253-006-0598-9

Bleakly S, Hayes M (2017) Review algal proteins: extraction, application, and challenges concerning production 1–34. https://doi.org/10.3390/foods6050033

Carvalho AP, Malcata FX (2005) Optimization of ω-3 fatty acid production by microalgae: crossover effects of CO2 and light intensity under batch and continuous cultivation modes. Mar Biotechnol 7:381–388. https://doi.org/10.1007/s10126-004-4047-4

Castro JDS, Assis LR De, Calijuri ML (2021) Technologies for improving microalgae biomass production coupled to effluent treatment: a life cycle approach 57. https://doi.org/10.1016/j.algal.2021.102346

Catone CM, Ripa M, Geremia E, Ulgiati S (2021) Bio-products from algae-based biorefinery on wastewater: a review. J Environ Manag 293:112792. https://doi.org/10.1016/j.jenvman.2021.112792

Chen L, Li P, Liu Z, Jiao Q (2009) The released polysaccharide of the cyanobacterium Aphanothece halophytica inhibits flocculation of the alga with ferric chloride. J Appl Phycol 21:327–331. https://doi.org/10.1007/s10811-008-9371-z

Cheng-Wu Z, Zmora O, Kopel R, Richmond A (2001) An industrial-size flat plate glass reactor for mass production of nannochloropsis sp. (Eustigmatophyceae). Aquaculture 195:35–49. https://doi.org/10.1016/S0044-8486(00)00533-0

Chini Zittelli G, Lavista F, Bastianini A et al (1999) Production of eicosapentaenoic acid by Nannochloropsis sp. cultures in outdoor tubular photobioreactors. In: Osinga R, Tramper J, Burgess JG, Wijffels RHBT-P in IM (eds) Marine bioprocess engineering. Elsevier, pp 299–312

Chisti Y (2007) Biodiesel from microalgae. Biotechnol Adv 25:294–306. https://doi.org/10.1016/j.biotechadv.2007.02.001

Couto AT, Cardador M, Santorio S et al (2022) Cultivable microalgae diversity from a freshwater aquaculture filtering system and its potential for polishing aquaculture-derived water streams. J Appl Microbiol 132:1543–1556. https://doi.org/10.1111/jam.15300

Cui X, Yang J, Cui M et al (2021) Comparative experiments of two novel tubular photobioreactors with an inner aerated tube for microalgal cultivation: enhanced mass transfer and improved biomass yield. Algal Res 58. https://doi.org/10.1016/j.algal.2021.102364

da Silva Ferreira V, Sant'Anna C (2016) Impact of culture conditions on the chlorophyll content of microalgae for biotechnological applications. World J Microbiol Biotechnol 33:20. https://doi.org/10.1007/s11274-016-2181-6

Debeni Devi N, Chaudhuri A, Goud VV (2022) Algae biofilm as a renewable resource for production of biofuel and value-added products: a review. Sustain Energy Technol Assess 53:102749. https://doi.org/10.1016/j.seta.2022.102749

Demirbas A (2010) Use of algae as biofuel sources. Energy Convers Manag 51:2738–2749. https://doi.org/10.1016/j.enconman.2010.06.010

Demirbas MF (2011) Biofuels from algae for sustainable development. Appl Energy 88:3473–3480. https://doi.org/10.1016/j.apenergy.2011.01.059

Divakaran R, Sivasankara Pillai VN (2002) Flocculation of algae using chitosan. J Appl Phycol 14:419–422. https://doi.org/10.1023/A:1022137023257

Duan Y, Shi F (2014) Bioreactor design for algal growth as a sustainable energy source. Elsevier B.V.

Farhan A, Udaiyappan M, Abu H, Sobri M (2017) A review of the potentials, challenges and current status of microalgae biomass applications in industrial wastewater treatment. J Water Process Eng 20:8–21. https://doi.org/10.1016/j.jwpe.2017.09.006

Felix C, Ubando A, Madrazo C et al (2019) Investigation of direct biodiesel production from wet microalgae using definitive screening design. Energy Procedia 158:1149–1154. https://doi.org/10.1016/j.egypro.2019.01.296

Fernández I, Acién FG, Berenguel M, Guzmán JL (2014) First principles model of a tubular photobioreactor for microalgal production. Ind Eng Chem Res 53:11121–11136. https://doi.org/10.1021/ie501438r

Gil-izquierdo A, Pedreño MA, Montoro-garcía S et al (2021) Science of the total environment a sustainable approach by using microalgae to minimize the eutrophication process of Mar Menor lagoon. Sci Total Environ 758:143613. https://doi.org/10.1016/j.scitotenv.2020.143613

Gonçalves AL, Pires JCM, Simões M (2017) A review on the use of microalgal consortia for wastewater treatment. Algal Res 24:403–415. https://doi.org/10.1016/j.algal.2016.11.008

Greenman J, Gajda I, Ieropoulos I (2019) Microbial fuel cells (MFC) and microalgae; photo microbial fuel cell (PMFC) as complete recycling machines. Sustain Energy Fuels 3:2546–2560. https://doi.org/10.1039/C9SE00354A

Han M, Zhang C, Li F, Ho SH (2022) Data-driven analysis on immobilized microalgae system: new upgrading trends for microalgal wastewater treatment. Sci Total Environ 852

Hernández D, Riaño B, Coca M, García-González MC (2015) Saccharification of carbohydrates in microalgal biomass by physical, chemical and enzymatic pre-treatments as a previous step for bioethanol production. Chem Eng J 262:939–945. https://doi.org/10.1016/j.cej.2014.10.049

Hon-Nami K (2006) A unique feature of hydrogen recovery in endogenous starch-to-alcohol fermentation of the marine microalga, Chlamydomonas perigranulata. Appl Biochem Biotechnol 131:808–828. https://doi.org/10.1385/ABAB:131:1:808

Hossain N, Zaini J, Indra Mahlia TM (2019) Life cycle assessment, energy balance and sensitivity analysis of bioethanol production from microalgae in a tropical country. Renew Sustain Energy Rev 115:109371. https://doi.org/10.1016/j.rser.2019.109371

Hosseinizand H, Lim CJ, Webb E, Sokhansanj S (2017) Economic analysis of drying microalgae Chlorella in a conveyor belt dryer with recycled heat from a power plant. Appl Therm Eng 124:525–532. https://doi.org/10.1016/J.APPLTHERMALENG.2017.06.047

Hu C, Li M, Li J et al (2008) Variation of lipid and fatty acid compositions of the marine microalga Pavlova viridis (Prymnesiophyceae) under laboratory and outdoor culture conditions. World J Microbiol Biotechnol 24:1209–1214. https://doi.org/10.1007/s11274-007-9595-0

Hu G, Li Y, Ye C et al (2019) Engineering microorganisms for enhanced CO2 sequestration. Trends Biotechnol 37:532–547

Huang J, Wang J, Huang Z et al (2023) Photothermal technique-enabled ambient production of microalgae biodiesel: Mechanism and life cycle assessment. Bioresour Technol 369:128390. https://doi.org/10.1016/j.biortech.2022.128390

Huang Q, Jiang F, Wang L, Yang C (2017) Design of photobioreactors for mass cultivation of photosynthetic organisms 3:318–329

Jacob-Lopes E, Franco TT (2013) From oil refinery to microalgal biorefinery. J CO2 Utiliz 2:1–7. https://doi.org/10.1016/j.jcou.2013.06.001

Jiang Y, Fan K-W, Tsz-Yeung Wong R, Chen F (2004) Fatty acid composition and squalene content of the marine microalga schizochytrium mangrovei. J Agric Food Chem 52:1196–1200. https://doi.org/10.1021/jf035004c

Johnson TJ, Anderson GA, Gibbons WR (2018) Photobioreactor cultivation strategies for microalgae and cyanobacteria photobioreactor cultivation strategies for microalgae and cyanobacteria. https://doi.org/10.1002/btpr.2628

Kholssi R, Vogelei P, Marks EAN et al (2021) Biocatalysis and agricultural biotechnology biotechnological uses of microalgae: a review on the state of the art and challenges for the circular economy. Biocatal Agric Biotechnol 36:102114. https://doi.org/10.1016/j.bcab.2021.102114

Kim M, Gwak Y, Jung W, Jin ES (2017) Identification and characterization of an isoform antifreeze protein from the antarctic marine diatom, chaetoceros neogracile and suggestion of the core region. Mar Drugs 15. https://doi.org/10.3390/md15100318

Knuckey RM, Brown MR, Robert R, Frampton DMF (2006) Production of microalgal concentrates by flocculation and their assessment as aquaculture feeds. Aquacult Eng 35. https://doi.org/10.1016/j.aquaeng.2006.04.001

Kooistra WHCF, Gersonde R, Medlin LK, Mann DG (2007) The origin and evolution of the diatoms: their adaptation to a planktonic existence. Evol Primary Prod Sea 207–249. https://doi.org/10.1016/B978-012370518-1/50012-6

Koutra E, Economou CN, Tsafrakidou P, Kornaros M (2018) Bio-based products from microalgae cultivated in digestates. Trends Biotechnol 36:819–833. https://doi.org/10.1016/j.tibtech.2018.02.015

Kumar P, Prajapati SK, Malik A, Vijay VK (2017) Cultivation of native algal consortium in semi-continuous pilot scale raceway pond for greywater treatment coupled with potential methane production. Biochem Pharmacol. https://doi.org/10.1016/j.jece.2017.10.044

Kumar G, Shekh A, Jakhu S et al (2020) Bioengineering of microalgae: recent advances, perspectives, and regulatory challenges for industrial application. Front Bioeng Biotechnol 8. https://doi.org/10.3389/fbioe.2020.00914

Liu R, Li S, Tu Y, Hao X (2021) Capabilities and mechanisms of microalgae on removing micropollutants from wastewater: a review. J Environ Manag 285:112149. https://doi.org/10.1016/j.jenvman.2021.112149

Logan BE, Call D, Cheng S et al (2008) Microbial electrolysis cells for high yield hydrogen gas production from organic matter. Environ Sci Technol 42:8630–8640. https://doi.org/10.1021/ES801553Z

Luo Y, Le-Clech P, Henderson RK (2017) Simultaneous microalgae cultivation and wastewater treatment in submerged membrane photobioreactors: a review. Algal Res 24:425–437. https://doi.org/10.1016/j.algal.2016.10.026

Ma Z, Cheah WY, Ng IS et al (2022) Microalgae-based biotechnological sequestration of carbon dioxide for net zero emissions. Trends Biotechnol. https://doi.org/10.1016/J.TIBTECH.2022.09.002

Mathuriya AS, Yakhmi J v. (2016) Microbial fuel cells—Applications for generation of electrical power and beyond. Crit Rev Microbiol 42:127–143. https://doi.org/10.3109/1040841X.2014.905513

Milledge JJ (2011) Commercial application of microalgae other than as biofuels: a brief review. Rev Environ Sci Biotechnol 10:31–41. https://doi.org/10.1007/s11157-010-9214-7

Mobin S, Alam F (2017) Some promising microalgal species for commercial applications: a review. Energy Procedia 110:510–517. https://doi.org/10.1016/j.egypro.2017.03.177

Morais WG, Gorgich M, Corrêa PS et al (2020) Microalgae for biotechnological applications: Cultivation, harvesting and biomass processing. Aquaculture 528:735562. https://doi.org/10.1016/j.aquaculture.2020.735562

Murthy KNC, Vanitha A, Rajesha J et al (2005) In vivo antioxidant activity of carotenoids from Dunaliella salina—A green microalga. Life Sci 76:1381–1390. https://doi.org/10.1016/j.lfs.2004.10.015

Mohan N, Sivasubramanian V (2010) Mass cultivation of chroococcus turgidus and Oscillatoria sp. and effective harvesting of biomass by low-cost methods. Nat Prec. https://doi.org/10.1038/npre.2010.4331.1

Nakas JP, Schaedle M, Parkinson CM et al (1983) System development for linked-fermentation production of solvents from Algal biomass. Appl Environ Microbiol 46:1017–1023. https://doi.org/10.1128/aem.46.5.1017-1023.1983

Nethravathy MU, Mehar JG, Mudliar SN, Shekh AY (2019) Recent advances in microalgal bioactives for food, feed, and healthcare products: commercial potential, market space, and sustainability 00:1–16. https://doi.org/10.1111/1541-4337.12500

Nie X, Mubashar M, Zhang S et al (2020) Current progress, challenges and perspectives in microalgae-based nutrient removal for aquaculture waste: a comprehensive review. J Clean Prod 277:124209. https://doi.org/10.1016/J.JCLEPRO.2020.124209

Nitsos C, Filali R, Taidi B, Lemaire J (2020) Current and novel approaches to downstream processing of microalgae: a review. Biotechnol Adv 45:107650. https://doi.org/10.1016/j.biotechadv.2020.107650

Nwoba EG, Parlevliet DA, Laird DW et al (2019) Light management technologies for increasing algal photobioreactor efficiency. Algal Res 39:101433. https://doi.org/10.1016/j.algal.2019.101433

Olabi AG, Shehata N, Sayed ET et al (2023) Role of microalgae in achieving sustainable development goals and circular economy. Sci Total Environ 854. https://doi.org/10.1016/J.SCITOTENV.2022.158689

Pagels F, Guedes AC, Amaro HM et al (2019) Phycobiliproteins from cyanobacteria: chemistry and biotechnological applications. Biotechnol Adv 37:422–443. https://doi.org/10.1016/j.biotechadv.2019.02.010

Pandey VC (2012) Phytoremediation of heavy metals from fly ash pond by Azolla caroliniana. Ecotoxicol Environ Saf 82:8–12. https://doi.org/10.1016/J.ECOENV.2012.05.002

Patil V, Källqvist T, Olsen E et al (2007) Fatty acid composition of 12 microalgae for possible use in aquaculture feed. Aquacult Int 15:1–9. https://doi.org/10.1007/s10499-006-9060-3

Peter AP, Khoo KS, Chew KW et al (2021) Microalgae for biofuels, wastewater treatment and environmental monitoring. Environ Chem Lett 19:2891–2904

Pisal DS, Lele SS (2005) Carotenoid production from microalga, Dunaliella salina. Indian J Biotechnol 4:476–483

Qiang X, Wang L, Niu J et al (2021) Phycobiliprotein as fluorescent probe and photosensitizer: a systematic review. Int J Biol Macromol 193:1910–1917. https://doi.org/10.1016/j.ijbiomac.2021.11.022

Ramos-Suárez JL, Cuadra FG, Acién FG, Carreras N (2014) Benefits of combining anaerobic digestion and amino acid extraction from microalgae. Chem Eng J 258:1–9. https://doi.org/10.1016/j.cej.2014.07.086

Riggio VA, Comino E, Rosso M (2015) Above ground part of common reed to enhance anaerobic co-digestion of farm biomasses: potential, monitoring and efficiency. Ecol Eng 79:35–41. https://doi.org/10.1016/J.ECOLENG.2015.01.015

Satya A, Harimawan A, Haryani GS (2017) Non-linear isotherm models, cadmium kinetics, and biosorption thermodynamics of dried biomass of native aphanothece sp. in a batch system 49:617–638. https://doi.org/10.5614/j.eng.technol.sci.2017.49.5.5

Satya A, Harimawan A, Haryani GS et al (2021a) Integrated treatment of submerged membrane and adsorption using dried Aphanothece sp for removing cadmium from synthetic wastewater. J Water Process Eng 41:102022. https://doi.org/10.1016/j.jwpe.2021.102022

Satya A, Harimawan A, Sri Haryani G et al (2021b) Fixed-bed adsorption performance and empirical modelingof cadmium removal using adsorbent prepared from the cyanobacterium Aphanothece sp cultivar. Environ Technol Innov 21. https://doi.org/10.1016/j.eti.2020.101194

Satya ADM, Cheah WY, Yazdi SK et al (2023) Progress on microalgae cultivation in wastewater for bioremediation and circular bioeconomy. Environ Res 218:114948. https://doi.org/10.1016/j.envres.2022.114948

Senatore V, Oliva G, Buonerba A et al (2022) An integrated algal membrane photobioreactor as a green-transition technology for the carbon capture and utilization. J Environ Chem Eng 10:107344. https://doi.org/10.1016/j.jece.2022.107344

Shao P, Darcovich K, McCracken T et al (2015) Algae-dewatering using rotary drum vacuum filters: process modeling, simulation and techno-economics. Chem Eng J 268:67–75. https://doi.org/10.1016/j.cej.2015.01.029

Sharma M, Salama ES, Zhang P et al (2022) Microalgae-assisted microbial fuel cells for electricity generation coupled with wastewater treatment: Biotechnological perspective. J Water Process Eng 49:102966. https://doi.org/10.1016/J.JWPE.2022.102966

Shen Y, Li H, Zhu W et al (2017) Microalgal-biochar immobilized complex: a novel efficient biosorbent for cadmium removal from aqueous solution. Bioresour Technol 244:1031–1038. https://doi.org/10.1016/J.BIORTECH.2017.08.085

Shukla M, Kumar S (2018) Algal growth in photosynthetic algal microbial fuel cell and its subsequent utilization for biofuels. Renew Sustain Energy Rev 82:402–414

Singh UB, Ahluwalia AS (2013) Microalgae: a promising tool for carbon sequestration. Mitig Adapt Strateg Glob Chang 18:73–95. https://doi.org/10.1007/s11027-012-9393-3

Sirohi R, Kumar Pandey A, Ranganathan P et al (2022) Design and applications of photobioreactors-a review. Bioresour Technol 349:126858. https://doi.org/10.1016/j.biortech.2022.126858

Sisman-Aydin G (2022) Comparative study on phycoremediation performance of three native microalgae for primary-treated municipal wastewater. Environ Technol Innov 28. https://doi.org/10.1016/j.eti.2022.102932

Slade R, Bauen A (2013) Micro-algae cultivation for biofuels: cost, energy balance, environmental impacts and future prospects. Biomass Bioenergy 53:29–38. https://doi.org/10.1016/j.biombioe.2012.12.019

Song C, Xie M, Qiu Y et al (2019) Integration of CO2 absorption with biological transformation via using rich ammonia solution as a nutrient source for microalgae cultivation. Energy 179:618–627. https://doi.org/10.1016/j.energy.2019.05.039

Srivastava A, Kalwani M, Chakdar H et al (2022) Biosynthesis and biotechnological interventions for commercial production of microalgal pigments: a review. Bioresour Technol 352:127071. https://doi.org/10.1016/j.biortech.2022.127071

Stirbet A, Lazár D, Papageorgiou GC, Govindjee (2018) Chlorophyll a fluorescence in cyanobacteria: relation to photosynthesis

Sukenik A, Shelef G (1984) Algal autoflocculation–verification and proposed mechanism. Biotechnol Bioeng 26:142–147. https://doi.org/10.1002/BIT.260260206

Sung YJ, Yoon HK, Lee JS et al (2022) Novel 3D-printed buoyant structures for improvement in flue gas CO2-derived microalgal biomass production by enhancing anti-biofouling on vertical polymeric photobioreactor. J Clean Prod 366. https://doi.org/10.1016/j.jclepro.2022.133030

Tijani H, Abdullah N, Yuzir A (2015) Integration of microalgae biomass in biomethanation systems. Renew Sustain Energy Rev 52:1610–1622. https://doi.org/10.1016/j.rser.2015.07.179

Ugoala E, Ndukwe G, Mustapha K, Ayo R (2012) Constraints to large scale algae biomass production and utilization. J Algal Biomass Utln 3:14–32

Vandamme D, Foubert I, Muylaert K (2013) Flocculation as a low-cost method for harvesting microalgae for bulk biomass production. Trends Biotechnol 31:233–239. https://doi.org/10.1016/j.tibtech.2012.12.005

Vazhappilly R, Chen F (1998) Eicosapentaenoic acid and docosahexaenoic acid production potential of microalgae and their heterotrophic growth. J Am Oil Chem Soc 75:393–397. https://doi.org/10.1007/s11746-998-0057-0

Veza I, Muhamad Said MF, Latiff ZA (2021) Recent advances in butanol production by acetone-butanol-ethanol (ABE) fermentation. Biomass Bioenergy 144:105919. https://doi.org/10.1016/j.biombioe.2020.105919

Vieira MV, Pastrana LM, Fuciños P (2020) Microalgae encapsulation systems for food, pharmaceutical and cosmetics applications. Mar Drugs 18

Vijayaraghavan K, Balasubramanian R (2015) Is biosorption suitable for decontamination of metal-bearing wastewaters? A critical review on the state-of-the-art of biosorption processes and future directions. J Environ Manag 160:283–296. https://doi.org/10.1016/j.jenvman.2015.06.030

Wang X, Jain A, Chen B et al (2022) Differential efficacy of water lily cultivars in phytoremediation of eutrophic water contaminated with phosphorus and nitrogen. Plant Physiol Biochem 171:139–146. https://doi.org/10.1016/J.PLAPHY.2021.12.001

Wu Z, Zhu Y, Huang W et al (2012) Evaluation of flocculation induced by pH increase for harvesting microalgae and reuse of flocculated medium. Bioresour Technol 110:496–502. https://doi.org/10.1016/j.biortech.2012.01.101

Wu W, Lin KH, Chang JS (2018) Economic and life-cycle greenhouse gas optimization of microalgae-to-biofuels chains. Bioresour Technol 267:550–559. https://doi.org/10.1016/j.biortech.2018.07.08

Wu W, Tan L, Chang H et al (2023) Advancements on process regulation for microalgae-based carbon neutrality and biodiesel production. Renew Sustain Energy Rev 171. https://doi.org/10.1016/j.rser.2022.112969

Xia L, Ge H, Zhou X et al (2013) Photoautotrophic outdoor two-stage cultivation for oleaginous microalgae Scenedesmus obtusus XJ-15. Bioresour Technol 144:261–267. https://doi.org/10.1016/j.biortech.2013.06.112

Yaashikaa PR, Keerthana Devi M, Senthil Kumar P et al (2022) A review on pretreatment methods, photobioreactor design and metabolic engineering approaches of algal biomass for enhanced biohydrogen production. Int J Hydrogen Energy. https://doi.org/10.1016/j.ijhydene.2022.10.092

Yadav G, Dash SK, Sen R (2019) A biorefinery for valorization of industrial waste-water and flue gas by microalgae for waste mitigation, carbon-dioxide sequestration and algal biomass production. Sci Total Environ 688:129–135. https://doi.org/10.1016/j.scitotenv.2019.06.024

Yaqoubnejad P, Rad HA, Taghavijeloudar M (2021) Development a novel hexagonal airlift flat plate photobioreactor for the improvement of microalgae growth that simultaneously enhance CO2 bio-fixation and wastewater treatment. J Environ Manag 298:113482. https://doi.org/10.1016/J.JENVMAN.2021.113482

An Overview on Managing Minute Duckweed (*Lemna Perpusilla* Torr) Cultivation for Fish Feed Purpose

Tjandra Chrismadha, Awalina Satya, Ika Atman Satya, Azalea Dyah Maysarah Satya, Sara Kazemi Yazdi, and Efri Mardawati

Abstract Feed is currently becoming a major problem in developing aquaculture sector. The use of aquatic plants has been pointed out for the resolution, as they can be easily produced while the nutrition value is reasonably good. One aquatic species, minute duckweed (*Lemna perpusilla* Torr.), is believed to be the solution. This species is abundantly distributed in tropical eutrophic waters. Although there has been some success of feeding fish on aquatic plants reported since a long time ago, little progress has been made for field implementation, as it has been hampered by the lack of knowledge on the proper and harmonious management of plant production to fulfill the fish requirement. This paper discusses a strategy to optimize the use of minute duckweed for feeding Nile tilapia fish. A scheme of integrated multitrophic aquaculture (IMTA) is proposed as a means of enhancing low-cost feed production, while simulations are performed to figure out the quantitative interconnection between the two commodities. Mass implementation of this integrated aquaculture scheme in Indonesia, however, faces a major problem of short land ownership, so institutional development is strongly needed, to encourage a segmentation farming activity as well as to set a fair play regulation that assures their business continuity.

Keywords Aquaculture · Alternative feed · Minute duckweed · *Lemna perpusilla Torr*

T. Chrismadha · A. Satya (✉)
Research Centre for Limnology and Water Resources, National Research and Innovation Agency, KST Soekarno, Cibinong, Indonesia
e-mail: awalina@brin.go.id

I. A. Satya
Research Centre for Computation, National Research and Innovation Agency, KST Soekarno, Cibinong, Indonesia

A. D. M. Satya · S. K. Yazdi
Department of Chemical and Environmental Engineering, Faculty of Science and Engineering, University of Nottingham Malaysia, Selangor Darul Ehsan, Jalan Broga, 43500 Semenyih, Malaysia

E. Mardawati
Department of Agro-Industrial Technology, Faculty of Agro-Industrial Technology, Padjadjaran University, Jatinangor, Indonesia

M. A. R. Lubis et al. (eds.), *Biomass Conversion and Sustainable Biorefinery*, Green Energy and Technology, https://doi.org/10.1007/978-981-99-7769-7_8

1 Introduction

The current major problem of fisheries aquaculture development is the high price of artificial feeds that lowers the benefit as well as the economic feasibility. It is mainly associated with a limited supply of protein sources while the price is considerably high, and up to now becomes the strategic issue that remains unresolved (Ogello et al. 2014; Journey W et al. 1993; Hartog and Plas 1970). Finding out alternative feed sources, particularly those related to low-cost sources of protein is very important to encourage the development of a sustainable and environmentally sound fisheries aquaculture (Ogello et al. 2014; Chrismadha et al. 2012; Pradhan et al. 2019).

Efforts to reduce the production cost can be directed by the utilization of any resources that cannot directly become human consumptions or marginal products with lower prices than the feed being produced (Arief et al. 2011; De et al. 2010). There have been many explorations works carried out, on some marginal resources with high biomass productivity and suitable nutrition value for alternative feeds (De et al. 2010; Satya et al. 2022; Said et al. 2022; Elangovan et al 2017). The most important fish feed purpose criteria are preferable and digestible, as well as good nutrition value, which generally assessed in terms of high protein substances with complete amino acids composition, while the fiber and antinutritional content are low (Helfrich and L a. 2002; Chakrabarti et al. 2018) One among resources that has many attentions for alternative feed is minute duckweed (*Lemna perpusilla* Torr.) (Chrismadha et al. 2012; Andriani et al. 2019).

2 Minute Duckweed (*Lemna Perpusilla Torr.*)

Minute duckweed belongs to the family of Lemnaceae (Goopy and Murray 2003; Les et al. 2002). There are two subfamilies in this group: Lemnoideae and Wolffiodeae are distinguished by the root existence, whereas the group of Wolffiodeae don't have any root. Lemnoideae consists of three genera: Lemna, Landoltia, and Spirodella which are all known as 'duckweeds', while Wolffiodeae consists of two genera: Wolffia and Wolffiella which are commonly known as water meals. Up to this time, there are 14 Lemna species identified worldwide (Les et al. 2002; Sree et al. 2016) and the species commonly found in Indonesia is *L. perpusilla*, which is recognized under local name of 'Matalele', while the global name is 'minute duckweed' (Goren et al. 2021; Ekperusi et al. 2019). This individual plant consists of only one flatted-oval form fake leaf, which is actually a composite of stem and leaf and is commonly called 'frond', and a root line prolongs from the frond base (Sree et al. 2016; Hartog and Plas 1970). The frond is composed of chlorenchymatous cells structure separated each other by aerial space providing buoyancy for floating on the surface of water.

The major components distinguishing minute duckweed from other Lemna species are the occurrence of three thickening lines spread out from the base near root to the opposite edge of frond which is called 'venae', and also the existence

Fig. 1 A clump of minute duckweed (left) and root sheath and venae as the specific marks (Chrismadha et al. 2012)

of triangle shaped membrane called 'root sheath' at the root base (Fig. 1). This species has a relatively small size, with maximum frond length of only 8 mm and get smaller under unfavorable growth condition. The root is very short under good growth condition and prolonged as long as 5 cm when suffering from nutrient limitation. One individual plant can produce up to 20 broods during 10 days of life span. The offspring frond usually emerges from the bud pockets located at the two opposing side just near the root base, and attach to the mother frond for several days, so the plant commonly looks like a clump of 2–3 fronds. Minute duckweed actually belongs to flowering plants (Spermatophytes), reproduces generatively by emerging flowers irregularly depending on the growth condition, and delivers seeds that are tolerant to long severe dry condition that will sprout up soon when the condition gets favorable.

Minute duckweed distributes broadly in fertile water of low- to middle-altitude tropical area (Acosta et al. 2021) Although it is also found in a relatively high inland water, such as in Samosir District of North Sumatera (>900 m above sea level), it possibly grows slower than that in the low land area. This tropical species has a close taxonomic relationship with common duckweed (*L minor*) which grows in cosmopolitan around the temperate area. They are only differentiated by the root sheath occurrence in minute duckweed, while no such root sheath is found in common duckweed (Pancho and Soerjani 1978). Those two species love highly fertile stagnant water, and are frequently found to form a green mat on the surface of organic contaminated waters, including ponds, swamps, irrigation and drainage channels, as well as slow moving river waters. It is commonly found to have a symbiotic growth with other floating aquatic plants, such as water meals, water ferns, and water hyacinth.

3 Integrated Multitrophic Aquaculture (IMTA)

Integrated multitrophic aquaculture (IMTA) is aquaculture activities involving various aquatic organisms set together in a harmonious way according to their trophic properties, and directed to enhance more effective space and water resources, as well as feed utilization. IMTA uses an ecological approach to encourage bio circular concept, which is to utilize wastes from aquaculture activities to produce some commodities that can provide added values while diminishing the negative environmental impacts in an integrated way (Barrington et al. 2009). IMTA has an objective to increase the rate as well as the efficiency of biomass productivity per unit of water column at any determined time period (e.g. kg/(m^3 year). IMTA consists of various components of cultivated organisms with different but complementary trophic state, arranged to be a uniting production system based on the inner circular mass and energy transfers to obtain internal ecological balance resulting in a higher production efficiency.

Research of IMTA in Research Center for Limnology and Water Resources—National Research and Innovation Agency of Republic of Indonesia (RCLWR-NRIA) is focused to utilize waste from cat fish (*Clarias sp.*) cultivation—trophic status of omnivore—carnivore—for growing floating aquatic plant minute duckweed (*L perpusilla*)—trophic status of autotrophic or producer—that can be used for feeding Nile tilapia fish (*Oreochromis nioloticus*)—trophic status of omnivore—herbivore (Fig. 2). The advantage value of this IMTA system is that the minute duckweed converts waste materials from fish culture into biomass that can be put back into fish culture as a valuable feed source. This implementation encourages the minute duckweed to deliver a double strategic function, as the water phytoremediation agent that keeps the water condition to suit the fish growth requirement, and as an additional feed source which reduces the cost for feeding expenses (Chrismadha et al. 2021). Accordingly, IMTA implementation will simultaneously enhance a better production efficiency and environmentally sound cultivation practice (Fig. 3).

4 Experimental for IMTA Implementation

Trials for implementation of IMTA scheme have been conducted in RCLWR-NRIA, including utilizing cat fish cultivation waste water to grow minute duckweed and using the biomass product for alternative feed in Nile tilapia culture. Observation on growth and biomass productivity of the plant based on cat fish cultivation waste water medium was carried out in a water closed recirculation ponds placed in an open space, consisting of a cat fish cultivation pond in which the water media was pumped in to three serial order minute duckweed cultivation ponds before it was recirculated back into the cat fish pond (Chrismadha and Said 2019a). The pond size was 2.4×5 m^2 each with 50 cm depth, while the fish density was 1,500 fishes/pond

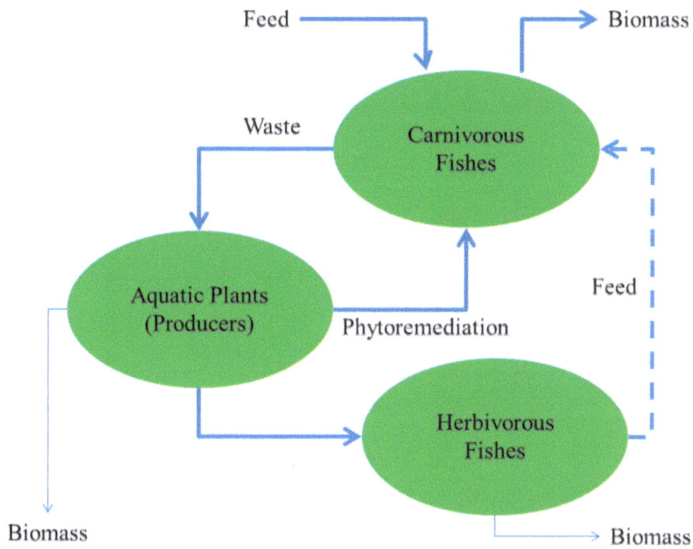

Fig. 2 Concept of integrated multitrophic aquaculture (IMTA) in RCLWR-NRIA

Fig. 3 IMTA trial ponds at RCLWR-NRIA, KST-Soekarno, Cibinong-West of Java

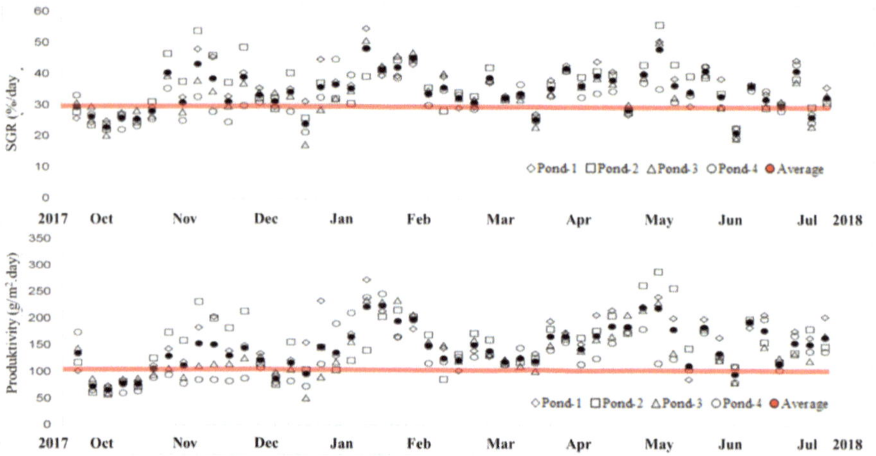

Fig. 4 Growth and biomass productivity of minute duckweed in IMTA ponds with cat fish aquaculture waste water as the growth medium (Chrismadha and Said 2019a)

(125 fishes/m^2), provided with commercial feed containing 30–33% protein at 3% of progressive grown body weight under assumption of 6% specific growth rate.

Continual four cultivation period observation from October 2017 to July 2018 showed the suitability of waste water from cat fish aquaculture for growth medium of minute duckweed. In the meantime, the plant can effectively control the water quality, so there was no water replacement required during the cultivation time, except in dry season when the evaporation rate was considerably high while there was no input of rain water. It saved the water requirement up to more than 80%. Almost all throughout the observation period the average duckweed biomass productivity was above 100 g/(m^2 day) or equal to 328 ton/(ha year) of the fresh weight (Fig. 4). The optimal harvesting period for the minute duckweed culture was 3–4 days.

Figure 4 also shows the fluctuation of the duckweed growth and biomass productivity in response to the cat fish growth period and the progressive amount of feed delivery. The waste oversaturation level was visible just close before the fish harvesting time, where the load of waste was too high and inhibited the minute duckweed growth. The recovery situation was there, however, several days after the fish was harvested and the fish cultivation was restarted back from the fingerling size. The cat fish culture productivity itself was in the range of 70–100 kg/(pond. harvest) or equal to 233–333 ton/(ha. year), with feed conversion value of 0.8–1.4.

Nutrition state of minute duckweed grown in IMTA ponds has been reported by Sutrisno et al. (2021), whichpointed out the advantage of high protein content up to 32.90% of the dry weight, while total lipid content was 9.73% DW, carbohydrate 8.75% DW, and ash 15.32%. Under a better growth condition, the plant has a higher protein content up to 38.10% DW or more (Ilyas et al. 2014; Hasan and Chakrabarti 2009). Sutrisno et al. (2021) also showed a complete amino acids composition in minute duckweed biomass. HPLC analysis with 15 reference amino acid standards revealed the existence of all the amino acids in the biomass (Table 1). Duckweeds are naturally widely known to have a complete amino acids composition, close to animal amino acids composition (Hasan and Chakrabarti 2009), so that it is considered to be very suitable for animal feed purpose, including for fishes. Chakrabarty et al. (2018) reported the occurrence of 20 amino acids species in common duckweed (*L minor*).

The main problem in utilizing minute duckweed for fish diet is the low energy density, which is mainly caused by limited lipid content in the minute duckweed biomass. Lipid is the most efficient energy source for fish, but the content in the minute duckweed biomass ranges from 4.15 to 9.73% DW, whereas the ideal lipid content in fish diet is 7–15% (Helfrich 2002). Beside its function as energy source, lipid also has an important role as media transport for fat dissolved vitamins, such as vitamin A, D, and E (Elangovan et al. 2017). Lipid in fish diet has also to contain appropriate content of essential fatty acids, particularly those belonging to omega-3 group. These polyunsaturated fatty acids (PUFAs) cannot be synthesized by fish itself, so the availability completely depends on the external supply from the diet, with an ideal portion of 0.5–2.0% of the total diet. There is no information yet available for fatty acid composition of the minute duckweed, but Chakrabarti et al. (2018) reported the omega-3 content of the closest related species, which is common duckweed was up to 46% of the total fatty acids.

Under the assumption that fatty acid content of minute duckweed is similar to that of common duckweed, the lipid supply can be directed for fulfilling the need of energy so that the available protein input can be more obtained for growth. It is mainly due to the carbohydrate content of minute duckweed that is also considerably low, which is 7.27–8.75% DW only. Carbohydrate component in fish diet is generally high of more than 20%, although it is not an efficient energy source for fish, carbohydrate has considerably lower price compared to other components. The important notes related

Table 1 Proximate and amino acids composition of minute duckweed (*L. perpusilla Torr*) grown in IMTA ponds

Proximate*		Amino acids**			
	% DW	Essentials	% BK	Non-essentials	% dry weight
Moisture	88.2–94.3	Histidine	0.54	Alanine	1.98
Ash	6.70–16.50	Isoleucine	1.53	Arginine	1.71
Protein	27.3–38.10	Leucine	2.61	Aspartic	2.97
Lipid	4.15–9.73	Lysine	1.53	Cysteine	
Carbohydrate	7.25–8.75	Methionine	0.54	Glutamic Acid	3.51
		Phenylalanine	1.53	Glycine	1.44
		Threonine	1.35	Proline	
		Tryptophan	NA	Serine	1.44
		Valine	1.98	Tyrosine	1.62

* Hasan and Chakrabarty (2009), Ilyas et al. (2014), ** Sutrisno et al. (2021)
NA not analyzed

to the duckweed carbohydrate content is the crude fiber fraction that has been reported to compose up to 44% DW (Ilyas et al. 2014), while in the group of duckweed the nitrogen free extract (NFE) which generally represent the nonstructural carbohydrate content, was 27.2–66.4% DW (Chakrabarti et al. 2018; Goopy and Murray 2003; Hasan and Chakrabarti 2009). Crude fibers are materials that cannot be digested by fish and become a source of inefficiency in terms of feed metabolism. Trials in RCLWS–NRIA, KST-Soekarno-Cibinong also show that minute duckweed biomass can be dried, to make it have a longer storage time as well as for a more flexible distribution purpose. This drying practice, however, might give a consequence of decreasing the nutrition value.

The next trial was utilization of the minute duckweed biomass for alternative feed in fish cultivation. In line with the IMTA conception, it was a direction to employ the fresh biomass for the feeding purpose. The fresh harvested biomass was delivered for fish feeding as soon as possible, so that it eliminates the requirement for any post-harvest processing as well as the storage places that might need some additional energy and cost expenditure. There have been some experiments to feed various fishes on fresh minute duckweed biomass, and the results show that almost all tested fishes are ready to forage on it. There is a group of fishes that strongly like to forage on minute duckweed, even when there is regular pellet available. This group consist of Nile tilapia, common carp, and cat fish. Other tested fishes, including gourami and pomfret fish consume minute duckweed when no other feed source is available.

Investigation to figure out fish growth performance fed on minute duckweed biomass was carried out in two serial trials of Nile tilapia culture. Firstly, it is conducted at a nursery level, involving new born fish of 1–2 weeks of age for the starter and grown up for about 8 weeks to size of 40–50 g/fish, that is ready for starter in the next round of grown-up cultivation. The second was the grown-up level, involving these starting fish of 15–17 g/fish size and cultivated up to about 100 g/fish size for around 4 months.

The nursery level trial was carried out in aquaria of a closed water recirculation system with water quality maintenance mainly directed for controlling the dissolved oxygen content >3 mg/L, with culture density of 100 fishes/m^2 (Sutrisno et al. 2021). The minute duckweed employed for feed was the fresh biomass obtained from the nearby IMTA ponds culture based on cat fish cultivation waste water medium. The observation result is summarized in Table 2, showing a great potential of minute duckweed biomass for the replacement of pellet commercial feed in the nursery-level fish cultivation activity. Although the fish fed on minute duckweed biomass had slower growth rate, but a simulation according to the specific growth rate values showed that the fish only needs as long as 12–15 days additional time to achieve the same size as that fed on commercial pellet. It can be concluded that minute duckweed can completely replace commercial feed for growing Nile tilapia fish with a consequence of a longer culture period.

This might become an interesting advantage when the minute duckweed biomass can be produced at significant lower cost than that of purchasing the commercial feed.

Table 2 Growth performance of Nile tilapia fish at nursery level. After Sutrisno et al. (2021)

	Time				
	Initial	Week-2	Week-4	Week-6	Week-8
Pellet					
Length (cm)	35,00 ± 0,00	57,03 ± 1,59	87,03 ± 0,76	123,27 ± 3,62	138,60 ± 4,88
Weight (g/fish)	0,86 ± 0,00	3,42 ± 0,18	11,56 ± 0,28	30,96 ± 4,56	47,91 ± 5,62
CF (g/cm)	0,24 ± 0,00	0,59 ± 0,02	1,30 ± 0,03	2,49 ± 0,23	3,40 ± 0,35
SGR (%/day)		9,89 ± 0,38	7,62 ± 0,48	6,54 ± 0,68	3,36 ± 0,28
WG (g/fish/day)		0,18 ± 0,03	0,51 ± 0,03	1,13 ± 0,16	1,30 ± 0,19
TF (g)		120,00 ± 0,00	730,00 ± 0,00	670,00 ± 0,00	990,00 ± 0,00
FCR		0,47 ± 0,03	0,89 ± 0,05	0,81 ± 0,12	1,19 ± 0,18
Minute duckweed					
Length (cm)	35 ± 0,00	51,07 ± 3,43	76,15 ± 2,68	106,40 ± 3,36	124,75 ± 3,65
Weight (g/fish)	0,86 ± 0,00	2,62 ± 0,24	8,42 ± 1,04	20,47 ± 2,31	32,18 ± 3,27
CF (g/cm)	0,24 ± 0,00	0,50 ± 0,03	1,07 ± 0,11	1,90 ± 0,16	2,57 ± 0,25
SGR (%day)		7,98 ± 0,64	7,27 ± 0,41	5,93 ± 0,25	3,48 ± 0,39
WG (g/fish/day)		0,13 ± 0,02	0,34 ± 0,03	0,69 ± 0,06	0,65 ± 0,06
TF (g)		191,67 ± 29,30	956,67 ± 40,41	983,33 ± 136,50	1100,00 ± 185,20
FCR		10,83 ± 0,30	17,56 ± 1,83	18,99 ± 0,95	27,78 ± 7,68

Notes CF = Condition Factor; SGR = *Specific Growth Rate*; WG = *Weight Gain*; TF = Total Feed; FCR = Food Conversion Ratio

Experiment of the grown-up level cultivation was carried out in larger vessels, which were 1 m^3 provided with a closed water recirculating system by means of submersible pumps to generate a water flow to maintain the dissolved oxygen level >3 mg/L (Chrismadha and Mulyana 2019b). There was a variation in culture density applied for the experimental treatment. At the lowest culture density, where less space and any other resources competition was occurred, a single feed of minute duckweed was able to stimulate fish growth up to 3%/day (Table 3). This growth rate decreased with the culture density.

Table 3 Growth performance of Nile tilapia fish at grown-up level. After Chrismadha and Mulyana (2019b)

Parameters	Stocking density (fishes/m^2)	Days			
		0	40	75	115
Length (cm)	5	9.80 ± 0.00	14.22 ± 0.25	16.55 ± 0.27	18.30 ± 0.14
	10	9.92 ± 0.00	13.65 ± 0.07	15.79 ± 0.86	17.24 ± 0.61
	20	10.28 ± 0.25	13.16 ± 0.48	24.53 ± 1.91	16.28 ± 0.73
Weight (g/fish)	5	16.57 ± 0.63	55.30 ± 2.91	87.91 ± 4.99	126.36 ± 5.29
	10	16.83 ± 0.13	49.55 ± 3.71	76.29 ± 7.25	107.09 ± 5.85
	20	17.34 ± 0.20	43.26 ± 5.78	63.63 ± 13.66	88.03 ± 11.88
Condition factor (g/cm)	5	1.69 ± 0.07	3.89 ± 0.14	5.31 ± 0.22	6.90 ± 0.24
	10	1.70 ± 0.01	3.63 ± 0.29	4.83 ± 0.20	6.21 ± 0.12
	20	1.69 ± 0.01	3.28 ± 0.32	4.35 ± 0,37	5.40 ± 0.49
Specific growth rate (%/day)	5		3.09 ± 0.04	1.32 ± 0.31	0.98 ± 0.04
	10		2.76 ± 0.21	1.23 ± 0.06	0.92 ± 0.11
	20		2.33 ± 0.39	1.08 ± 0.24	0.90 ± 0.22
Weight gain (g/fish/day)	5		0.99 ± 0.06	0.84 ± 0.20	0.99 ± 0.01
	10		0.84 ± 0.10	0.86 ± 0.09	0.79 ± 0.04
	20		0.66 ± 0.16	0.52 ± 0.20	0.63 ± 0.05
Total feed (g)	5		3,501 ± 634	4,500 ± 1061	7,450 ± 354
	10		6,550 ± 2192	8,800 ± 1556	12,100 ± 1131
	20		12,225 ± 884	13,800 ± 2687	18,400 ± 424
Food conversion ratio	5		18.02 ± 2.22	27.62 ± 0.19	38.76 ± 2.14
	10		20.56 ± 9.11	32.81 ± 1.47	39.41 ± 5.46
	20		24.04 ± 3.94	35.25 ± 7.05	37.78 ± 1.90

The growth rate in this trial was found to be lower than previously reported by Rejeki et al. (2013). But it was in the range of that reported by Khalil et al. (2011). which was 0.31–1.44%/day for the same fish species grown in floating cages in reservoir water. Fish generally has a higher growth rate when cultivated in floating net placed in reservoir waters compared to that cultivated in inland pond water. As it has a better water circulation obtained from big water volume of the reservoir. Accordingly, the growth performance of Nile tilapia in this experiment, which used vessels of closed water recirculation system can be considered to be suitably good and can reflect the ability of duckweed biomass to support fish growth even when it is delivered as a single died.

From this experiment, it can also be figured out the pattern of fish consumption rate on the duckweed biomass, which tends to progressively follow the body weight development with the rate ranging from 38.91 to 40.50% of the body weight in which the efficiency of feed conversion ratio was tend to decrease to form a logistic model on the body weight development. Calculation of the minute duckweed biomass requirement to grow fish up to a certain size can be conducted by plotting the progressive daily FCR values along the body weight development and multiply this daily FCR value with the daily weight gain. It is estimated that to produce a fish of 250 g size the required fresh biomass of minute duckweed is as much as 12.591–12.946 g which is equal to the FCR value of 50.08–50.60. According to this FCR value, it can be estimated that the production rate of 328 ton/(ha year) as mentioned above is suitable to fulfill the feed requirement for producing 250 g size fish as much as 7–8 ton.

5 IMTA Criterions

IMTA has many types of implementations as its design is tailored to the environmental and ecological aspects of the location it is implemented. It usually involves the integration of many species consisting of fed species (usually fish) which provide waste or detritus to be consumed by the extractable species such as marine invertebrates and/or algae (Alexander et al. 2016). The synergistic culture will attribute to the trophic level which showcased the relationship between the components within the pond. IMTA is an evolution from the concept of polyculture which has a long route back to 2300-1700 BC in Southern China, long time ago in a form of ponds for water storage during the Hans dynasty (Li 1994). Meanwhile, the modern type of polyculture with a quantitative and scientific approach close to the current system stems from the work which evaluated the bioremediation work of organisms (shellfish) and simple phytae consisting of microalgae and seaweeds in a systematic municipal water treatment (Ryther et al. 1975). It demonstrated the combined tertiary sewage treatment using an aquaculture system where the solid wastes from shellfish were then fed to small invertebrates and other small creatures which were regarded as high-quality foods to lobsters, flounders, and other commercially valuable secondary crops. While the dissolved waste excreted was then used as nutrients for unicellular algae applied as the final polishing step. The system gave a final effluent free of inorganic nitrogen and did not contribute to eutrophication. However, the work was considered to not be quantitative and practical enough thus it was further revised by the addition of phytoplankton as a biofilter which further supports the uptake of nutrients done by algae, pioneered by Gordin et al. (1981) and worked on continuously by Harlin et al. (1982). Advances toward ecological engineering would only be done to further improve IMTA. The pioneering work of Troell et al. enforced this by stressing and eventually landing the basis of IMTA where it should factor in the sustainability of aquaculture. The sustainability can also be measured by counting in the economic, societal, and environmental benefits thus including recycling of waste nutrients from higher trophic-level species into commercially valued lower trophic-level crops (Troell et al. 2009). IMTA can either be land-based or open water-based with species configurations. The basic set-up of this system is by utilizing ponds. The typical set-up known to date is based on varied species systems run in open water as first developed by Troell et al. (2003). The interconnected ponds are set up as two structures placed close to each other. These also include protection from the fed aquaculture and the extractive components to the existing environmental conditions by the use of a caging system. The use of IMTA is additionally referred to as a provider of 'rich food for impoverished people' as it provided a good environment for fish with particularly high proteins, Lipids, Minerals, and vitamins (Ye et al. 2017). In 2022, rapid fisheries and aquaculture production was observed with a record-breaking of 214 million tons of production with approximately 112.6 million tons attributed to aquaculture. As sludge of aquaculture is high in nutrients, removal efficiencies must be paid attention to avoid environmental stress one of which is by recirculating water (Couto et al. 2022). Closed water circulation is part of IMTA as it ensured the system will not affect the environment uncontrollably. In terms of setting up this closed system, careful consideration of choosing IMTA types has to consider the following choices provided in Table 4. Table 4 shows that the selection criteria for IMTA all fit well with the implementation of Duckweed.

6 The Economic Potential

The main objective of IMTA development is enhancing fishery aquaculture production efficiency. In the implementation, IMTA involves alternative feeds that can become a new commodity and needs a further direction for its cultivation practice in the field. The important thing to encourage minute duckweed culture development is the feature of its economic value. Referring to the FCR value as mentioned above ($\approx$50), it means that to produce 1 kg Nile tilapia fish it requires about 50 kg fresh duckweed biomass. If it makes equal to the tilapia feed price, where the current market price is IDR 13,500/kg and the commercial feed conversion value is 1.5, the selling price of the fresh minute duckweed biomass can be calculated as much as IDR 405/kg. Accordingly, the potential of increasing land value by using it for minute duckweed cultivation can be determined.

Table 4 Criteria of selecting species for IMTA

Species selection criteria for IMTA	Definition	In our trial IMTA implementation
Complementary co-cultured species	This is to ensure that the extractive species can be eaten by fed aquaculture to ensure improvement in water equality and good economic output	Duckweed presented biomass that supports fish growth in the pond
Native species	The choice of native species will avoid invasion and save acclimatization time and cost too	Duckweed is native to the natural system where catfish grows
Consistency of waste material	Consistent waste material is to ensure that the system will be able to run sustainably	The growth of duckweed and catfish can be seen to grow side by side in previous studies
Choice of the bio-filtering organism	In choosing a biofilter organism. It should have a high growth potential which can be periodically harvested. This is also to avoid IMTA's biological bottlenecks which include the lack of seed, long periods of production, lack of knowledge in cultivating, Biofouling, and predation between the feed and bio-filtering organisms	Duckweed also serves as a bio-filtering organism and even can live in consortia for improving water quality. The growth of this organism is also quite fast
Established market species	In developing IMTA, the economic value should be co-cultured with buyers to invite larger investments	Although duckweed is not well-established in the market. The need for this species will grow with the growing food demands for aquaculture
Commercial hindrances	In choosing a species for IMTA, other similar species must be considered to avoid commercialization hindrances	Microalgae and duckweed have almost the same growth and simpler structure. Further research can be done in comparing these two or even consider them as co-culture
Socio-ecological and political compatibility	In ensuring a sustainable set-up, socio-ecological and political compatibility must be well-thought before implementation. This includes the IMTA bottlenecks such as licensing and regulations	Indonesia, for example, is one of the largest island countries with high diversity in their waters. Therefore the use of IMTA with duckweed and local catfish can improve the currently existing aquaculture businesses

Source Nissar et al. (2023), Chen et al. (2018)

Food and Agricultural Organization (FAO) of United Nations Organization (UNO) uses the parameter of land equivalent ratio (LER) for the evaluation of land use efficiency in polyculture farming activities. In principle, the value of land used for a certain purpose can be directly compared between two commodities or indirectly by referring to a certain main reference commodity. This approach can be used for the evaluation of potential land use for minute duckweed cultivation from the perspective of economic development. Taking paddy field cultivation land as the standard reference with an assumption of productivity level 24 tons/(ha year) and the selling price of unhauled rice is IDR 5,000/kg while the selling price of minute duckweed refers to tilapia feed value as described above (IDR 13,500/kg and FCR = 1.5) the land use value for minute duckweed cultivation is considerably higher than that of paddy field purpose (Table 5). Therefore minute duckweed cultivation can be considered to be more beneficial and potentially becomes an alternative activity for earning life, particularly in the locations where highly fertile water is available.

The important note regarding the above economic evaluation is that the minute duckweed selling price is presumed to be equal to the feeding cost in Nile tilapia cultivation activity. The minute duckweed selling value has actually not yet been performed and will be self-revealed to follow the market mechanism when the mass cultivation has been developed. As an example, the minute duckweed selling price determination refers to its utilization for fertilizer purpose will possibly lower down the value compared to the above description for Nile tilapia fish feed. However, the value will increase if it refers to the purpose for feeding cat fish. Lower price will stimulate the segment of consumer but becomes a disincentive for the producers and on the contrary, higher price will encourage the producer but lowering the consumer's enthusiasm. The current duckweed selling price on online sites ranged from IDR 20,000–40,000 per kg and is more directed for the purpose of ornamental fish feed.

As an additional note, trials in RCLWS–NRIA have also revealed that duckweed biomass can be utilized for additional feed in cat fish cultivation to reduce

Table 5 Land equivalent ratio of land used for various aquaculture commodities

	Units	Nile Tilapia	Cat fish	Minute duckweed	Rice field
Wide	m^2	100	100	100	Land productivity assumption of 24 ton/Ha/year
stocking Density	Fishes	2000	15000	30	
Survival rate	%	90	90		
Specific growth rate	%/day	1.76	4.47	30	
Culture period	day	157	63	4	
Harvesting quantity	$kg/m^2/day$	0.029	0.214	0.1	0.0066
Unit price	IDR/kg	22.000	18.000	405	5000
Land values	IDR/m^2day	630.5732	3857.143	40.5	33

the commercial pellet, need up to 30% without significant harm to the fish growth (Chrismadha et al. 2021).

7 Challenges

It has been pointed out that challenges in minute duckweed culture development are mainly associated with its position as a new commodity where the technical aspect needs to be adapted to landscape suitability and socio-cultural of the local farmers (Journey et al. 1993). Minute duckweed culture needs a large more intensive maintenance compared to rice planting or even fishery aquaculture. This is mainly due to the production cycle of minute duckweed that can be counted only in several days while that of rice field is in 2–4 months. This is so for fisheries aquaculture even though there is a need to have daily feeding activity, the production cycle is generally more than two months. Beside the need for more intensive work a short harvesting period also makes the farmer to have a close interaction with the users to market due to the minute duckweed biomass property that contains high level of protein to make it more susceptible to damage. In the tropical condition it is relatively warm and fresh biomass of minute duckweed cannot sustain for more than 24 h. Whereas as a new commodity the market place for minute duckweed has not yet been established so that for the initiation minute duckweed culture development is only feasible when it is conducted in an integrated mode. And this integrated mode must be done with any fisheries aquaculture so that the resulting minute duckweed biomass can be directly distributed as fish feed such as under the scheme of IMTA.

Implementation of integrated cultivation of minute duckweed and fish however requires a considerably wide land area. Meanwhile, the land ownership level of Indonesian farmers is relatively short. By average in the year of 2007, the land ownership level of Indonesian farmers was 0.36 ha (Susilowati and Maulana 2012). As has been mentioned above minute duckweed biomass conversion ratio to be Nile tilapia flesh is $\approx$50, which means that to support Nile tilapia fish production it needs an extensive minute duckweed production system.

A simulation of the integrated aquaculture ponds as shown in Table 5 shows that an optimal utilization of minute duckweed biomass can be enhanced by the arrangement of Nile tilapia fish cultures in a time serial segmentation in which a regular seeding and harvesting are performed to keep the total population in a constant number so that the feen requirement can also be managed in a steady amount. The extent of segmentation which then determines the demand on the pond number depends on the fish growth rate, while the pond large of the ponds has to follow the tonnage target. As shown in Table 6, there are 14 culture time serial segmentation required to accommodate an optimal minute duckweed feeding scheme of Nile tilapia fish grow at a rate of 2%/day and the target harvesting size of 250 g/fish. To install a fish production capacity of 90 kg/week with the minute duckweed productivity of 0.1 kg/m^2/day there will be a need of totally 3,058 m^2 pond area consisting of 458 m^2 fish ponds and 2,600 m^2 minute duckweed ponds. Considering the relative shortness of Indonesian farmers land ownership the implementation of IMTA has to bring up a communal activity by forming a group of fish farmers with an integrated farming segmentation controlled by a predefined production target. According to this predefined target the farming activities will divide into two groups of minute duckweed and Nile tilapia fish cultivations in which each group will have a guarantee for their business continuity where the fish farmer will obtain an assurance of feed supply that at the same time gives a commitment to definite market for the product of the minute duckweed farmers.

Regarding to this issue an institutional means has to be developed to organize all the components of the involved production system including farmers, infrastructures, science and technology, and markets. An authority has to be established to collect farmers and divide them into segments of farming activities with accountable roles and targets and gives a financial commitment as well as supply of any production requirements to encourage the success of cultivation works.

Some important issues regarding this institutional aspect are as follows:

(1) Minute duckweed is a new farming commodity that is not yet widely known. In this case. The emerged institution has to have a capacity to provide a bridge for science and technology to enter and have an adaption to the local farmer with their available resources. The new institution has also to have a close interaction with the relevant R&D institution so that all the newest technology can be received and implemented. At the same time all the actual problems in the field can be communicated and have a appropriate assistance for the resolutions.
(2) As a new commodity minute duckweed has not been well-defined in the market. So that the emerged institution has to have a capacity to create an integrated relationship between producers and users to guarantee the market availability for the minute duckweed biomass product. The simplest pattern is a combination of minute duckweed cultivation and fish farming in which a farming segmentation is designed to define the fish production target and its feed requirement. This market issue is very important particularly from the minute duckweed cultivation side as it has to be conducted in an intensive way with harvesting time every 3–4 days and biomass product susceptible to damage so that the product has to be delivered as soon as possible to the users.
(3) Fair price making that gives incentives to all the parties. In this case the institution takes a role as a mediator for price determination that gives advantages to both the minute duckweed producer and the fish farmers.
(4) Incentive assurances of payment and financial capital credit. The institution has to be able to deliver financial services providing payment upon the delivered minute duckweed product as soon as possible, which is credited to the users that will be billed at the harvesting time. This direct cash payment is believed to become a good incentive for minute duckweed farmers to carry on the cultivation activity.
(5) To become a feed buffer agency at the time of minute duckweed cultivation fails to provide suitable feed quantity. This failure moment has to be considered as the minute duckweed is an autotrophic organism and the growth is largely affected

Table 6 Simulation on time serial segmentation of Nile tilapia fish culture for harmonizing use of minute duckweed culture

Weeks	Weight (W)	Weight gain (WG)	Daily feed intake (F)	Weekly feed requirement														
	g/fish	(W * G)	(WG * FCR)	(F * P/1,000)														
		g/fish/day	g/fish/day	kg/day														
				Pond numbers														Total
				1	2	3	4	5	6	7	8	9	10	11	12	13	14	
1	42.05	0.84	15.14	0.84														0.84
2	47.36	0.95	17.05	0.95	0.84													1.79
3	54.40	1.09	19.58	1.09	0.95	0.84												2.88
4	62.49	1.25	22.50	1.25	1.09	0.95	0.84											4.13
5	71.78	1.44	25.84	1.44	1.25	1.09	0.95	0.84										5.56
6	82.45	1.65	29.68	1.65	1.44	1.25	1.09	0.95	0.84									7.21
7	94.71	1.89	34.10	1.89	1.65	1.44	1.25	1.09	0.95	0.84								9.11
8	108.80	2.18	39.17	2.18	1.89	1.65	1.44	1.25	1.09	0.95	0.84							11.28
9	124.97	2.50	44.99	2.50	2.18	1.89	1.65	1.44	1.25	1.09	0.95	0.84						13.78
10	143.56	2.87	51.68	2.87	2.50	2.18	1.89	1.65	1.44	1.25	1.09	0.95	0.84					16.65
11	164.90	3.30	59.36	3.30	2.87	2.50	2.18	1.89	1.65	1.44	1.25	1.09	0.95	0.84				19.95
12	189.42	3.79	68.19	3.79	3.30	2.87	2.50	2.18	1.89	1.65	1.44	1.25	1.09	0.95	0.84			23.74
13	217.58	4.35	78.33	4.35	3.79	3.30	2.87	2.50	2.18	1.89	1.65	1.44	1.25	1.09	0.95	0.84		28.09
14	244.95	4.90	88.18	4.90	4.35	3.79	3.30	2.87	2.50	2.18	1.89	1.65	1.44	1.25	1.09	0.95	0.84	32.99

(continued)

Table 6 (continued)

Weeks	Weight (W)	Weight gain (WG)	Daily feed intake (F)	Weekly feed requirement														
	g/fish	(W * G)	(WG * FCR)	(F * P/1,000)														
		g/fish/ day	g/fish/ day	kg/day														
				Pond numbers														Total
				1	2	3	4	5	6	7	8	9	10	11	12	13	14	
15	42.05	0.84	15.14	0.84	4.90	4.35	3.79	3.30	2.87	2.50	2.18	1.89	1.65	1.44	1.25	1.09	0.95	32.99
16	47.36	0.95	17.05	0.95	0.84	4.90	4.35	3.79	3.30	2.87	2.50	2.18	1.89	1.65	1.44	1.25	1.09	32.99

Notes By segmentation of 14 pond units which are sequentially harvested one pond every week. there will be obtained a constant feed requirement avter the week of by the weather so that under extreme weather condition this plant cultivation might face a failure and it needs some days to a recover.

(6) The institution has also to be capable of collaborations with the government agencies in case of relevant regulations and development program implementation.

8 Conclusion

Involvement of minute duckweed in fisheries aquaculture can potentially become an alternative way to resolve the current problem of high-cost feed. An appropriate cultivation system has to be developed to enable this potential into real implementation. An integrated culture practice is suggested where the minute duckweed production cycle is synchronized with fish farming production target in an optimal way. Fish culture has to be divided into certain time serial segmentation of stocking and harvest to let a constant requirement of feed quantity so the minute duckweed target production can be conveniently set up. Developing a harmonious interaction is very important in order to make this integrated culture practices become profitable and sustainable.

Acknowledgements IMTA research was conducted in RCLWR-BRIN under various financial schemes. including LIPI Competitive Research Program (2012). LIPI Science and Technology Park (2019) and DIPA of previous RC Limnology-LIPI (2010–2014).

Author Contributionship Tjandra Chrismadha and Awalina Satya. were the main contributors (data curation. writing original draft); others were supporting contributors: Ika Atman Satya (review). Azalea Dyah Maysarah Satya (review). Sara Kazemi Yazdi (review). and Efri Mardawati (review).

References

Acosta K, Appenroth KJ, Borisjuk L et al (2021) Return of the Lemnaceae: duckweed as a model plant system in the genomics and postgenomics era. Plant Cell 33:3207–3234. https://doi.org/10.1093/plcell/koab189

Alexander KA, Angel D, Freeman S et al (2016) Improving sustainability of aquaculture in Europe: Stakeholder dialogues on Integrated multi-trophic aquaculture (IMTA). Environ Sci Policy 55:96–106. https://doi.org/10.1016/J.ENVSCI.2015.09.006

Andriani Y, Irawan B, Iskandar et al (2019) Short communication: diversity of duckweed (Araceae-Lemnoideae) morphological characteristics and its potentials as food sources for herbivorous fishes in West Java. Indonesia. Biodiversitas 20:1617–1623. https://doi.org/10.13057/biodiv/d200618

Arief M, Pertiwi DK, Cahyoko Y (2011) Pengaruh pemberian pakan buatan. pakan alami. dan kombinasinya terhadap pertumbuhan. rasio konservasi pakan dan tingkat kelulushidupan Ikan Sidat (Anguilla bicolor). Jurnal Ilmiah Perikanan Dan Kelautan 3:61–66

Barrington K, Chopin T, Robinson S (2009) Integrated multi-trophic aquaculture (IMTA) in marine temperate waters. In: Integrated mariculture—A global review—FAO fisheries and aquaculture technical Paper No. 529 7–46. https://doi.org/10.1016/S0044-8486(03)00469-1

Chakrabarti R, Clark WD, Sharma JG et al (2018) Mass production of Lemna minor and its amino acid and fatty acid profiles. Front Chem 6:1–16. https://doi.org/10.3389/fchem.2018.00479

Chen G, Fang Y, Huang J et al (2018) Duckweed systems for eutrophic water puri fi cation through converting wastewater nutrients to high- starch biomass: comparative evaluation of three different genera (Spirodela polyrhiza. Lemna Minor Polycult 17927–17937. https://doi.org/10.1039/C8RA01856A

Chrismadha T, Sulawesty F, Satya A et al (2012) Use of duckweed (Lemna perpusilla Torr.) for natural feed and fitoremedial agent in aquaculture: profit improvement and sustainability enhancement. In: International seminar on new paradigm and innovation on natural science and its application (ISNPINSA), pp 1–5

Chrismadha T, Said D (2019a) Pengembangan alternatif budidaya perikanan jaring apung: Meningkatkan efisiensi produksi kolam perikanan darat dengan tumbuhan mata lele (Lemna perpusilla Torr.). Laporan Teknis Teknologi Penyehatan Danau untuk Mendukung Budidaya Perikanan dan Pariwisata

Chrismadha T, Mulyana E (2019b) Laju Konsumsi Tumbuhan Air Mata Lele (Lemna perpusilla) oleh Ikan Nila (Oreochromis sp.) dengan Padat Tebar Berbeda. Limnotek: Perairan Darat Tropis Di Indonesia 26:39–46

Chrismadha T, Tanjung LR, Sutrisno (2021) Use of minute duckweed (Lemna perpusilla) for supplemental feed in catfish (Clarias sp.) culture: determination of the optimal proportion using powersim simulation. E3S Web Conf 322:02015. https://doi.org/10.1051/e3sconf/202132202015

Couto AT, Cardador M, Santorio S et al (2022) Cultivable microalgae diversity from a freshwater aquaculture filtering system and its potential for polishing aquaculture-derived water streams. J Appl Microbiol 132:1543–1556. https://doi.org/10.1111/jam.15300

Craig S, Helfrich LA (2002) Understanding fish nutrition feeds and feeding. Virginia Cooperative Extension 1–18

de Almeida Tavares F, Tavares A, Lapolli F et al (2010) Use of domestic effluent through duckweeds and red tilapia farming in integrated system. Panam J Aquat Sci 5:1–10

Ekperusi AO, Sikoki FD, Nwachukwu EO (2019) Chemosphere application of common duckweed (Lemna minor) in phytoremediation of chemicals in the environment: state and future perspective 223

Elangovan P, Nathan F, Prabu E et al (2017) An overview on significance of fish nutrition in aquaculture industry NADP-development of cost efficient feed for enhancing freshwater fish production view project an overview on significance of fish nutrition in aquaculture industry. Int J Fish Aquat Stud 5

Goopy JP, Murray PJ (2003) A review on the role of duckweed in nutrient reclamation and as a source of animal feed. Asian-Australas J Anim Sci 16:297–305. https://doi.org/10.5713/ajas.2003.297

Gordin H, Motzkin F, Hughes-Games A, Porter C (1981) Seawater mariculture pond—An integrated system. Eur Aquac Spec Publ 6:1–13

Goren AY, Okten HE, Sofuoglu SC (2021) Environmental technology & innovation biosorption of methylene blue from water by live Lemna minor. Environ Technol Innov 22:101432. https://doi.org/10.1016/j.eti.2021.101432

Harlin MM, Thorne-Miller B, Boothroyd JC (1982) Seagrass-sediment dynamics of a flood-tidal delta in Rhode Island (U.S.A.). Aquat Bot 14:127–138. https://doi.org/10.1016/0304-3770(82)90092-4

Hartog CD, van der Plas F (1970) A synopsis of the lemnaceae. Blumea XVIII(2)

Hasan MR, Chakrabarti R (2009) Use of algae and aquatic macrophytes as feed in small-scale aquaculture

Ilyas AP, Nirmala K, Widiyanto T (2014) Pemanfaatan Lemna perpusilla sebagai pakan kombinasi untuk ikan nila (Oreochromis niloticus) pada sistem resirkulasi Anita. Limnotek 2014:193–201

Journey W, Skillicorn P, Spira W (1993) A new aquatic farming system for developing countries. The World Bank Group 76:76

Khalil F, Mehrim A, El-Shebly A, Abdelaal M (2011) An assessment study of tilapia polyculture in floating net cages. J Anim Poultry Prod 2:75–92. https://doi.org/10.21608/jappmu.2011.83344
Les DH, Crawford DJ, Landolt E et al (2002) Phylogeny and systematics of Lemnaceae. The duckweed family. Syst Bot 27:221–240
Li S (1994) Characteristics of freshwater fish resources. Dev Aquac Fish Sci 28:27–50
Nissar S, Bakhtiyar Y, Arafat MY et al (2023) The evolution of integrated multi-trophic aquaculture in context of its design and components paving way to valorization via optimization and diversification. Aquaculture 565:739074. https://doi.org/10.1016/j.aquaculture.2022.739074
Ogello EO, Munguti JM, Sakakura Y, Hagiwara A (2014) Complete replacement of fish meal in the diet of Nile Tilapia (Oreochromis niloticus L.) grow-out with alternative protein sources. A review. Int J Adv Res (indore) 2:962–978
Pancho JV, Soerjani M (1978) Aquatic weeds of Southeast Asia: a systematic account of common Southeast Asian acquatic weeds. National Pub. Cooperative
Pradhan D, Sukla LB, Mishra BB, Devi N (2019) Biosorption for removal of hexavalent chromium using microalgae. J Clean Prod 209:617–629. https://doi.org/10.1016/j.jclepro.2018.10.288
Rejeki S, Hastuti S, Elfitasari T (2013) Uji Coba Budidaya Nila Larasati Di Karamba Jaring Apung Dengan Padat Tebar Berbeda. J Saintek Perikanan 9:29–39
Ryther JH, Goldman JC, Gifford CE et al (1975) Physical models of integrated waste recycling-marine polyculture systems. Elsevier Scientific Publishing Company
Said DS, Chrismadha T, Mayasari N et al (2022) Nutrition value and growth ability of aquatic Weed Wolffia globosa as alternative feed sources for aquaculture system. IOP Conf Ser Earth Environ Sci 950:0–9. https://doi.org/10.1088/1755-1315/950/1/012044
Satya A, Chrismadha T, Satya ADM, Satya IA (2022) Optimizing crude protein production from minute duckweed (Lemna perpusilla Torr) grown in varied NPK based medium. IOP Conf Ser Earth Environ Sci 1062:12008. https://doi.org/10.1088/1755-1315/1062/1/012008
Sree KS, Bog M, Appenroth KJ (2016) Taxonomy of duckweeds (Lemnaceae). potential new crop plants. Emir J Food Agric 28:291–302. https://doi.org/10.9755/ejfa.2016-01-038
Susilowati SH, Maulana M (2012) Eksistensi Petani Gurem dan Urgensi Kebijakan Reforma Agraria Farm Business Land Size. Analisis Kebijakan Pertanian 10:17–30. https://doi.org/10.21082/akp.v10n1.2012.17-30
Sutrisno S, Chrismadha T, Mayasari N (2021) Potensi Wolffia globosa dan Lemna perpusilla (Lemnaceae) sebagai Pakan Ikan Nila (Oreochromis niloticus). Limnotek: perairan darat tropis di Indonesia 28:1–11. https://doi.org/10.14203/limnotek.v28i1.346
Troell M, Halling C, Neori A et al (2003) Integrated mariculture: asking the right questions. Aquaculture 226:69–90. https://doi.org/10.1016/S0044-8486(03)00469-1
Troell M, Joyce A, Chopin T et al (2009) Ecological engineering in aquaculture—Potential for integrated multi-trophic aquaculture (IMTA) in marine offshore systems. Aquaculture 297:1–9. https://doi.org/10.1016/J.AQUACULTURE.2009.09.010
Ye Y, Barange M, Beveridge M et al (2017) FAO's statistic data and sustainability of fisheries and aquaculture: comments on Pauly and Zeller. Mar Policy 81:401–405. https://doi.org/10.1016/J.MARPOL.2017.03.012

Structured Lipids Based on Palm Oil

Siti Nurhasanah and S. Joni Munarso

Abstract Palm oil is usually used for the needs of food, chemical industry, and cosmetic industry. The basic processing of palm fruit can produce two types of oil namely crude palm oil (CPO) which is produced from the extraction process of the mesocarp part of the oil palm fruit and palm kernel oil (PKO) as an extract of the palm kernel part. Naturally, oils and fats have specific characteristics, and the development of food processing and technology causes these characteristics to not able to meet all the expected needs to obtain products with certain functional properties such as: lipids for sufferers of coronary heart disease, type 2 diabetes sufferers, patients in the post-operative recovery period, patients who suffer from allergies or digestive problems, and consumers who are controlling their weight low-calorie products. The dominant fatty acids in palm oil are palmitic, oleic acid, and linoleic acid. CPO also contains minor components such as squalene, sterols, and carotenoids. Structured lipids (SLs) are the result of modification or restructuring of triacylglycerols, which can be obtained by chemical or enzymatic interesterification of triacylglycerols containing short, medium, and/or long chain fatty acids. SLs are the result of modification or restructuring of triacylglycerols, which can be obtained by chemical or enzymatic interesterification of triacylglycerols. SLs can be sourced from animal or vegetable fats, or genetic engineering. SLs are synthesized for the purpose of obtaining functional lipids or nutraceuticals, which can improve or modify the physical, chemical, and rheological characteristics of oils and fats, and changing or enhancing nutrition properties of food, giving a certain health benefit. Palm oil has special fatty acids and other minor components, making it possible to be used as a raw material for the manufacture of SLs so that their bioavailability increases. Functional oil and fat production can be catalyzed by lipase. Fats/oils can improve physicochemical and nutritional properties using a lipase catalyst. Palm oil has special fatty acids and other minor components, making it possible to be used as a raw material for the manufacture of SLs so that their bioavailability increases. Functional oil and fat

S. Nurhasanah (✉)
Department of Food and Technology, Faculty of Agro-Industrial Technology, Padjadjaran University, Bandung 45363, Indonesia
e-mail: siti.nurhasanah@unpad.ac.id

S. J. Munarso
National Research and Innovation Agency, Bogor 16124, Indonesia

M. A. R. Lubis et al. (eds.), *Biomass Conversion and Sustainable Biorefinery*, Green Energy and Technology, https://doi.org/10.1007/978-981-99-7769-7_9

production can be catalyzed by lipase. Fats/oils can improve physicochemical and nutritional properties using a lipase catalyst.

Keywords Fatty acid · Interesterification · Palm oil · Structured lipids

1 Introduction

Palm oil is a vegetable oil that can be obtained from the mesocarp of the fruit of the oil palm tree, generally of the species Elaeis guineensis. Naturally, palm oil is red in color due to its high beta-carotene content. This oil is a common cooking ingredient in tropical countries including Africa, Southeast Asia, and parts of Brazil. The results of the palm oil industry are not only cooking oil but can also be used as a basic ingredient for other industries such as the food and cosmetics industries. Half of all fats and vegetable oils consumed come from palm oil (Murgianto et al. 2021).

Palm oil is usually used for the needs of food (popular components in baking, processed meals, snacks, frozen foods, and chocolate due to their neutral taste, texture, and practicality), chemical industry, and cosmetic industry. The largest demand for palm oil is used for food ingredients such as cooking oil, margarine, shortening, emulsifiers, cocoa butter substitutes, and other derivative products. The basic processing of palm fruit can produce two types of oil namely crude palm oil (CPO) which is produced from the extraction process of the mesocarp part of the oil palm fruit and palm kernel oil (PKO) as an extract of the palm kernel part (Ali et al. 2014). Commercial cooking oil is made from palm olein, whereas palm stearin is mostly used for shortening and margarine products. PKO has a sharp melting profile because high in lauric acid therefore it is suitable for confectioneries and other specialty fats. PKO is more widely used in a variety of non-edible products, such as cosmetics, detergents, surfactants, plastics, herbicides, and various other industrial and agricultural chemicals. The widespread use of palm oil makes this oil an important commodity that plays a role in food security in the world (Dian et al. 2017; Chew et al. 2021).

Naturally, oils and fats have specific characteristics. The development of food processing and technology causes these characteristics to be not able to meet all the expected needs. For example, lipids for sufferers of coronary heart disease, type 2 diabetes sufferers, patients in the post-operative recovery period, patients who suffer from allergies or digestive problems, and consumers who are controlling their weight low-calorie products and to obtain products with certain functional properties. Therefore, it is necessary to modify the fatty acids that make up the triacylglycerol (TAG) into structured lipids (SLs).

SLs can be carried out by interesterification, either chemically or enzymatically. The advantages of enzymatic interesterification are that the reaction is more specific, the reaction conditions (pH, temperature, pressure) are milder and the waste produced is minimal, thereby reducing environmental pollution. The composition of the TAG after enzymatic interesterification will depend on the reaction conditions such as the type of enzyme, reaction time and temperature, water activity of the enzyme, substrate ratio, and the type of substrate used. The interesterification process results in changes in the chemical, physical, and oxidative properties of the substrate of origin. With the superiority of fatty acids in palm oil, it is possible to serve as SLs which can be used as raw materials for functional food.

2 Palm Oil

2.1 General

Palm (*Elaeis guineensis*) is a plant originating from the West African region and classified as a tropical plant. Palm fruit has an oval shape and is attached to bunches and can weigh up to 10–40 kg. The palm fruit consists of 2 main parts, namely the pericarp and the seed. Oil palm fruit bunches are composed of fruit flesh (60%), bunch skin (29%), and fruit seeds/palm kernel (11%). The pericarp consists of the exocarp, mesocarp, and endocarp, while the seed portion consists of the endosperm (palm kernel/kernel), embryo, and seed coat (Ezechi and Muda 2019; Mahlia et al. 2019).

CPO obtained from extraction or from the pressing process of the flesh of the palm fruit and has not been refined. CPO refinery to make cooking oil always begins with heating, threshing, boiling, stirring, and pressing then proceed with filtering, purification, and phase separation (fractionation). Fractional crystallization, which can be classified into dry and solvent fractionation, is used to separate oils and fats into two or more components. The first stage of fractionation is crystallization, which is followed by the separation of solid and liquid fractions (Hasibuan et al. 2021). Palm oil is very useful for edible oil to make food products, such as frying or cooking oil including margarine, shortenings, red palm oil, and also specialty fats. Palm olein has no unpleasant odor, no trans-fatty acid, and has high resistance to oxidation. Therefore, it is suitable for frying and applications and it is commonly used for cooking, either with pan frying and deep frying and can be reused several times in restaurants, fast food restaurants, and also in the manufacture of snack foods and instant noodles (Purnama et al. 2020).

2.2 Palm Oil Processing

CPO still contains unwanted impurities. Components in the form of phospholipids, proteins, residues, and carbohydrates in the oil need to be removed through the degumming process. The principle of degumming is to separate the phosphatides into the aqueous phase so that they can be separated by precipitation, filtering, or centrifugation. Processing of oil palm bunches begins with heating, threshing, boiling, stirring, and pressing then followed by filtering, purifying, and separating phases (fractionation). Oil refining consists of several stages: gum separation, neutralization, bleaching, and deodorization. Bleaching process reduces the amount of carotene to get clear cooking oil (Andarwulan et al. 2018).

The liquid fraction of palm oil that has been refined at the deodorizing stage to remove free fatty acids and clarification to remove color and deodorize and fractionate is known as refined, bleached, and deodorized palm olein (RBDP Olein). CPO processing by modifying the refining process, without bleaching, to maintain the carotene content in palm oil can produce red palm oil or Red Palm Oil (RPO) (Sumarna et al. 2022). The content of carotene, especially α-carotene and β-carotene, is a precursor of vitamin A in the body. Before use, CPO extracted from fresh fruit bunches needs to be purified because CPO still contains unwanted impurities. Components in the form of phospholipids, proteins, residues, and carbohydrates in the oil need to be removed through the degumming process. The principle of degumming is to separate the phosphatides into the aqueous phase so that they can be separated by precipitation, filtering, or centrifugation.

Degumming process can be carried out in various ways. Wet degumming is able by adding phosphoric acid, and citric acid. Dry degumming uses the same steps as wet degumming but without adding water. Of the two methods performed, the wet degumming method produces better quality than the other method of dry degumming (Putri et al. 2019; Mayalibit et al. 2020). Neutralization process is carried out to separate the free fatty acids. Generally neutralization is done by reacting NaOH with acid free fat contained in the oil to form soap that can be separated (Musyaroh and Hidayat 2018). The process of deodorization of oil is carried out as one of the important steps that must be taken in order to reduce unwanted aromas or tastes. Both of these parameters are usually caused by damage to the oil component in the oil during processing, such as unsaturated hydrocarbons, terpenes, sterols, and tocopherols (Ramlah and Sampebarra 2018). Deodorization also affects the reduction of free fatty acid levels in oil, although not as much as degumming, neutralization, and bleaching processes (Riyadi et al. 2016).

The fractionation process in oil is carried out to separate the olein fraction (liquid phase) and the stearin fraction (solid phase) of the oil. The olein fraction is often used as cooking oil while the stearin fraction is often used as an ingredient raw margarine or shortening. Fractionation is accomplished by winterization, crystallization (hexane, acetone, or isopropyl alcohol), and detergent processes. Winterization can result in oil loss which is abundant whereas organic solvents are flammable and expensive in the process. The most commonly used is dry fractionation separated by cooling first so that the stearin freezes and the olein can be filtered.

Palm oil is naturally semi-solid at room temperature, meaning there is no need for it to be hydrogenated and therefore it contains no trans fats. Replacing trans fats with palm oil may reduce heart disease risk markers and improve blood lipids. In the hydrogenation process, unsaturated triglycerides are attacked more quickly than di-oleo-glycerides, and finally mono-oleo compounds. The concept of fatty acid selectivity indicates that the rate of hydrogenation of various unsaturated fatty acids depends on the concentration of hydrogen. Consequently the reaction rates of two different unsaturated fatty acids, for example, linoleic acid and oleic acid, are only a function of the concentrations of these acids and their rate constants, so that the concentrations of these acids can be determined experimentally, while the ratio of the hydrogenation rate constant can be calculated (Gunstone and Norris 2013). During the hydrogenation process, double bonds are not only saturated, but they may also shift their position along the fatty acid chain (positional isomerization) and/or undergo *cis-trans* isomerization. Because *trans* isomers strongly affect the physical properties of the triglycerides, the extent of the formation of these isomers has to be controlled. It is generally expressed as the "*trans* selectivity" or as the "isomerization index," which are defined as the increase in *trans* content (expressed as % elaidic) divided by the decrease in iodine value or by the decrease in double bond content, respectively, as observed in the early stages of the hydrogenation process. Elaidic acid, $C_{18}H_{34}O_2$, has an essentially linear alkyl chain. The double bond is twisted across the mean direction of the alkyl chain. In the crystal structure, the molecules form centrosymmetric O–HO hydrogen-bonded dimers.

2.3 *Composition Palm Oil*

The RBDPO fractionation will produce two unique fractions, namely the olein fraction which has a high iodine number and low melting point and the stearin fraction which has a lower iodine number and a high melting point (Tables 1 and 2).

CPO contains unsaturated fatty acids and saturated fatty acids. Saturated fatty acids have only single bonds between their carbon atoms, whereas unsaturated fatty acids have at least one double bond between their constituent atoms. Saturated fatty acids are more stable or less visible than unsaturated fatty acids. Double bonds in unsaturated fatty acids easily react with oxygen or are easily oxidized. Composition

Table 1 Comparison of fatty acid composition of palm oil

Fatty acids	CPO (%)	Oleic (%)	Stearic (%)
Lauric (C12)	0.10–0.40 (0.24)	0.20–0.40 (0.27)	0.10–0.30 (0.18)
Miristic (C14)	1.00–1.40 (1.11)	0.90–1.20 (1.09)	1.10–1.70 (1.27)
Palmitic (C16)	40.90–47.50 (44.14)	36.80–43.20 (40.93)	49.80–68.10 (56.79)
Stearic (C18)	3.80–4.80 (4.44)	3.70–4.80 (4.18)	3.90–5.60 (4.93)
Oleic (C18:1)	36.40–41.20 (39.04)	39.80–44.60 (41.51)	20.40–34.40 (29.00)
Linoleic (C18:2)	9.20–11.60 (10.57)	10.40–12.90 (11.64)	5.00–8.90 (7.23)
Linolenic (C18:3)	0.05–0.60 (0.37)	0.10–0.60 (0.40)	0.00–0.50 (0.09)
Arachidic (C20:0)	0.20–0.70 (0.38)	0.30–0.50 (0.37)	0.00–0.50 (0.24)

Gee (2007)

Table 2 Physical characteristics and chemical composition

Characteristics	Typical	Range
Specific gravity, 30 °C	0.918	0.951–0.920
Index refraction, 25 °C	–	1.470–1.474
Iodin value	124.0	118.0–128.0
Saponification	–	187–193
Unsaponification	–	1.3–2.3
Melting point (°C)	–	−12 to −10
Solid point (°C)	–	1.0–20.0
Stability AOM (h)	19	16–19
α-Tocopherol (ppm)	152	116–172
β-Tocopherol (ppm)	12	0–22
γ-Tocopherol (ppm)	1276	119–1401
δ-Tocopherol (ppm)	61	59–65

O'Brien (2008)

of saturated and unsaturated fatty acids in palm oil is balanced. The dominant fatty acids consist of palmitic (44–45%), oleic acid (39–40%), and linoleic acid (10–11%). This allows it to be fractionated into two main fractions: liquid oil (65–70%), palm olein (mp 18–20 °C), and solid fraction (30–35%), stearin (mp 48–50 °C). CPO also contains minor components (1%) such as squalene (200–600 ppm), sterols (250–620 ppm), and carotenoids (500–700 ppm), pigments responsible for the reddish-orange color and the richest source of tocotrienols in the world (de Almeida et al. 2019).

Beta carotene is a minor component found in palm oil and is a secondary metabolite that belongs to the group of carotenoid compounds synthesized by plants, algae, and some microorganisms. This type of carotenoid is an organic compound that is a chromophore and has eleven conjugated double bonds in its structure. The beta carotene molecule has two beta-ionic rings which theoretically will experience chain termination at $-C15 = C15'-$ so beta carotene will be converted into two retinol molecules (Bogacz-Radomska and Harasym 2018).

Beta carotene is a carotenoid compound that has high bioactivity. The bioactivity properties of beta carotene are useful as a source of provitamin A which can support the growth of embryo in pregnant women, the growth of children, and influence for eye health. Apart from that, beta carotene as well has high anticancer and antioxidant properties so it can ward off cancer, the bad influence of free radicals, increase body immunity, prevent aging in early childhood, and the risk of cardiovascular disease (Langi et al. 2018).

2.4 *Deterioration of Palm Oil*

The ratio of oil degradation mainly depends on fatty acid composition, type, and quality of the oil. The selection of oils should be based on the optimization of the process with regard to culinary aspects as well as nutritional, physiological, and technological requirements. The application of high temperatures to the oil results in a change in the composition of the fatty acids and produces monoglycerides, diglycerides, free fatty acids, primary and secondary oxidation products through processes such as oxidation, polymerization, and hydrolysis (Jadhav et al. 2022). Free fatty acid (FFA %) content is the most widely used criterion to determine the quality of palm oil. Codex Alimentarius standard for maximum concentration of FFA to 5.0% for CPO and up to 0.3% for RPO in oleic acid. Thus, according to this norm, all fresh oil (zero month storage) is within the established limits (de Almeida et al. 2019).

Lipid oxidation is an important quality criterion for the food industry. Oxidation is a reaction between unsaturated fats and oxygen which is accelerated by heat, light, and metals. Acid value gives an indicator of free fatty acids present in the sample of oil. Peroxide value is an indicator of primary oxidation products (hydroperoxides) formed due to oxidation of lipids in food which can be decomposed into aldehydes and ketones. Formation of secondary oxidation products like aldehyde and ketones by degradation of peroxide can be measured by p-Anisidine value. Lipid oxidation not only produces a rancid flavor but can also reduce nutritional quality and safety, namely the formation of oxidation products which cause toxic products in the presence of peroxide decomposition to produce secondary reaction products, and provide other physiological and pathological effects (Nurhasanah et al. 2019; Jadhav et al. 2022). Thermoxidative changes in heated oil with total polar components (TPC), anisidine (AV) values, formation of color components, and changes in the composition of fatty acids and tocopherols.

3 Structured Lipids

3.1 *Definition*

SLs are the result of modification or restructuring of triacylglycerols, which can be obtained by chemical or enzymatic interesterification of triacylglycerols containing short, medium, and/or long chain fatty acids. SLs can be sourced from animal or vegetable fats, or genetic engineering. SLs are synthesized for the purpose of obtaining functional lipids or nutraceuticals, which can improve or modify the physical, chemical, and rheological characteristics of oils and fats, and changing or enhancing nutrition properties of food, giving a certain health benefit. The properties of fatty acids based on physical, biological, and nutritional are largely determined by the position, number, and configuration of their double bonds. These determine the shape of the molecules, the way molecules can pack together in solid phases, monolayers, bilayers, and how individual molecules can interact with enzymes and receptors. Changes in the composition and position of the fatty acids in triacylglycerols are caused by the interesterification reaction causing changes in several properties such as solid fat content, crystallization behavior, physical properties, chemical properties, thermal properties, and consistency when compared to native lipids. The changes are verified through physical, chemical, and functional analysis due to the fact that SLs can present triacylglycerols. SLs can provide essential fatty acids such as linoleic (18:2n-6), oleic acid (18:1n-9), and linolenic acid (18:3n-3) as found in many vegetable oils, such as soybean oil, olive oil, palm oil, coconut oil. These fatty acids are essential for growth and development throughout the human life cycle, as well as the promotion of improvement in health, and plays an important role in reducing risk of metabolic syndrome (Moreira et al. 2017).

One form of lipid structure is specialty fats. Specialty fats are type of fat that has a special function, so that it has the potential for special applications such as to make confectionery fat, usually used to replace all or part of cocoa butter and dairy butter. Among specialty fats, cocoa butter alternatives represent perhaps the most diverse and widely developed specialty fats. Cocoa butter alternatives are designed to provide an alternative, both economically and functionally, to a high-value ingredient, cocoa butter. These fats are formulated or modified from palm oil, since the cocoa butter and dairy butter are expensive and their supply unreliable. The role of these fats is to provide specific texture and richness of taste. Furthermore, the confectionery fats should have a sharp melting behavior to melt easily in the month (Talbot 2015; Ramadan 2019).

SLs are beneficial for human nutrition because they can be tailor-made to target specific diseases and metabolic conditions, and reduce calories by specifically positioning certain fatty acids in the glycerol backbone. Such low-calorie fats are usually designed to take advantage of the limited absorption of long-chain saturated fatty acids or the lower caloric density of short-chain saturated fatty acids. SLs also lower cholesterol, LDL cholesterol, and triglycerides, given a normal diet as well as an atherogenic diet. In addition, lipid accumulation in the arteries was also significantly reduced. Thus low-calorie structured fat has the added benefit of reducing serum and liver lipids which are considered risk factors for cardiovascular disease (Kanjilal et al. 2016).

Lipase is more promising when certain positional modifications of triacylglycerols are addressed, in addition to resulting in less residue in SL production. In addition, enzymes can be reused many times, minor lipids, among other compounds with bioactive functions, are preserved due to milder reactions compared to chemical interesterification. After the interesterification reaction, the number of triacylglycerols is higher than diacylglycerol and monoacylglycerol content, hydrolysis followed by esterification leaving a small amount of diacylglycerol and monoacylglycerol, either by using lipase Lipozyme TL IM or by Rhizopus sp. (Moreira et al. 2017).

3.2 *Structured Lipid Synthesis*

3.2.1 Chemical

Chemically catalyzed is the most common interesterification process which is described as a “reshuffling”. Accidental and limited interesterification can occur when the oil is heated to above 200 °C, as illustrated by the characteristic change in crystallization of confectionery fat after deodorization (Gunstone and Norris 2013). In general, the chemical interesterification process takes place with three kinds of reactions at once, namely: (1) Alcoholysis, (2) Acidolysis, and (3) Transesterification. For the interesterification reaction, use under low pressure in a water bath at 80–85 °C for 30 min. After drying, 1% (m/m) sodium methoxide powder (Oliveira et al. 2017).

The advantages of chemical interesterification are relatively faster methods, one of which is for the synthesis of tripalmitin. Palmitic acid is not easily oxidized even at relatively high temperatures. The final product contained is 97.60% PPP, 1.46% dipalmitin, 0.08 and 0.26% monopalmitin. The disadvantage of this method is that the alkaline catalyst in liquid form mixes perfectly with the product so that purification of the product from the catalyst is relatively difficult. In addition, the use of alkaline catalysts results in side reactions that are very disturbing, namely the occurrence of saponification reactions to form unwanted by-products thereby reducing yields (Wei et al. 2015).

3.2.2 Enzymatic

In addition to chemically catalyzed randomization, there are also directed interesterification processes in which the equilibrium associated with complete randomization is disturbed either by distillation of the most volatile components, such as FAME (fatty acid methyl esters) (Gunstone and Norris 2013).

Enzymes are well-known as one of the biocatalysts of a wide variety of processes that are highly effective and efficient catalysts characterized by high activity and selectivity to accelerate biochemical reactions. The conversion of substrate catalyzed by lipase into products is carried out by reducing the energy of the activation reaction, carried out under mild conditions (pH and temperature), very good selectivity, and substrate activity. Many commercially essential processes, especially in various food, cosmetic, and pharmaceutical industries use lipases as a natural catalyst. Lipases are a powerful tool for biotransformation on a broad substrate range (Robinson 2015; Liu and Dong 2020; Pandey et al. 2020). Catalysts are widely used in the chemical industry sector: in basic chemistry, in polymerization chemistry, and in refining such as refining including coconut oil-based products, as well as in liquid reactions industrial processes (pharmacy, food, cosmetic, etc.) (Bedade et al. 2019).

The enzymes that can be used for the above purposes are Lipozyme TL IM and Novozyme 435. Lipozyme TL IM is a type of commercial lipase enzyme immobilized using silica gel from *Thermomyces lanuginosus* and is widely used for various esterification reactions. This enzyme has positional specificity of the TAG molecule, namely at the primary position (sn-1,3) with a mild optimal temperature (25–60 °C). Meanwhile, Novozyme 435 is a type of non-specific immobilized commercial lipase enzyme using macroporous acrylic resin beads with an optimum temperature of 40–60 °C (Ortiz et al. 2019).

Modification of the simple mixture by these two lipases causes a change in thermal profile, which causes a delayed crystallization process, as well as the decrease in enthalpy, indicating that it is interesterified the sample releases less energy during crystallization. Energy measured in this process refers to the rearrangement of the liquid phase molecules, which release energy and reformulate in the solid state. In addition, the crystallization curve also reveals that the higher degree of unsaturation fatty acids in the TAGs lowers the end temperature and enthalpy of crystallization. This phenomenon can be explained by the fact that enthalpies are calculated by the area of each peak and according to the number of crystals formed during cooling indicated crystallization curve of the SLs mixture difference performances presented by lipases, synthesized SLs with Rhizopus sp. lipase showed three crystallization peaks, due to incomplete restructuring of the tri- and saturated TAG formed presenting a higher enthalpy value for each peak compared to samples catalyzed by Lipozyme TL IM (Moreira et al. 2017).

Bioavailability of fatty acids is not only determined by composition, but also determined by the position of each type of fatty acid on the glycerol backbone. Unsaturated long chain fatty acids in position sn-2 can improve bioavailability, because pancreatic lipase confers less activity on these fatty acids when esterified to the sn-1 and sn-3 positions. The absorption of long chain fatty acids and MCFA will be higher if they are present in the sn-2 position of the TAG. The TAG will be converted into 2 monoglycerides which are more water soluble by pancreatic lipase in the body. At position sn-1,3 can support low lipid absorption, without compromising the fatty acids located in sn-2. Based on this perspective, this type of SL synthesis requires modification at certain positions on the glycerol backbone, which can be obtained by enzymatic interesterification (Moreira et al. 2017).

Lipases can be produced by plants, animals, and microorganisms, and microbial lipases are receiving more attention from the industry because of their ability to remain active at extreme temperatures, organic solvents, pH, exhibit high selectivity, wide substrate specificity, and do not require cofactors. The advantages of lipases can reduce the number of hazardous solvents needed; the total reaction steps make the process cheaper and more environmentally friendly. Other benefits of using lipases are mild reaction conditions, low energy consumption, biodegradability, and yields of a pure product. The lipase from *Thermomyces lanuginosus* expressed in *Aspergillus oryzae* is the first commercialized recombinant lipase. Lipase is an enzyme that can work reversibly, catalyzes the hydrolysis of triacylglycerol to glycerol and free fatty acids and, or partial hydrolysis to diacylglycerols (DAGs) and monoacylglycerols (MAGs) (Subroto et al. 2019).

Enzymatic interesterification process system can be done with the feedbatch system and system continuous. (Wei et al. 2015) observed *Thermomyces lanuginosus* lipase activity during batch interesterification process. Activity enzymes decreased rapidly after 6 times (equal to 24 h) reaction on batch system.

4 Structured Lipids Based on Palm Oil

Palm oil has special fatty acids and other minor components, making it possible to be used as a raw material for the manufacture of SLs so that their bioavailability increases. Functional oil and fat production can be catalyzed by lipase. Fats/oils can improve physicochemical and nutritional properties using a lipase catalyst. Unsaturated fatty acids in triglycerides are mostly in the sn-2 position. The lipase specificity of sn-1,3 can be used to catalyze the transesterification reaction while maintaining sn-2 fatty acids. Strategies for Human Milk Fat Subtitute (HMFS) have been developed to mimic the fat composition and distribution of human milk. HMFS is used in infant formula to mimic the fat of breast milk. Breast milk is the main choice for newborns, infant formula that most closely resembles breast milk is a good substitute for baby nutrition when breastfeeding is insufficient or cannot be done. Fatty acid components in human milk fat are oleic acid, palmitic acid, linoleic acid, followed by stearic acid, myristic acid, and lauric acid. The distribution position of breast milk fatty acids is 70% palmitic acid is in the sn-2 position and UFA (oleic acid, linoleic acid linolenic acid, etc.) is in the sn-1.3 position. This characteristic makes breast milk fat different from most vegetable oils in that most of the UFA is in the sn-2 position and the SFA especially palmitic acid is in the sn-1,3 position (Qin et al. 2014; Hasibuan et al. 2021).

Cocoa butter (CB) is a very important ingredient that contributes to the textural and sensory properties of confectionery products, particularly chocolate products (up to 32% CB in chocolate formulations). CB is hard and brittle under room temperature, but when eaten, it melts perfectly in the mouth with a soft creamy texture and a cold sensation. The CB polymorphism has a major influence on the physical properties of chocolate products, such as gloss, snap, contraction, heat resistance, fast and sharp melting in the mouth, and bloom resistance. The special nature of CB is not followed by supply, price, use in hot climates, and consistency of quality between regions (Zhang et al. 2020). SL plays a role in the development of bakery products. Based on the SL1 melting profile, it is suitable for the manufacture of biscuits and cakes in terms of sensory and organoleptic properties. The organoleptic characteristics and quality of biscuits and cakes made with SL1 were indistinguishable from those prepared with traditional bread tallow. This suggests that bakery fats can be fully replaced by SLs studied to prepare trans-free low-calorie cakes and biscuits. Thus the use of low-calorie fat can be applied to bakery products, cakes and biscuits are not only trans-free but also have added value for health such as low calories and hypocholesterolemic properties (Kanjilal et al. 2016). Binary mixture of PKO and interesterified fats was dominant in β′ crystal. The chocolate showed consistent texture before and after tempering process (Zhang et al. 2020).

PKO with other vegetable oils can be used as raw material for SL. PKO contains a wide variety of fatty acids (C6−C20) and has more β′ polymorphs than β. Margarine made with SL from a mixture of PKO, canola oil, and stearin fractions has a hardness, stickiness, or compactness similar to commercial margarine. Therefore, the SL is suitable for the formulation of trans-free margarines with low atherogenicity and desirable textural properties (Kim 2008).

5 Conclusion

Palm oil is produced from the extraction process of the mesocarp part of the palm fruit (CPO) and palm kernel oil (PKO). Naturally, oils and fats have specific characteristics, developments in food processing and technology have made these characteristics unable to meet all the expected requirements for obtaining products with certain functional properties by making SLs by modifying or restructuring triacylglycerols, which can be obtained by chemical or enzymatic interesterification of triacylglycerols. SLs can be sourced from animal or vegetable fats, or genetic engineering. SL is synthesized with the aim of obtaining functional lipids or nutraceuticals, which can improve or modify the characteristics of oils and fats, and change or enhance the nutritional properties of foods, providing certain health benefits. Palm oil has special fatty acids and other minor components, making it possible to use it as a raw material for SL so that its bioavailability increases.

Acknowledgements If any.

References

Ali FS, Shamsudin R, Yunus R (2014) The effect of storage time of chopped oil palm fruit bunches on the palm oil quality. Agric Agric Sci Procedia 2:165–172. https://doi.org/10.1016/j.aaspro.2014.11.024

Andarwulan N, Hariyadi P, Mursalin (2018) Engineering process of deodorization to improve product quality of red palm oil with rich of carotene. Indones Food Sci Technol J 1:20–26. https://doi.org/10.22437/ifstj.v1i1.5014

Bedade DK, Sutar YB, Singhal RS (2019) Chitosan coated calcium alginate beads for covalent immobilization of acrylamidase: process parameters and removal of acrylamide from coffee. Food Chem 275:95–104. https://doi.org/10.1016/j.foodchem.2018.09.090

Bogacz-Radomska L, Harasym J (2018) β-carotene-properties and production methods. Food Qual Saf 2:69–74. https://doi.org/10.1093/fqsafe/fyy004

Chew CL, Ng CY, Hong WO et al (2021) Improving sustainability of palm oil production by increasing oil extraction rate: a review. Food Bioprocess Technol 14:573–586 de Almeida DT, Viana TV, Costa MM et al (2019) Effects of different storage conditions on the oxidative stability of crude and refined palm oil, olein and stearin (Elaeis guineensis). Food Sci Technol 39:211–217. https://doi.org/10.1590/fst.43317

Dian NLHM, Hamid RA, Kanagaratnam S et al (2017) Palm oil and palm kernel oil: versatile ingredients for food applications. J Oil Palm Res 29:487–511. https://doi.org/10.21894/jopr.2017.00014

Ezechi EH, Muda K (2019) Overview of trends in crude palm oil production and economic impact in Malaysia. Sriwij J Environ 4:19–26. https://doi.org/10.22135/sje.2019.4.1.19

Gee PT (2007) Analytical characteristics of crude and refined palm oil and fractions. Eur J Lipid Sci Technol 109:373–379. https://doi.org/10.1002/ejlt.200600264

Gunstone FD, Norris FA (2013) Lipids in foods: chemistry, biochemistry and technology. Elsevier Science

Hasibuan HA, Sitanggang AB, Andarwulan N, Hariyadi P (2021) Solvent fractionation of hard palm stearin to increase the concentration of tripalmitoylglycerol and dipalmitoyl-stearoyl-glycerol as substrates for synthesis of human milk fat substitute. Int J Food Sci Technol 56:4549–4558. https://doi.org/10.1111/ijfs.15206

Jadhav HB, Gogate PR, Waghmare JT, Annapure US (2022) Comparative assessment of thermo-oxidative stability of palm oil designer lipid and palm oil blends as frying medium. Appl Food Res 2:100039. https://doi.org/10.1016/j.afres.2021.100039

Kanjilal S, Shanker KS, Rao BVSK et al (2016) Application of low calorie hypocholesterolemic structured lipid as potential bakery fat. Int Food Res J 23:854–859

Kim D (2008) Demand and pricing in the US margarine industry. J Agri Food Ind Organ 6(1)

Langi P, Kiokias S, Varzakas T, Proestos C (2018) Carotenoids: from plants to food and feed industries. Methods Mol Biol 1852:57–71. https://doi.org/10.1007/978-1-4939-8742-9_3

Liu DM, Dong C (2020) Recent advances in nano-carrier immobilized enzymes and their applications. Process Biochem 92:464–475. https://doi.org/10.1016/j.procbio.2020.02.005

Mahlia TMI, Ismail N, Hossain N et al (2019) Palm oil and its wastes as bioenergy sources: a comprehensive review. Environ Sci Pollut Res. https://doi.org/10.1007/s11356-019-04563-x

Mayalibit AP, Sarungallo ZL, Paiki SN (2020) Pengaruh proses degumming menggunakan asam sitrat terhadap kualitas minyak buah merah (Pandanus conoideus Lamk). Agritechnology 2:23. https://doi.org/10.51310/agritechnology.v2i1.25

Moreira DKT, Ract JNR, Ribeiro APB, Macedo GA (2017) Production and characterization of structured lipids with antiobesity potential and as a source of essential fatty acids. Food Res Int 99:713–719. https://doi.org/10.1016/j.foodres.2017.06.034

Murgianto F, Edyson E, Ardiyanto A et al (2021) Potential content of palm oil at various levels of loose fruit in oil palm circle. J Agro Ind Perkeb 9:91–98. https://doi.org/10.25181/jaip.v9i2.2161

Musyaroh, Hidayat N (2018) The effect of stirring length time and NaOH concentration on the neutralization process of superworm cooking oil purification. Ind J Teknol dan Manaj Agroindustri 7:81–88. https://doi.org/10.21776/ub.industria.2018.007.02.2

Nurhasanah S, Wulandari N, Joni Munarso S, Hariyadi P (2019) Production of structured lipids rich in triacylglycerols containing medium-chain fatty acids and unsaturated fatty acids at the Sn-2 position through enzymatic interesterification. Int J Adv Sci Eng Inf Technol 9:1624–1630. https://doi.org/10.18517/ijaseit.9.5.10076

O'Brien RD (2008) Fats and oils: formulating and processing for applications, 3rd edn. CRC Press

Oliveira PD, Rodrigues AMC, Bezerra CV, Silva LHM (2017) Chemical interesterification of blends with palm stearin and patawa oil. Food Chem 215:369–376. https://doi.org/10.1016/j.foodchem.2016.07.165

Ortiz C, Ferreira ML, Barbosa O et al (2019) Novozym 435: the "perfect" lipase immobilized biocatalyst? Catal Sci Technol 9:2380–2420. https://doi.org/10.1039/c9cy00415g

Pandey D, Daverey A, Arunachalam K (2020) Biochar: production, properties and emerging role as a support for enzyme immobilization. J Clean Prod 255:120267. https://doi.org/10.1016/j.jclepro.2020.120267

Purnama KO, Setyaningsih D, Hambali E, Taniwiryono D (2020) Processing, characteristics, and potential application of red palm oil—A review. Int J Oil Palm 3:40–55. https://doi.org/10.35876/ijop.v3i2.47
Putri DO, Mardawati E, Putri SH, Frank D (2019) Comparison of the CPO (crude palm oil) degumming method on the characteristics of the lecithin produced. J Ind Pertan 1:88–94
Qin XL, Zhong JF, Wang YH et al (2014) 1,3-Dioleoyl-2-palmitoylglycerol-rich human milk fat substitutes: production, purification, characterization and modeling of the formulation. Eur J Lipid Sci Technol 116:282–290. https://doi.org/10.1002/ejlt.201300343
Ramadan MF (2019) Chemistry and functionality of fruit oils: an introduction. Springer
Ramlah S, Sampebarra AL (2018) Karakteristik dan Citarasa Cokelat Putih dari Lemak Kakao Non Deodorisasi dan Deodorisasi (Characteristics and flavor of white chocolate from non deodorized and deodorized cocoa butter). J Ind Has Perkeb 13:117–128
Riyadi AH, Muchtadi TR, Andarwulan N, Haryati T (2016) Pilot plant study of red palm oil deodorization using moderate temperature. Agric Agric Sci Procedia 9:209–216. https://doi.org/10.1016/j.aaspro.2016.02.129
Robinson PK (2015) Enzymes: principles and biotechnological applications. Essays Biochem 59:1–41. https://doi.org/10.1042/BSE0590001
Subroto E, Supriyanto UT, Hidayat C (2019) Enzymatic glycerolysis–interesterification of palm stearin–olein blend for synthesis structured lipid containing high mono- and diacylglycerol. Food Sci Biotechnol 28:511–517. https://doi.org/10.1007/s10068-018-0462-6
Sumarna D, Sumarlan SH, Wijaya S, Hidayat N (2022) Processing of olein fraction red palm oil with minimal refining method and optimization of deodorization process. Proc Int Conf Trop Agrifood Feed Fuel (ICTAFF 2021) 17:167–175. https://doi.org/10.2991/absr.k.220102.026
Talbot G (2015) Specialty oils and fats in food and nutrition: properties, processing and applications. Woodhead Publishing
Wei W, Feng Y, Zhang X et al (2015) Synthesis of structured lipid 1,3-dioleoyl-2-palmitoylglycerol in both solvent and solvent-free system. Lwt 60:1187–1194. https://doi.org/10.1016/j.lwt.2014.09.013
Zhang Z, Song J, Lee WJ et al (2020) Characterization of enzymatically interesterified palm oil-based fats and its potential application as cocoa butter substitute. Food Chem 318:126518. https://doi.org/10.1016/j.foodchem.2020.126518

Recent Developments of the Agroindustry Byproducts Utilization in Bacterial Cellulose Production and Its Medical Devices Applications

Khatarina Meldawati Pasaribu, I Putu Mahendra, Novia Amalia Sholeha, Nanang Masruchin, Efri Mardawati, and Myrtha Karina

Abstract Bacterial cellulose (BC) is a renewable material which is currently playing a central role in medical device applications due to its biocompatibility and capability to be structurally, chemically, and morphologically modified at macro, micro, and nano scales. In addition, BC also has high water content, mechanical strength, and purity which are also excellent properties for use in biomedical applications. Despite the numerous advantages of BC properties for biomedical applications, its use for commercialization is still a challenge due to the high expense of the carbon and nitrogen sources required for BC synthesis. This study will provide an overview of numerous alternate sources of carbon and nitrogen from agricultural byproducts for BC synthesis that have been investigated and the potential of BC to be used for medical devices.

K. M. Pasaribu (✉) · N. A. Sholeha · N. Masruchin · M. Karina
Research Center for Biomass and Bioproducts, National Research and Innovation Agency of Indonesia (BRIN), Cibinong 16911, Indonesia
e-mail: khat004@brin.go.id

K. M. Pasaribu · N. A. Sholeha · N. Masruchin · E. Mardawati · M. Karina
Research Collaboration Center for Biomass and Biorefinery, BRIN – UNPAD, Jatinangor 45363, Indonesia

I. P. Mahendra
Institut Teknologi Sumatera, Jalan Terusan Ryacudu, Way Hui, Jati Agung, Lampung Selatan, Lampung 35365, Indonesia

K. M. Pasaribu · N. Masruchin · M. Karina
Research Collaboration Center for Nanocellulose, BRIN - UNAND, Padang 25163, Indonesia

E. Mardawati
Faculty of Agroindustrial Technology, Universitas Padjadjaran, Jatinangor, Bandung 45363, Indonesia

N. A. Sholeha
School of Vocational Studies, Bogor Agricultural University (IPB University), Jalan Kumbang No. 14, Bogor 16151, Indonesia

M. A. R. Lubis et al. (eds.), *Biomass Conversion and Sustainable Biorefinery*, Green Energy and Technology, https://doi.org/10.1007/978-981-99-7769-7_10

Keywords Bacterial cellulose · Production · Substrate · Agroindustry byproduct · Medical devices

1 Introduction

In the biomedical field, synthetic or natural polymers are used for different purposes (Sionkowska 2011; Tian et al. 2012). Although the application of synthetic polymers in medical devices has advanced significantly, it is known that using these polymer products in biomedicine poses irritation risk issues for tissue due to their low biocompatibility (Gunatillake et al. 2003; Bilgi et al. 2016). The higher biocompatibility properties of natural polymers make these polymers the most widely recommended alternatives as synthetic polymers substitutes in medical applications (Aravamudhan et al. 2014). Currently, plant-derived cellulose is the most abundant natural polymer produced on earth and is used extensively in the textile and food industries. However, its use in the medical and cosmetic fields is limited due to its impurities such as lignin, hemicellulose, and pectin (Klemm et al. 2005; Moran-Mirabal and Cranston 2015).

Apart from being sourced from plants, there is cellulose which is sourced from microbial synthesis called bacterial cellulose (BC). In contrast to plant-derived cellulose, BC is a cellulose with a high purity, thus it has found extensive use in the biomedical sector, including as polymer scaffolding for bone and cartilage repair, wound dressings to restore burned skin, a membrane for skin drug delivery, artificial blood vessels for microsurgery, and wound dressings to treat burned skin (Trovatti et al. 2011; Gomes et al. 2013). Bacterial cellulose is a natural extracellular polymer with the molecular formula of $(C_6H_{10}O_5)_n$ which is synthesized extensively by *Gluconacetobacter* strains via linear coupling of glucopyranose sugar monomers (Shoda and Sugano 2005). The strain that was employed to synthesize this polymer is non-pathogenic, extensively distributed in fruits and their products, and simple to grow in a lab (Klemm et al. 2009; Moosavi-Nasab and Yousefi 2011). This polymer is a biomaterial with superior properties due to its nanofibrous network structure (50–120 nm), 100% purity, high surface area, high crystallinity, high degree of polymerization, high capacity to absorb and hold water, high wet tensile strength, biocompatible, and easily degradable (Chen et al. 2013; Dhar et al. 2019).

However, the use of BC for industrial-scale applications for medical devices is constrained by its high production costs. For industrial-scale application, production parameters including temperature, pH, surface area to volume ratio of culture medium air–liquid interface (S/V), inoculum ratio and incubation time should be optimized for high quality, cost-effective, and high-yield BC production (Bilgi et al. 2016; Gea et al. 2018). Among the factors that affect the production costs, the sources of carbon and nitrogen used in the BC production are one of the factors that significantly determine the manufacturing costs, which can amount to 65% of the total cost (Chen et al. 2013; Sudying et al. 2019). The traditional source of carbon for BC fermentation is sugars such as glucose, fructose, and sucrose. Coconut water is currently the most affordable and sustainable raw material utilized in BC industrial production. However, coconuts are only grown in the tropics countries such as the Philippines, Indonesia, and other South and Southeast Asian (Cao et al. 2018). In most countries in the world, coconut production is extremely low and import-dependent. Thus, the use of this alternative substrate cannot be a solution for reducing BC production costs in many countries.

Meanwhile, the demand for BC, especially in the food and renewable material industry, continues to increase and it is predicted that demand for BC will eventually outpace supply in the future (Çakar et al. 2014; Padmanaban et al. 2015; Gea et al. 2022). Adding to this, in 2016, BC market was valued at US$207 million, and it is anticipated to reach US$700 million in 2026 (Calderón-Toledo et al. 2022).

In recent years, many researchers have focused on efforts to produce BC by developing cost-effective carbon and nitrogen feedstocks from local agroindustry product residues (Hong et al. 2011, 2012). Agricultural waste is known to contain high amounts of lipids, carbohydrates, including mono-, oligo- and polysaccharides, and proteins which can be converted into renewable energy sources with high added value (Calderón-Toledo et al. 2022). The advantage of using agricultural residues as raw materials in BC production is to reduce the cost of raw materials so that they can be mass-produced and commercialized and can be used as a basis for developing advanced materials, especially in their applications as biomedical devices (Chen et al. 2013). In addition, effective utilization of agricultural byproducts will be a good mode of recycling biomass which will simultaneously reduce the burden of waste treatment (Cao et al. 2018).

2 Bacterial Cellulose

Bacterial Cellulose is an extracellular polymer produced by bacteria from the genera *Sarcina*, *Gluconacetobacter*, and *Agrobacterium* through oxidative fermentation in non-synthetic and synthetic media (Esa et al. 2014; Huang et al. 2014). Among the various types of bacteria, *Gluconacetobacter xylinus* is the producer that receives the most attention for commercial-level production because it can digest a variety of carbohydrates in liquid media to produce BC in larger quantities (El-Gendi et al. 2022). In the BC synthesis, the bacterial body will releases glucose chains through the tiny holes in their cell envelopes, and the hydroxyl groups in the polymer chains form microfibrils, which subsequently group to form random nanofibers, will create 3D nanopore networks (Calderón-Toledo et al. 2022). The main structure of this fibril is formed of β-1,4-glycosidic linear polysaccharide structure that is connected by hydrogen bonds (Ul-Islam et al. 2012). Because it is secreted in the form of 2–4 nm fibrils, which are 100 times smaller than plant cellulose fibrils, this fiber is categorized as a nanoscale network (Gayathry and Gopalaswamy 2014; Costa et al. 2017).

In its natural state, BC possesses good hydration and high water holding capacity which is up to 100 times of its initial weight (Rani and Appaiah 2013). In addition, BC has a high degree of polymerization, crystallinity, and biological adaptability. BC also offers biocompatibility, biodegradability, and renewable capabilities (Costa et al. 2017). Therefore, BC is a desirable material for future use in a variety of industries, including electronics, paper, and food as well as the biomedical industry (bone and cartilage reconstruction, tissue engineering, wound dressings, implants, corneal restorations, orthodontics, artificial blood vessels), pharmaceuticals, veterinary medicine, and the leather industry (Andritsou et al. 2018; Popa et al. 2022).

3 Nutritional Requirements for Bacterial Cellulose Production

Currently, the use of bacteria to produce high-purity cellulose is in great demand because this cellulose has good nanometric, thermal, and mechanical properties. Moreover, it does not require aggressive treatment for purification and is environmentally friendly (Lin et al. 2013; Shi et al. 2014). Since this sort of cellulose is produced by bacteria, the availability of nutrient-rich elements in the growth media is a key factor for the production process as illustrated in Fig. 1 although there are also several things that need to be considered, such as, culture type (static vs. dynamic, surface vs. submerged), pH, shear force, and level of oxygenation support (Fernandes et al. 2020; Popa et al. 2022). Therefore, carbon sources include sugars such as fructose, glucose, sucrose; polysaccharides such as amylose or starch; nitrogen such as peptones, yeast extract, and casein hydrolyzates; inorganic salts such as ammonium sulfate; organic acids such as citric acid; some low molecular weight hydroxylated compounds such as glycerol, mannitol, and ethanol; are among the most important in the production of BC.

Currently, synthetic media, such as Hestrin-Schramm (HS) media, are utilized extensively in laboratory-scale BC production. However, the use of this medium in a large-scale BC production is not recommended due to its high cost, low yield, time consumption, and intensive labor required (Calderón-Toledo et al. 2022; Popa et al. 2022). These facts present a significant barrier and restrict the commercial use of BC products. Consequently, finding inexpensive substrates becomes crucial (Cavka et al. 2013; Huang et al. 2014). According to estimations, fermented media consumes up to 65% of the whole production budget (Chen et al. 2013; Sudying et al. 2019). Thus, the main challenge in bacterial cellulose production is to find an abundant, suitable, inexpensive, and non-competitive carbon source with food production (Dórame-Miranda et al. 2019). In recent years, agroindustrial wastes have emerged as a potential affordable substitute medium for BC production (El-Gendi et al. 2022). This method is advantageous because it is able to produce high-quality

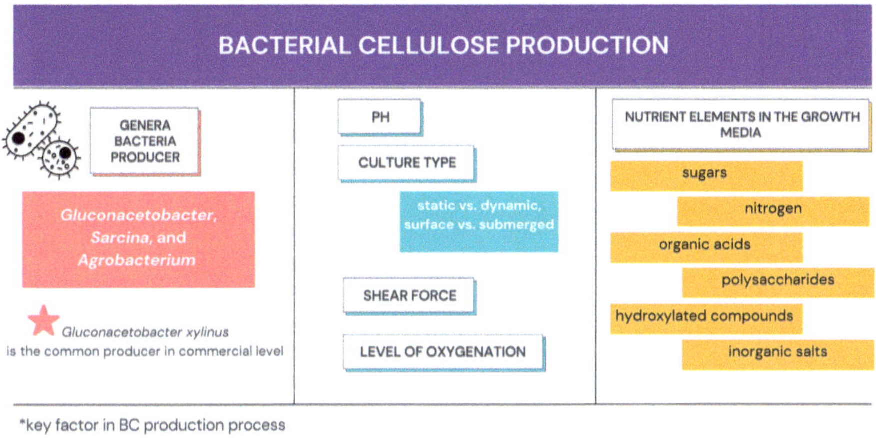

Fig. 1 Key factor in BC production process

BC and reduces waste disposal by converting waste water into a cheap fermentation promoter (Popa et al. 2022).

4 Agroindustry Byproducts that Used as Bacterial Cellulose Medium Culture

Given that BC has demonstrated excellent potential as a useful biopolymer in many applications, numerous studies have been conducted to identify an efficient method for generating BC at a reasonable price. The most notable is the utilization of agricultural byproducts as a replacement of carbon and protein for BC production. Each of the agroindustry byproducts and the bacterial strain that have been studied in BC production is shown in Table 1. This table also includes information of BC yields and highlights of BC produced from each agro-industrial byproduct used.

4.1 Wine Pomace

Pomace grapes are the main solid byproduct of the wine industry consisting of grape skins and seeds. These byproducts contain lignified fiber and soluble compounds such as sugars, phenolics, and a number of alcohols (Muhlack et al. 2018; Troncozo et al. 2019). As a result, grape pomace and its hydrolyzate have the potential to be used as a low-cost bioactive substrate to produce BC which has antioxidant and antibacterial activity. The pomace substrate pretreatment process was carried out by hydrolysis using commercial enzymes, namely pectinase and cellulase (Li et al. 2021).

Table 1 The utilization of agroindustry byproducts as a carbon and nitrogen source replacement in BC synthesis

Byproduct source	Bacterial strain	Yield	Highlight	References
Wine pomace	*Komagataeibacter rhaeticus*	4.28 ± 0.21 g/L in 10 day	Possess antioxidant and antibacterial activity	Li et al. (2021)
Carob-haricot bean extract	*Gluconacetobacter xylinus*	1.8 g/L in 9 days	Predict and optimize BC production yield	Bilgi et al. (2016)
Mango (*Mangifera indica*) waste extract	*Komagataeibacter rhaeticus*	25.34 g/L in 21 days	Provide a high yield with a reasonable price	Calderón-Toledo et al. (2022)
Citrus pulp	*Gluconacetobacter hansenii* and *Gluconacetobacter xylinum*	8.77 g/L in 10 days	Almost achieve industrial level	Cao et al. (2018)
Corn stalk	*Acetobacter xylinum*	2.86 g/L in 10 days	Matched with the idea of a biorefinery because it is affordable, green, and sustainable	Chen et al. (2013)
Sugarcane straw	*Komagataeibacter xylinus*	2.96 g/L in 12 days	Provides a different strategy to meet the commercial demand	Dhar et al. (2019)
Pecan nutshell (*Carya illinoinensis*)	*Gluconacetobacter entanii*	2.816 g/L in 28 days	Chemical functionalization of BC becomes methylcellulose and reduces its crystallinity	Dórame-Miranda et al. (2019)
Dry olive mill residue	*Gluconacetobacter sacchari*	0.85 g/L in 4 days	Demonstrate positive outcomes to combat high production costs in BC	Gomes et al. (2013)
Cashew tree residue	*Komagataeibacter rhaeticus*	2.8 g L^{-1} in 7 days	Up to a 33% of cost reduction	Pacheco et al. (2017)

(continued)

Table 1 (continued)

Byproduct source	Bacterial strain	Yield	Highlight	References
Coffee cherry husk	*Gluconacetobacter hansenii*	8.2 g/L in 14 days	Provides a more affordable alternative substrate and offers solution for toxic agricultural waste disposal	Rani and Appaiah (2013)
Rice washing drainage	*Komagataeibacter nataicola*	0.20 g/L in 13 days	A potential alternate carbon source that does not require any pretreatment	Sudying et al. (2019)
Tobacco waste extract (TWE)	*Acetobacter xylinum*	5.2 g/L in 7 days	Nicotine in TWE, which is known to inhibit BC production, can be removed by steam distillation treatment	Ye et al. (2019)

4.2 Carob and Haricot Bean Extracts

Carob (*Ceratonia siliqua* L.) production in Turkey is known to reach 15 × 10^3 tons/year with a total production of more than 400 × 10^3 tons/year in the world. Carob contains about 50% sugar (75% of sucrose and the rest is fructose, maltose, and glucose), 8% protein, and a number of important minerals (AYAZ et al. 2007; Bilgi et al. 2016). Meanwhile, haricot beans (*Phaseolus vulgaris*) are known to contain protein of 18.5–22%, a number of minerals (such as magnesium and calcium) and carbohydrates (de Almeida Costa et al. 2006; Shimelis et al. 2006). Therefore, the mixture of haricot nuts and carob has the potential to be employed in BC production as sources of carbon and nitrogen, respectively. According to the studies that have been published, Plackett-Burman and Central Composite Design techniques were used to prepare carob and haricot bean extracts as alternative growth media for *G. xylinus* in the development of cost-effective BC production methods. While, the pretreatment process used in this medium is by hydrolyzation process of the extract with distilled water using an autoclave (Bilgi et al. 2016).

4.3 Mango Variety (Mangifera Indica)

The Food and Agriculture Organization estimates that around 55% of mango produced globally, equivalent to 1.1 million tons, is wasted due to rotting during the transport, packaging, and storage (García-Sánchez et al. 2020). Mango extract (*Mangifera indica*) is known to contain 13.7–15.0% of sugar in the form of glucose, fructose, and sucrose and 1.5–5.5% of protein, making mango waste potentially can be used as an alternative low-cost substrate source for BC production (Prasanna et al. 2003; Maldonado-Celis et al. 2019). Moreover, mango fermentation with yeast and bacteria has been widely used in producing probiotic juice, wine, vinegar, yeast lipase, etc. (Li et al. 2012; da Pereira et al. 2019). Previous studies stated that it is necessary to use mangoes as a medium for BC production by first applying a pretreatment in the form of a hydrolysis process followed by a sugars reversion reaction with acid hydrolysis in 1 M HCl (Calderón-Toledo et al. 2022).

4.4 Citrus Pulp Waste

Oranges are one of the fruits that are produced in very big quantities worldwide, with an annual production reaching 27.1 million tons where 40–60% of its total weight is a byproduct in the form of peels and pulp (Fan et al. 2016). Currently, oranges have been used as vitamins by ingesting their active constituents or as a beverage ingredient in ethanol, vinegar, liquor, lactic acid beverages, mushrooms, and high-protein feed (Shan 2016). In China, peel of the orange has been used to produce traditional medicine by extracting its pectin, refined oil, and flavonoids content. However, orange pulp manufacturing has not yet been optimized, resulting in an accumulation of this byproduct. While it is known that citrus pulp is used as a production medium in BC by first being pretreated with enzymatic hydrolysis using cellulase and pectinase enzymes to reduce the viscosity of citrus pulp-based medium and reduce sugar so that mono/disaccharides could be used by the microorganism to synthesize BC (Cao et al. 2018).

4.5 Corn Stalk

Corn stalks are a byproduct of the annual production of corn agriculture, which is widely available worldwide (Luo et al. 2017). Thus far, this byproduct has only been treated inefficiently as compost (Boufi and Chaker 2016). Recently, the utilization of this biomass has been developed to produce bioethanol with biorefinery process (Shen and Wyman 2011; Cheng et al. 2017). Furthermore, according to a number of research, corn stalk prehydrolyzate contains high sugar content that can be employed in the fermentation process, offering another option for maximizing the use of these agricultural byproducts (Boufi and Chaker 2016; Esteves Costa et al. 2016). However, it is known that the utilization of the prehydrolyzate content in corncobs does not optimally support the production of BC. Therefore, its utilization as a carbon source to cultivate *Acetobacter xylinum* required a pretreatment using acetic acid and followed by pre-hydrolysis liquor (PHL) detoxification using activated carbon and ion exchange resin (Cai et al. 2016; Jiang et al. 2016).

4.6 Sugarcane Straw

Sugarcane is an agricultural commodity that is produced in large quantities every year. Brazil, is one of the largest sugarcane producing countries in the world, estimated to produce 93.3 $\times$ 10^6 tons of sugarcane waste each year in the form of bagasse and straw (dos Santos Rocha et al. 2017). It is estimated that every 1000 kg of sugar cane that processed in agroindustrial, will result in residues of 176 kg and 231 kg in the form of straw and dregs respectively. However, this residue is still not managed adequately, and as a result, the primary method of this waste treatment is burning. Recently, bagasse has been utilized as a component in bioethanol, biobutanol, or for bioenergy production for power plants, however the utilization of straw as waste is still rare (Dhar et al. 2019). It is known that sugarcane straw contains cellulose, hemicellulose, and lignin fraction ranging from 35–45%, 25–35%, to 10–25% respectively, which can be used as a medium for BC synthesis. Several pretreatment techniques such as enzymatic hydrolysis, acid hydrolysis, or hydrothermal treatment can be used to optimize the ultilization of this byproduct as a medium for BC production. Interestingly, the utilization of biomass as the carbon source was reported to produce BC with higher yield compared to pure glucose (Costa et al. 2017).

4.7 Pecan Nutshell

Pecan nut (*Carya illinoinensis*) is an agricultural product in which its seed is the part that is most widely used. The Food and Agriculture Organization (FAOSTAT) estimates that 460 $\times$ 10^3 tons of Pecan nut are produced annually while 40–50% is hazelnut shells which is a byproduct of this agroindustry (do Prado et al. 2014; Hilbig et al. 2018; Dórame-Miranda et al. 2019). Pecan nut shells are recognized to have the potential to be used as a source of nutrition for bacteria that specialize in producing BC because they contain significant levels of crude fiber (particularly lignin and cellulose) and carbohydrates (~90%) (Flores-Córdova et al. 2016). The utilization of hazelnut shells as a substrate in the production of BC is also reported does not require special pretreatment (Dórame-Miranda et al. 2019).

4.8 Dry Olive Mill Residue

Olive oil industry is one of the significant economic activity in several nations, including Portugal (Trichopoulou and Critselis 2004; Sieri et al. 2004). An estimated

40 × 103 tons of two-phase olive pomace (OP), which is made up of the pulp, skin, and stone fragments of the olives, are produced in Portugal each year as a result of the industrial extraction of olive oil (Fernández-Bolaños et al. 2006). OP has been used to make OP oil with a yield of 9.2%, therefore this method creates a byproduct known as dry olive mill residue (DOR), which can reach up to 35% of the mass of the original dry OP (Vlyssides et al. 2004; Sánchez Moral and Ruiz Méndez 2006). Currently, DOR has been utilized to produce electricity, organic fertilizers, and additives for animal feed (Martín García et al. 2003; López-Piñeiro et al. 2007). The utilization development of DOR continues to be studied and it is reported that DOR contains a substrate rich in sugars monomer as a source of carbon and nutrients for BC production after pretreatment in the form of mild acid hydrolysis (Gomes et al. 2013).

4.9 Cashew Tree Residues

Cashew (*Anacardium occidentale*) is an agricultural product whose main product is cashew nuts and cashew juice and has byproducts in the form of cashew pulp and exudate. Since cashew nuts are the most valuable output of the cashew tree, the exudate from this tree trunk must regularly be removed in order to stimulate the production of cashew nuts. Each of these trees produces approximately 700 g of exudate each year, which becomes a waste (Pacheco et al. 2017). Cashew tree exudate contains of arabinogalactan proteins, mono- and oligosaccharides, mineral salts, and 70% of branched heteropolysaccharides, called as cashew gum (Pereira-Netto et al. 2007; Silva et al. 2010). The polymer chains are mainly composed of D-galactopyranose units, as the primary building block, which are joined by β-(1 → 4) glycosidic bonds. Glucose, rhamnose, glucuronic acid, and arabinose are additional sugars that can incorporate branched cashew gum chemical structures (Pacheco et al. 2017). At least 68×10^3 and 48×10^3 tons of cashew gum and tree exudate respectively, are produced annually (Pacheco et al. 2017). Large quantities of this byproduct made an opportunity to carry out a research in efforts to treat this waste for the benefit of the economy and environmental sustainability. This residue has been extensively utilized in the food, biotechnology, and pharmaceutical industries (Kumar et al. 2012). The usage of cashew tree exudate and cashew gum as an alternative carbon source in BC production has been examined and claimed to exhibit good chances to be used in lowering the production cost (Pacheco et al. 2017).

4.10 Coffee Cherry Husk

Coffee cherry husk is one of the most prevalent byproduct that is produced in coffee cherries agroindustrial. The amount of this byproduct is almost 18% of the total coffee cherries that are processed (Rani and Appaiah 2013). This husk is rich in polyphenols, minerals, proteins, and carbohydrates. However, their use in agriculture has been restricted leading to a significant pollution issue at coffee cherries processing due to the existence of undesirable compounds like tannins, caffeine, and other polyphenols. While, according to studies, coffee cherry husk has the potential to be utilized in bioprocesses as an alternative substrate which is affordable. Coffee cherry husk can be added straight to the medium without going through any extra processing as a carbon source (Rani and Appaiah 2013).

4.11 Rice Washing Drainage

The global average of rice consumption per year in 2016 was 478.38 $\times$ 10^6 tons (Sudying et al. 2019). According to this data, almost all industries that are engaged in rice processing washed the rice with clean water before being cooked for consumption. In this rice washing drainage, numerous amino acids, saccharides, vitamins, and other nutrients are contained. Previous study reported the utilization of rice washing drainage as a source of carbon for BC synthesis can be used as cost reduction alternative because does not need special pretreatment (Sudying et al. 2019).

4.12 Tobacco Waste Extract

Tobacco (*Nicotiana*) is a plant with high economic value that is cultivated worldwide because it is the main ingredient in the cigarette production industry (Wang et al. 2015). Every year, it is estimated that about half (50%) of the tobacco used in cigarette production ends up as waste (Liu et al. 2015). This waste is found in the form of tobacco stems, tobacco leaves, and unwanted waste (Zhong et al. 2010; Wang et al. 2013). This waste has a high toxic nicotine content so it tends not to be managed but only thrown away or burned (Zhang et al. 2013; Okunola et al. 2014). This certainly endangers human health and contributes to environmental pollution. Recently, tobacco waste has been widely studied for its use as a substrate for producing fertilizers, pectinases, and some drug precursors (Wang et al. 2015; Zheng et al. 2017). Furthermore, the potential of tobacco waste as a substrate for BC production was also studied because it has a high sugar content in the form of glucose, sucrose, fructose, and other polysaccharides. In its utilization as a substrate for BC pretreatment production in the form of a steam distillation process used to remove nicotine which can inhibit microorganisms in producing BC. TWE is reported to be an ideal substrate for lower cost BC production (Ye et al. 2019).

5 Bacterial Cellulose in Medical Application

The unique physico-mechanical properties of BC, especially its biocompatibility properties make it widely used for direct applications in biomedical fields such as tissue engineering, wound healing, and drug delivery (Choi et al. 2022; El-Gendi et al. 2022). In this section, the superior qualities of BC will be described, along with the rationale for its use and research development in medical device.

5.1 *Wound Dressings*

In the biomedical sector to date, BC has played a significant role in the development of dressings for various types of skin trauma such as chronic skin ulcers, burns, surgical incisions, and other trauma (Popa et al. 2022). In general, the wound healing process is a complex process which is divided into four stages, namely: hemostasis, inflammation, proliferation, and maturation. The type of dressing material used to treat the wound is known to affect how quickly and effectively each of these stages progresses (Kushwaha et al. 2022). According to the scientific method of wound care, a bandage must have a number of qualities in order to support and hasten the wound healing process. These qualities include the ability to maintain moisture, absorb exudate, support angiogenesis, enable gas exchange, create thermal insulation in the wound area, prevent microbial infection, and be non-toxic, non-sticky, and non-allergenic (Niculescu and Grumezescu 2022).

In this case, BC is one of the greatest materials for creating a good wound dressing because BC has a high water holding capacity which enables it to absorb wound exudate, maintain a moist environment at the injury site, and stimulate the acceleration of the re-epithelialization process (Hajmohammadi et al. 2020; Pasaribu et al. 2020a, b). Additionally, the high water content of BC can also prevent pain and secondary trauma in patients during dressing removal (Weyell et al. 2019). Numerous hydroxyl groups in BC can form hydrogen bonds with water to produce flexibility, which makes it flexible enough, especially for contoured skin surface, to act as wound physical barrier from the outside environment (Swingler et al. 2021). However, from a biofunctional perspective, BC lacks the antibacterial and antimicrobial properties which are helpful in preventing infection throughout wound healing phase. In order to maximize the contribution of BC wound dressings to expedite the wound healing process, a combination with other substances is required (Choi et al. 2022). Numerous studies show that BC works well for wound healing when mixed with other compounds or materials such nanoparticles, benzalkonium chloride, hydroxyapatite, Aloe vera, and vaccarin (Picheth et al. 2017; Hasibuan et al. 2021). The effectiveness of BC as a wound dressing is also proven by the availability of several BC commercial wound dressing products such as BioFillTM (Curitiba, Brazil) and DermafillTM (Londrina, Brazil) which are used to treat burns and boils, Membracel® (Curitiba, Brazil) for ulcers and leg vein lacerations, xCell® (New York, NY, USA) for venous leg ulcers, and EpiProtect® (Royal Wootton Bassett, UK) for burns (Cielecka et al. 2019).

5.2 Bone Tissue Engineering

Tissue engineering has recently emerged as a viable solution for the replacement of damaged tissue (Swetha et al. 2010; Zhou et al. 2014). Therefore, research engaged in the manufacture of extracellular matrix (ECM) scaffolds that imitate the composition and architecture of natural ECM of target tissues is widely carried out, especially in bone tissue engineering (Mano et al. 2007; Stevens 2008; Swetha et al. 2010). Bone, which is a component of the skeletal system, is produced through a formation process known as osteogenesis that developed during the prenatal until adulthood phase of each individual. Thus bone has the capacity for restoration and regeneration to repair minor injuries and mechanical damage caused by normal trauma as long as it is not a birth defect. Bone tissue engineering using scaffolds is currently widely recommended in the treatment of bone repair by stimulating bone regeneration through a complementary combination of cells, biomaterials, and factor therapy (Li et al. 2013; Vadaye Kheiry et al. 2018).

The selection of biomaterial for scaffolding in bone tissue engineering is crucial because in this approach the biomaterial acts as a structural and/or functional supporting template for the cell regeneration process (Sill and von Recum 2008). Bone tissue engineering requires unique biomaterials with characteristics such as strong mechanical stress resistance and tunable biodegradability (Atila et al. 2019). Bacterial cellulose (BC), is one of the biomaterials that have received extensive study for Bone tissue engineering scaffolding production due to its high compatibility, although on the other hand, these polymers do not have the appropriate mechanical properties (Sell et al. 2010; Huang et al. 2014). Due to its superior purity, tensile strength, modulus, and elasticity, BC is used in regenerative medicine more frequently than plant cellulose. Moreover, BC also has biofunctionality and is biocompatible (Khan et al. 2015). BC-based scaffolds also have porosity and 3D network structures that support cell growth (Wan et al. 2007). BC was also reported to have a structure similar to bone collagen and increase cell proliferation in-vitro (Chen et al. 2009; García-Gareta et al. 2015). When compared to animal-derived biomaterials like collagen, the usage of BC scaffolds has also been shown to enhance tissue and bone regeneration and decrease the potential danger of cross-infection (Kong et al. 2004; Popa et al. 2022).

5.3 Dental Implants

In dental clinical practice, dental implants are a common operation, however this practice is frequently constrained because the maxillary region lacks the necessary bone height for the treatment (de Oliveira Barud et al. 2021). The integration of dental implants into the surrounding tissue is a significant concern. Additionally, for bone regeneration, osseointegration between the implant and bone must be complete (Choi et al. 2022). Despite their excellent application in tissue engineering and biomedical devices, BC-based materials are still under-explored in dentistry. Whereas the use of BC for commercial purposes in dental applications is very profitable because of its good absorption capacity, volume retention, and mechanical strength (Mensah et al. 2022). In addition, research suggest that BC can preserve graft space, enhance bone structure, and be used for dental implant insertion when applied to guided tissue regeneration techniques for the treatment of periodontal disease (de Oliveira Barud et al. 2021).

5.4 Vascular Grafts and Artificial Blood Vessels

Vascular Grafts and Artificial Blood Vessels is a method used in replacing blood vessels by cutting damaged or diseased blood vessels. In this treatment, the development of intimal hyperplasia is severely hampered by the material incompatibility of artificial blood vessels (Choi et al. 2022). Currently, ePTFE, Dacron, and polyurethane are the most commonly utilized materials for artificial blood vessels. Comparative studies have shown that BC is superior to PET and ePTFE products for usage as a vascular graft material (Picheth et al. 2017). BC is a promising new material for application in artificial blood vessels because BC nanofiber may avoid blood clots by displaying delayed thrombin production on the surface (Fink et al. 2010). In general, the use of pure BC nanofibrous scaffold for tissue regeneration does have limitations due to the presence of nanopores which can inhibit cell infiltration and vascularization of the 3D scaffold. However, several modifications made to the BC reported the potential for the BC to adapt to mechanical properties similar to those of small diameter vessels (<5 mm) (Picheth et al. 2017). Development of BC tubes also shows better resemblance to the human saphenous vein (4.27×10–2% per mmHg for 30–120 mmHg) than commercially available Dacron saphenous vein products and ePTFE (Choi et al. 2022).

5.5 Delivery of Drug and Bioactive Agents

In recent years, numerous natural biopolymer-based hydrogels have been extensively researched for drug delivery applications (Dasari et al. 2022). Drugs are manufactured into drug delivery dosage forms because they are typically administered in numerous doses, possess fluctuating plasma concentrations, and have shorter half-lives. BC-based hydrogel scaffolds have recently been employed in drug delivery applications due to their potential in terms of high reactive surface, fine tissue structure, and high porosity (Swingler et al. 2021). Moreover, BC can be easily modified, blended, and impregnated with nanoparticles to change how receptive it is to drug release (Choi et al. 2022). Lyophilization followed by immersion is the most typical technique used to load the drug into the membrane of BC (Swingler et al. 2019). The drugs most commonly incorporated into bacterial cellulose are anti-inflammatory drugs, such as diclofenac and ibuprofen, and antimicrobial drugs (Ao et al. 2020; Bernardelli de Mattos et al. 2020; Junka et al. 2020). The effectiveness of BC as a drug delivery material can be boosted by utilizing the tensile strength and water absorption of BC to load it with antimicrobial substances such as antibiotics, to provide new features and functionalities (Gupta et al. 2019, 2020; Swingler et al. 2021). Researchers also frequently use BC-based as controlled-release drug delivery agents, for instance, the application of BC powder as paracetamol tablets coat via a spray-coating approach, demonstrates that the thickness of the BC film is used to affect the *in-vitro* drug release rate (Amin et al. 2012). Due to the lack of barrier interference and erythema, BC has also been widely researched as a transdermal medication delivery agent. It has been discovered that the skin is well-tolerated and moisturized because BC porosity can manage hydrophilicity of environment and also alter the release rate of the drug (Almeida et al. 2014; Ullah et al. 2016).

6 Conclusion and Future Perspective

This study focuses on the review of alternative substrates that sourced from the agroindustry byproducts in order to replace carbon and nitrogen and the advantages of the basic properties of bacterial cellulose for application as medical devices. Bacterial cellulose is a polysaccharide synthesized by various non-pathogenic bacteria under specific cultivation conditions. This appealing biopolymer has a number of physicochemical, mechanical, and biological qualities, including: environmental friendliness, biodegradability, biocompatibility, non-toxicity, optimal viscoelasticity, a 3D porous structure, high tensile strength, easy to modify, sufficient capacity to retain large amounts of water, and higher crystallinity and purity than plant cellulose. Either by itself or in combination with other biopolymers and bioactive substances BC has been reported to have therapeutic effects on various body areas of humans when used as medical devices. However, the development of BC as the main component in commercial medical device applications is still constrained due to the high-cost production of BC. Therefore, further research to find alternative synthesis media to reduce the BC production costs will continue to develop in the future.

Acknowledgements Khatarina Meldawati Pasaribu expresses her sincere gratitude to the National Research and Innovation Agency of Indonesia (BRIN) for providing the necessary resources to conduct her postdoctoral work at the Nanocellulose research group at the Research Center for Biomass and Bioproducts from July 1, 2022, to June 30, 2023.

References

Almeida IF, Pereira T, Silva NHCS et al (2014) Bacterial cellulose membranes as drug delivery systems: an in vivo skin compatibility study. Eur J Pharm Biopharm 86:332–336. https://doi.org/10.1016/j.ejpb.2013.08.008

Amin MCIM, Abadi AG, Ahmad N et al (2012) Bacterial cellulose film coating as drug delivery system: physicochemical, thermal and drug release properties. Sains Malaysiana 41:561–568

Andritsou V, De Melo EM, Tsouko E et al (2018) Synthesis and characterization of bacterial cellulose from citrus-based sustainable resources. ACS Omega 3:10365–10373. https://doi.org/10.1021/acsomega.8b01315

Ao H, Jiang W, Nie Y et al (2020) Engineering quaternized chitosan in the 3D bacterial cellulose structure for antibacterial wound dressings. Polym Test 86:106490. https://doi.org/10.1016/j.polymertesting.2020.106490

Aravamudhan A, Ramos DM, Nada AA, Kumbar SG (2014) Chapter 4—Natural polymers: polysaccharides and their derivatives for biomedical applications. In: Kumbar SG, Laurencin CT, Deng MBT-N and SBP (eds). Elsevier, Oxford, pp 67–89

Atila D, Karataş A, Evcin A et al (2019) Bacterial cellulose-reinforced boron-doped hydroxyapatite/gelatin scaffolds for bone tissue engineering. Cellulose 26:9765–9785. https://doi.org/10.1007/s10570-019-02741-1

Ayaz FA, Torun H, Ayaz S et al (2007) Determination of chemical composition of anatolian carob pod (ceratonia siliqua l.): sugars, amino and organic acids, minerals and phenolic compounds. J Food Qual 30:1040–1055. https://doi.org/10.1111/j.1745-4557.2007.00176.x

Bernardelli de Mattos I, Nischwitz SP, Tuca A-C et al (2020) Delivery of antiseptic solutions by a bacterial cellulose wound dressing: Uptake, release and antibacterial efficacy of octenidine and povidone-iodine. Burns 46:918–927. https://doi.org/10.1016/j.burns.2019.10.006

Bilgi E, Bayir E, Sendemir-Urkmez A, Hames EE (2016) Optimization of bacterial cellulose production by Gluconacetobacter xylinus using carob and haricot bean. Int J Biol Macromol 90:2–10. https://doi.org/10.1016/j.ijbiomac.2016.02.052

Boufi S, Chaker A (2016) Easy production of cellulose nanofibrils from corn stalk by a conventional high speed blender. Ind Crops Prod 93:39–47. https://doi.org/10.1016/j.indcrop.2016.05.030

Cai D, Li P, Luo Z et al (2016) Effect of dilute alkaline pretreatment on the conversion of different parts of corn stalk to fermentable sugars and its application in acetone–butanol–ethanol fermentation. Bioresour Technol 211:117–124. https://doi.org/10.1016/j.biortech.2016.03.076

Çakar F, Katı A, Özer I et al (2014) Newly developed medium and strategy for bacterial cellulose production. Biochem Eng J 92:35–40. https://doi.org/10.1016/j.bej.2014.07.002

Calderón-Toledo S, Horue M, Alvarez VA et al (2022) Isolation and partial characterization of Komagataeibacter sp. SU12 and optimization of bacterial cellulose production using Mangifera indica extracts. J Chem Technol Biotechnol 97:1482–1493. https://doi.org/10.1002/jctb.6839

Cao Y, Lu S, Yang Y (2018) Production of bacterial cellulose from byproduct of citrus juice processing (citrus pulp) by Gluconacetobacter hansenii. Cellulose 25:6977–6988. https://doi.org/10.1007/s10570-018-2056-0

Cavka A, Guo X, Tang S-J et al (2013) Production of bacterial cellulose and enzyme from waste fiber sludge. Biotechnol Biofuels 6:25. https://doi.org/10.1186/1754-6834-6-25

Chen YM, Xi T, Zheng Y et al (2009) In vitro cytotoxicity of bacterial cellulose scaffolds used for tissue-engineered bone. J Bioact Compat Polym 24:137–145. https://doi.org/10.1177/0883911509102710

Chen L, Hong F, Yang X xia, Han S (2013) Biotransformation of wheat straw to bacterial cellulose and its mechanism. Bioresour Technol 135:464–468. https://doi.org/10.1016/j.biortech.2012.10.029

Cheng Z, Yang R, Liu X et al (2017) Green synthesis of bacterial cellulose via acetic acid pre-hydrolysis liquor of agricultural corn stalk used as carbon source. Bioresour Technol 234:8–14. https://doi.org/10.1016/j.biortech.2017.02.131

Choi SM, Rao KM, Zo SM et al (2022) Bacterial cellulose and its applications. Polymers (Basel) 14
Cielecka I, Szustak M, Kalinowska H et al (2019) Glycerol-plasticized bacterial nanocellulose-based composites with enhanced flexibility and liquid sorption capacity. Cellulose 26:5409–5426. https://doi.org/10.1007/s10570-019-02501-1
Costa AFS, Almeida FCG, Vinhas GM, Sarubbo LA (2017) Production of bacterial cellulose by Gluconacetobacter hansenii using corn steep liquor as nutrient sources. Front Microbiol 8:1–12. https://doi.org/10.3389/fmicb.2017.02027
da Pereira AS, Fontes-Sant'Ana GC, Amaral PFF (2019) Mango agro-industrial wastes for lipase production from Yarrowia lipolytica and the potential of the fermented solid as a biocatalyst. Food Bioprod Process 115:68–77. https://doi.org/10.1016/j.fbp.2019.02.002
Dasari S, Njiki S, Mbemi A et al (2022) Pharmacological effects of cisplatin combination with natural products in cancer chemotherapy. Int J Mol Sci 23
de Almeida Costa GE, da Silva Queiroz-Monici K, Pissini Machado Reis SM, de Oliveira AC (2006) Chemical composition, dietary fibre and resistant starch contents of raw and cooked pea, common bean, chickpea and lentil legumes. Food Chem 94:327–330. https://doi.org/10.1016/j.foodchem.2004.11.020
de Fernandes IAA, Pedro AC, Ribeiro VR et al (2020) Bacterial cellulose: from production optimization to new applications. Int J Biol Macromol 164:2598–2611. https://doi.org/10.1016/j.ijbiomac.2020.07.255
de Oliveira Barud HG, da Silva RR, Borges MAC et al (2021) Bacterial nanocellulose in dentistry: perspectives and challenges. Molecules 26. https://doi.org/10.3390/MOLECULES26010049
Dhar P, Pratto B, Gonçalves Cruz AJ, Bankar S (2019) Valorization of sugarcane straw to produce highly conductive bacterial cellulose/graphene nanocomposite films through in situ fermentation: Kinetic analysis and property evaluation. J Clean Prod 238:117859. https://doi.org/10.1016/j.jclepro.2019.117859
do Prado ACP, da Silva HS, da Silveira SM et al (2014) Effect of the extraction process on the phenolic compounds profile and the antioxidant and antimicrobial activity of extracts of pecan nut [Carya illinoinensis (Wangenh) C. Koch] shell. Ind Crops Prod 52:552–561. https://doi.org/10.1016/j.indcrop.2013.11.031
Dórame-Miranda RF, Gámez-Meza N, Medina-Juárez L et al (2019) Bacterial cellulose production by Gluconacetobacter entanii using pecan nutshell as carbon source and its chemical functionalization. Carbohydr Polym 207:91–99. https://doi.org/10.1016/j.carbpol.2018.11.067
dos Santos Rocha MSR, Pratto B, de Sousa R et al (2017) A kinetic model for hydrothermal pretreatment of sugarcane straw. Bioresour Technol 228:176–185. https://doi.org/10.1016/j.biortech.2016.12.087
El-Gendi H, Taha TH, Ray JB, Saleh AK (2022) Recent advances in bacterial cellulose: a low-cost effective production media, optimization strategies and applications. Springer, Netherlands
Esa F, Tasirin SM, Rahman NA (2014) Overview of bacterial cellulose production and application. Agric Agric Sci Procedia 2:113–119. https://doi.org/10.1016/j.aaspro.2014.11.017
Esteves Costa CA, Coleman W, Dube M et al (2016) Assessment of key features of lignin from lignocellulosic crops: stalks and roots of corn, cotton, sugarcane, and tobacco. Ind Crops Prod 92:136–148. https://doi.org/10.1016/j.indcrop.2016.07.032
Fan X, Gao Y, He W et al (2016) Production of nano bacterial cellulose from beverage industrial waste of citrus peel and pomace using Komagataeibacter xylinus. Carbohydr Polym 151:1068–1072. https://doi.org/10.1016/j.carbpol.2016.06.062
Fernández-Bolaños J, Rodríguez G, Rodríguez R et al (2006) Extraction of interesting organic compounds from olive oil waste. Grasas Aceites 57:95–106. https://doi.org/10.3989/gya.2006.v57.i1.25
Fink H, Faxälv L, Molnár GF et al (2010) Real-time measurements of coagulation on bacterial cellulose and conventional vascular graft materials. Acta Biomater 6:1125–1130. https://doi.org/10.1016/j.actbio.2009.09.019

Flores-Córdova MA, Sánchez Chávez E, Chávez-Mendoza C et al (2016) Bioactive compounds and phytonutrients in edible part and nutshell of pecan (Carya illinoinensis). Cogent Food Agric 2:1262936. https://doi.org/10.1080/23311932.2016.1262936

García-Gareta E, Coathup MJ, Blunn GW (2015) Osteoinduction of bone grafting materials for bone repair and regeneration. Bone 81:112–121. https://doi.org/10.1016/j.bone.2015.07.007

García-Sánchez ME, Robledo-Ortíz JR, Jiménez-Palomar I et al (2020) Production of bacterial cellulose by Komagataeibacter xylinus using mango waste as alternative culture medium. Rev Mex Ing Quim 19:851–865. https://doi.org/10.24275/rmiq/Bio743

Gayathry G, Gopalaswamy G (2014) Production and characterisation of microbial cellulosic fibre from Acetobacter xylinum. Indian J Fibre Text Res 39:93–96

Gea S, Pasaribu KM, Sebayang K et al (2018) Enhancing the quality of nata de coco starter by channeling the oxygen into the bioreactor through agitation method. AIP Conf Proc 2049. https://doi.org/10.1063/1.5082469

Gea S, Pasaribu KM, Sarumaha AA, Rahayu S (2022) Cassava starch/bacterial cellulose-based bioplastics with Zanthoxylum acanthopodium. Biodiversitas 23:2601–2608. https://doi.org/10.13057/biodiv/d230542

Gomes FP, Silva NHCS, Trovatti E et al (2013) Production of bacterial cellulose by Gluconacetobacter sacchari using dry olive mill residue. Biomass Bioenerg 55:205–211. https://doi.org/10.1016/j.biombioe.2013.02.004

Gunatillake PA, Adhikari R, Gadegaard N (2003) Biodegradable synthetic polymers for tissue engineering. Eur Cells Mater 5:1–16. https://doi.org/10.22203/eCM.v005a01

Gupta A, Keddie DJ, Kannappan V et al (2019) Production and characterisation of bacterial cellulose hydrogels loaded with curcumin encapsulated in cyclodextrins as wound dressings. Eur Polym J 118:437–450. https://doi.org/10.1016/j.eurpolymj.2019.06.018

Gupta A, Briffa SM, Swingler S et al (2020) Synthesis of silver nanoparticles using curcumin-cyclodextrins loaded into bacterial cellulose-based hydrogels for wound dressing applications. Biomacromolecules 21:1802–1811. https://doi.org/10.1021/acs.biomac.9b01724

Hajmohammadi K, Esmaeili Zabihi R, Akbarzadeh K, Parizad N (2020) Using a combination therapy to combat scalp necrosis: a case report. J Med Case Rep 14:132. https://doi.org/10.1186/s13256-020-02450-5

Hasibuan PAZ, Yuandani TM et al (2021) Antimicrobial and antihemolytic properties of a CNF/AgNP-chitosan film: a potential wound dressing material. Heliyon 7:e08197. https://doi.org/10.1016/j.heliyon.2021.e08197

Hilbig J, de Britto Policarpi P, de Souza Grinevicius VM, Mota NS et al (2018) Aqueous extract from pecan nut [Carya illinoinensis (Wangenh) C. Koch] shell show activity against breast cancer cell line MCF-7 and Ehrlich ascites tumor in Balb-C mice. J Ethnopharmacol 211:256–266. https://doi.org/10.1016/j.jep.2017.08.012

Hong F, Zhu YX, Yang G, Yang XX (2011) Wheat straw acid hydrolysate as a potential cost-effective feedstock for production of bacterial cellulose. J Chem Technol Biotechnol 86:675–680. https://doi.org/10.1002/jctb.2567

Hong F, Guo X, Zhang S et al (2012) Bacterial cellulose production from cotton-based waste textiles: enzymatic saccharification enhanced by ionic liquid pretreatment. Bioresour Technol 104:503–508. https://doi.org/10.1016/j.biortech.2011.11.028

Huang Y, Zhu C, Yang J et al (2014) Recent advances in bacterial cellulose. Cellulose 21:1–30. https://doi.org/10.1007/s10570-013-0088-z

Jiang D, Ge X, Zhang T et al (2016) Photo-fermentative hydrogen production from enzymatic hydrolysate of corn stalk pith with a photosynthetic consortium. Int J Hydrogen Energy 41:16778–16785. https://doi.org/10.1016/j.ijhydene.2016.07.129

Junka A, Bartoszewicz M, Dziadas M et al (2020) Application of bacterial cellulose experimental dressings saturated with gentamycin for management of bone biofilm in vitro and ex vivo. J Biomed Mater Res Part B Appl Biomater 108:30–37. https://doi.org/10.1002/jbm.b.34362

Khan S, Ul-Islam M, Ullah MW et al (2015) Engineered regenerated bacterial cellulose scaffolds for application in in vitro tissue regeneration. RSC Adv 5:84565–84573. https://doi.org/10.1039/C5RA16985B

Klemm D, Heublein B, Fink H-P, Bohn A (2005) Cellulose: fascinating biopolymer and sustainable raw material. Angew Chemie Int Ed 44:3358–3393. https://doi.org/10.1002/anie.200460587

Klemm D, Schumann D, Kramer F et al (2009) Nanocellulose materials—Different cellulose, different functionality. Macromol Symp 280:60–71. https://doi.org/10.1002/masy.200950608

Kong XD, Cui FZ, Wang XM et al (2004) Silk fibroin regulated mineralization of hydroxyapatite nanocrystals. J Cryst Growth 270:197–202. https://doi.org/10.1016/j.jcrysgro.2004.06.007

Kumar A, Moin A, Ahmed A et al (2012) Cashew gum a versatile hydrophyllic polymer: a review. Curr Drug ther 7:2–12. https://doi.org/10.2174/157488512800389146

Kushwaha A, Goswami L, Kim BS (2022) Nanomaterial-based therapy for wound healing. Nanomaterials 12

Li X, Yu B, Curran P, Liu S-Q (2012) Impact of two Williopsis yeast strains on the volatile composition of mango wine. Int J Food Sci Technol 47:808–815. https://doi.org/10.1111/j.1365-2621.2011.02912.x

Li X, Wang L, Fan Y et al (2013) Nanostructured scaffolds for bone tissue engineering. J Biomed Mater Res Part A 101A:2424–2435. https://doi.org/10.1002/jbm.a.34539

Li Z, Azi F, Dong J et al (2021) Green and efficient in-situ biosynthesis of antioxidant and antibacterial bacterial cellulose using wine pomace. Int J Biol Macromol 193:2183–2191. https://doi.org/10.1016/j.ijbiomac.2021.11.049

Lin SP, Loira Calvar I, Catchmark JM et al (2013) Biosynthesis, production and applications of bacterial cellulose. Cellulose 20:2191–2219. https://doi.org/10.1007/s10570-013-9994-3

Liu Y, Dong J, Liu G et al (2015) Co-digestion of tobacco waste with different agricultural biomass feedstocks and the inhibition of tobacco viruses by anaerobic digestion. Bioresour Technol 189:210–216. https://doi.org/10.1016/j.biortech.2015.04.003

López-Piñeiro A, Murillo S, Barreto C et al (2007) Changes in organic matter and residual effect of amendment with two-phase olive-mill waste on degraded agricultural soils. Sci Total Environ 378:84–89. https://doi.org/10.1016/j.scitotenv.2007.01.018

Luo Z, Li P, Cai D et al (2017) Comparison of performances of corn fiber plastic composites made from different parts of corn stalk. Ind Crops Prod 95:521–527. https://doi.org/10.1016/j.indcrop.2016.11.005

Maldonado-Celis ME, Yahia EM, Bedoya R et al (2019) Chemical composition of mango (Mangifera indica L.) fruit: nutritional and phytochemical compounds. Front Plant Sci 10:1–21. https://doi.org/10.3389/fpls.2019.01073

Mano JF, Silva GA, Azevedo HS et al (2007) Natural origin biodegradable systems in tissue engineering and regenerative medicine: present status and some moving trends. J R Soc Interface 4:999–1030. https://doi.org/10.1098/rsif.2007.0220

Martín García AI, Moumen A, Yáñez Ruiz DR, Molina Alcaide E (2003) Chemical composition and nutrients availability for goats and sheep of two-stage olive cake and olive leaves. Anim Feed Sci Technol 107:61–74. https://doi.org/10.1016/S0377-8401(03)00066-X

Mensah A, Chen Y, Christopher N, Wei Q (2022) Membrane technological pathways and inherent structure of bacterial cellulose composites for drug delivery. Bioengineering 9

Moosavi-Nasab M, Yousefi A (2011) Biotechnological production of cellulose by Gluconacetobacter xylinus from agricultural waste. Iran J Biotechnol 9:94–101

Moran-Mirabal JM, Cranston ED (2015) Cellulose nanotechnology on the rise. Ind Biotechnol 11:14–15. https://doi.org/10.1089/ind.2015.1501

Muhlack RA, Potumarthi R, Jeffery DW (2018) Sustainable wineries through waste valorisation: a review of grape marc utilisation for value-added products. Waste Manag 72:99–118. https://doi.org/10.1016/j.wasman.2017.11.011

Niculescu A-G, Grumezescu AM (2022) An up-to-date review of biomaterials application in wound management. Polymers (Basel) 14

Okunola AA, Babatunde EE, Chinwe D et al (2014) Mutagenicity of automobile workshop soil leachate and tobacco industry wastewater using the Ames Salmonella fluctuation and the SOS chromotests. Toxicol Ind Health 32:1086–1096. https://doi.org/10.1177/0748233714547535

Pacheco G, Nogueira CR, Meneguin AB et al (2017) Development and characterization of bacterial cellulose produced by cashew tree residues as alternative carbon source. Ind Crops Prod 107:13–19. https://doi.org/10.1016/j.indcrop.2017.05.026

Padmanaban S, Balaji N, Muthukumaran C, Tamilarasan K (2015) Statistical optimization of process parameters for exopolysaccharide production by Aureobasidium pullulans using sweet potato based medium. 3 Biotech 5:1067–1073. https://doi.org/10.1007/s13205-015-0308-3

Pasaribu KM, Gea S, Ilyas S et al (2020a) Fabrication and in-vivo study of micro-colloidal Zanthoxylum acanthopodium-loaded bacterial cellulose as a burn wound dressing. Polymers (Basel) 12. https://doi.org/10.3390/polym12071436

Pasaribu KM, Gea S, Ilyas S et al (2020b) Characterization of bacterial cellulose-based wound dressing in different order impregnation of chitosan and collagen. Biomolecules 10:1–15. https://doi.org/10.3390/biom10111511

Pereira-Netto AB, Pettolino F, Cruz-Silva CTA et al (2007) Cashew-nut tree exudate gum: identification of an arabinogalactan-protein as a constituent of the gum and use on the stimulation of somatic embryogenesis. Plant Sci 173:468–477. https://doi.org/10.1016/j.plantsci.2007.07.008

Picheth GF, Pirich CL, Sierakowski MR et al (2017) Bacterial cellulose in biomedical applications: a review. Int J Biol Macromol 104:97–106. https://doi.org/10.1016/j.ijbiomac.2017.05.171

Popa L, Ghica MV, Tudoroiu EE et al (2022) Bacterial cellulose—A remarkable polymer as a source for biomaterials tailoring. Materials (Basel) 15. https://doi.org/10.3390/ma15031054

Prasanna V, Yashoda HM, Prabha TN, Tharanathan RN (2003) Pectic polysaccharides during ripening of mango (Mangifera indica L). J Sci Food Agric 83:1182–1186. https://doi.org/10.1002/jsfa.1522

Rani MU, Appaiah KAA (2013) Production of bacterial cellulose by Gluconacetobacter hansenii UAC09 using coffee cherry husk. J Food Sci Technol 50:755–762. https://doi.org/10.1007/s13197-011-0401-5

Sánchez Moral P, Ruiz Méndez MV (2006) Production of pomace olive oil. Grasas Aceites 57:47–55. https://doi.org/10.3989/gya.2006.v57.i1.21

Sell SA, Wolfe PS, Garg K et al (2010) The use of natural polymers in tissue engineering: a focus on electrospun extracellular matrix analogues. Polymers (Basel) 2:522–553

Shan Y (2016) Chapter 7—Biotransformation of citrus peel. In: Shan YBT-CU of CB-P (ed). Academic Press, pp 85–90

Shen J, Wyman CE (2011) A novel mechanism and kinetic model to explain enhanced xylose yields from dilute sulfuric acid compared to hydrothermal pretreatment of corn stover. Bioresour Technol 102:9111–9120. https://doi.org/10.1016/j.biortech.2011.04.001

Shi Z, Zhang Y, Phillips GO, Yang G (2014) Utilization of bacterial cellulose in food. Food Hydrocoll 35:539–545. https://doi.org/10.1016/j.foodhyd.2013.07.012

Shimelis EA, Meaza M, Rakshit SK, Ababa A (2006) Physico-chemical properties, pasting behavior and functional characteristics of flours and starches from improved bean (Phaseolus vulgaris L.) varieties grown in East Africa. Agric Eng VIII:1–19

Shoda M, Sugano Y (2005) Recent advances in bacterial cellulose production. Biotechnol Bioprocess Eng 10:1. https://doi.org/10.1007/BF02931175

Sieri S, Krogh V, Pala V et al (2004) Dietary patterns and risk of breast cancer in the ORDET cohort. Cancer Epidemiol Biomarkers Prev 13:567–572. https://doi.org/10.1158/1055-9965.567.13.4

Sill TJ, von Recum HA (2008) Electrospinning: applications in drug delivery and tissue engineering. Biomaterials 29:1989–2006. https://doi.org/10.1016/j.biomaterials.2008.01.011

Silva TM, Santiago PO, Purcena LLA, Fernandes KF (2010) Study of the cashew gum polysaccharide for the horseradish peroxidase immobilization—Structural characteristics, stability and recovery. Mater Sci Eng C 30:526–530. https://doi.org/10.1016/j.msec.2010.01.016

Sionkowska A (2011) Current research on the blends of natural and synthetic polymers as new biomaterials: review. Prog Polym Sci 36:1254–1276. https://doi.org/10.1016/j.progpolymsci.2011.05.003
Stevens MM (2008) Biomaterials for bone tissue engineering. Mater Today 11:18–25. https://doi.org/10.1016/S1369-7021(08)70086-5
Sudying P, Laingaumnuay N, Jaturapiree P (2019) Production and characterization of bacterial cellulose from rice washing drainage (Rwd) by komagataeibacter nataicola lil. Key Eng Mater (KEM) 824:30–37. https://doi.org/10.4028/www.scientific.net/KEM.824.30
Swetha M, Sahithi K, Moorthi A et al (2010) Biocomposites containing natural polymers and hydroxyapatite for bone tissue engineering. Int J Biol Macromol 47:1–4. https://doi.org/10.1016/j.ijbiomac.2010.03.015
Swingler S, Gupta A, Heaselgrave W et al (2019) An investigation into the anti-microbial properties of bacterial cellulose wound dressings loaded with curcumin: hydroxypropyl-β-cyclodextrin supramolecular inclusion complex an investigation into the anti-microbial properties of bacterial cellulose wound. Access Microbiol 1. https://doi.org/10.1099/acmi.amrmeds2019.po0012
Swingler S, Gupta A, Gibson H et al (2021) Recent advances and applications of bacterial cellulose in biomedicine. Polymers (Basel) 13:1–29. https://doi.org/10.3390/polym13030412
Tian H, Tang Z, Zhuang X et al (2012) Biodegradable synthetic polymers: preparation, functionalization and biomedical application. Prog Polym Sci 37:237–280. https://doi.org/10.1016/j.progpolymsci.2011.06.004
Trichopoulou A, Critselis E (2004) Mediterranean diet and longevity. Eur J Cancer Prev 13
Troncozo MI, Lješević M, Beškoski VP et al (2019) Fungal transformation and reduction of phytotoxicity of grape pomace waste. Chemosphere 237:124458. https://doi.org/10.1016/j.chemosphere.2019.124458
Trovatti E, Silva NHCS, Duarte IF et al (2011) Biocellulose membranes as supports for dermal release of lidocaine. Biomacromolecules 12:4162–4168. https://doi.org/10.1021/bm201303r
Ul-Islam M, Khan T, Park JK (2012) Water holding and release properties of bacterial cellulose obtained by in situ and ex situ modification. Carbohydr Polym 88:596–603. https://doi.org/10.1016/j.carbpol.2012.01.006
Ullah H, Santos HA, Khan T (2016) Applications of bacterial cellulose in food, cosmetics and drug delivery. Cellulose 23:2291–2314. https://doi.org/10.1007/s10570-016-0986-y
Vadaye Kheiry E, Parivar K, Baharara J et al (2018) The osteogenesis of bacterial cellulose scaffold loaded with fisetin. Iran J Basic Med Sci 21:965–971. https://doi.org/10.22038/ijbms.2018.25465.6296
Vlyssides AG, Loizides M, Karlis PK (2004) Integrated strategic approach for reusing olive oil extraction by-products. J Clean Prod 12:603–611. https://doi.org/10.1016/S0959-6526(03)00078-7
Wan YZ, Huang Y, Yuan CD et al (2007) Biomimetic synthesis of hydroxyapatite/bacterial cellulose nanocomposites for biomedical applications. Mater Sci Eng C 27:855–864. https://doi.org/10.1016/j.msec.2006.10.002
Wang J, He H, Wang M et al (2013) Bioaugmentation of activated sludge with Acinetobacter sp. TW enhances nicotine degradation in a synthetic tobacco wastewater treatment system. Bioresour Technol 142:445–453. https://doi.org/10.1016/j.biortech.2013.05.067
Wang W, Xu P, Tang H (2015) Sustainable production of valuable compound 3-succinoyl-pyridine by genetically engineering pseudomonas putida using the tobacco waste. Sci Rep 5:16411. https://doi.org/10.1038/srep16411
Weyell P, Beekmann U, Küpper C et al (2019) Tailor-made material characteristics of bacterial cellulose for drug delivery applications in dentistry. Carbohydr Polym 207:1–10. https://doi.org/10.1016/j.carbpol.2018.11.061
Ye J, Zheng S, Zhang Z et al (2019) Bacterial cellulose production by Acetobacter xylinum ATCC 23767 using tobacco waste extract as culture medium. Bioresour Technol 274:518–524. https://doi.org/10.1016/j.biortech.2018.12.028

Zhang K, Zhang K, Cao Y, Pan W (2013) Co-combustion characteristics and blending optimization of tobacco stem and high-sulfur bituminous coal based on thermogravimetric and mass spectrometry analyses. Bioresour Technol 131:325–332. https://doi.org/10.1016/j.biortech.2012.12.163

Zheng Y, Wang Y, Pan J et al (2017) Semi-continuous production of high-activity pectinases by immobilized Rhizopus oryzae using tobacco wastewater as substrate and their utilization in the hydrolysis of pectin-containing lignocellulosic biomass at high solid content. Bioresour Technol 241:1138–1144. https://doi.org/10.1016/j.biortech.2017.06.066

Zhong W, Zhu C, Shu M et al (2010) Degradation of nicotine in tobacco waste extract by newly isolated Pseudomonas sp. ZUTSKD. Bioresour Technol 101:6935–6941. https://doi.org/10.1016/j.biortech.2010.03.142

Zhou C, Ye X, Fan Y et al (2014) Biomimetic fabrication of a three-level hierarchical calcium phosphate/collagen/hydroxyapatite scaffold for bone tissue engineering. Biofabrication 6:35013. https://doi.org/10.1088/1758-5082/6/3/035013

Extraction, Isolation, Purification, and Potential Application of Xylose and Xylooligosaccharides from Lignocellulosic Biomass

Euis Hermiati, Hans Wijaya, and Dwi Ajias Pramasari

Abstract Xylose and XOS become products of interest and have good markets. Xylose and XOS are derived from xylans, which are parts of hemicellulose fraction of lignocellulosic biomass. The demand tends to increase due to depletion of fossil resources and a new paradigm shift in consumer preferences for healthier and natural products. There are different extraction methods or fractionation processes to extract xylose and XOS from lignocellulosic biomass feedstocks, including autohydrolysis and hydrolysis using acid, alkaline, solvent, and inorganic salts. The hydrolysis usually involves high temperature and pressure. It is important to find the most suitable, effective, and affordable method to first fractionate biomass major chemical components and achieve the practical applications of the method. There are some unwanted substances and oligosaccharides of various degree of polymerization (DP) produced during the manufacture of XOS and xylose. These substances should be removed to obtain xylose and XOS with high purity. Some purification methods such as solvent extraction, adsorption separation, chromatographic separation, and membrane filtration, or combinations of those methods could be applied. Xylose can be utilized for a variety of purposes, either directly as xylose or as a feedstock for the subsequent production of a variety of products, including furfural, furfuryl alcohol, xylitol, levulinic acid, ethanol, butanol, and hydrogen through chemical or biological conversion. XOS can be used as antioxidant, prebiotic, gelling agent, cosmetics, plant

E. Hermiati (✉) · D. A. Pramasari
Research Center for Biomass and Bioproducts, National Research and Innovation Agency (BRIN), Jl. Raya Bogor Km 46, Cibinong, Bogor 16911, Indonesia
e-mail: euis.hermiati@brin.go.id

D. A. Pramasari
e-mail: dwi.ajias.pramasari@brin.go.id

E. Hermiati
Research Collaboration Center for Biomass and Biorefinery Between BRIN, Universitas Padjadjaran, Jatinangor, Sumedang 45363, Indonesia

H. Wijaya
Research Center for Applied Microbiology, National Research and Innovation Agency (BRIN), Jl. Raya Bogor Km 46, Cibinong, Bogor 16911, Indonesia
e-mail: hans.wijaya@brin.go.id

M. A. R. Lubis et al. (eds.), *Biomass Conversion and Sustainable Biorefinery*, Green Energy and Technology, https://doi.org/10.1007/978-981-99-7769-7_11

growth regulator, treatment of diabetes, arteriosclerosis, and colon cancer, and are commercially interesting to be used as animal feed, food, beverage, and pharmaceutical ingredients. The production of xylose and XOS from lignocellulosic biomass still has some challenges regarding the technology to produce the products that are feasible commercially, but it has good prospects in the future as the increasing awareness to use renewable resources to produce healthier and environmentally friendlier products.

Keywords Lignocellulosic biorefinery · Oligosaccharides · Separation · Value-added products · Xylan derivatives

1 Introduction

The depletion of fossil resources has triggered research on the use of renewable resources for producing fuels and other compounds that are now obtained from fossils. Lignocellulosic biomass is among potential feedstocks for the production of fuels and chemicals, since it is abundantly available as by-products of agriculture and forestry industries. Lignocellulosic biomass can be used for the production of cellulosic ethanol, which can substitute the use of gasoline. Unfortunately, the commercialization of cellulosic ethanol or the second generation of bioethanol is still hampered by the high production cost, which makes the price of cellulosic ethanol high and hard to compete with gasoline. Efforts to produce economical cellulosic ethanol have been conducted through the improvement of technology, starting from pretreatment up to fermentation. Another effort that could be done is by applying biorefinery concept. In biorefinery, the transformation of lignocellulosic into energy, especially cellulosic ethanol, and other chemicals or bioproducts is carried out as an integrated technology, so that each component of lignocellulosic biomass could be converted and utilized efficiently. By doing this, the whole process would be economically more feasible.

Lignocellulosic biomass is made up of three major components: cellulose, hemicellulose, and lignin. In the production of cellulosic ethanol, the focus is to obtain cellulose that will be converted to glucose, which will further fermented to ethanol. Lignin is a challenge and becomes a recalcitrance in the production of cellulosic ethanol, but it is beneficial for producing different kinds of aromatic compounds or monomers that could substitute fossil-derived monomers in polymer industries. Hemicellulose is easily hydrolyzed to its oligomers or monomers, for example through water, acid, alkali, or enzymatic hydrolysis, and then separated from lignin and cellulose. Unlike cellulose, which has orderly packed structure, hemicellulose has amorphous characteristics with different kinds of branches. After being recovered and purified from lignocellulosic biomass, hemicellulose monomers can undergo a variety of transformations that result in compounds with added value, hence increasing the profit margin for the relevant businesses. Before completely

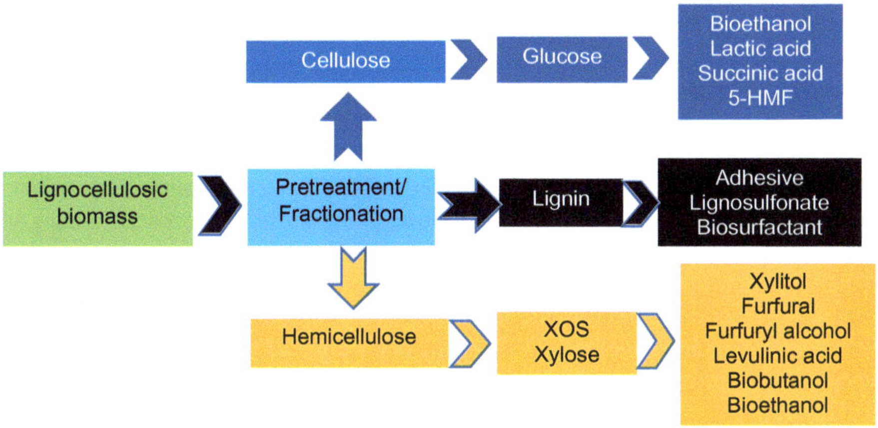

Fig. 1 Schematic diagram of lignocellulosic biomass fractionation

hydrolyzed to its monomers, the hemicellulose could also be hydrolyzed to oligosaccharides. As an integrated process, it is important to take advantage of every step in the lignocellulosic biomass conversion to ethanol, for example optimizing the pretreatment process to produce both cellulosic ethanol and sugar monomers or oligosaccharides. Schematic diagram of lignocellulosic biomass fractionation is shown in Fig. 1.

Xylose and XOS become products of interest and have good markets. According to Data Bridge Market Research, during the projection period of 2022–2029, the global xylose market is projected to increase at a CAGR of 5.4% from its value of USD 1,797.80 million in 2021 to USD 2,738.21 million by 2029. The price of XOS is dependent on its purity (70–95%), ranging from USD 25 to 50/kg, and the global market is anticipated to reach USD 130 million by 2025 (Santibáñez et al. 2021; Singh et al. 2018; Brenelli et al. 2022). The global XOS market was worth USD 88.09 million in 2016 and is anticipated to reach USD 119.62 million by the end of 2022, expanding at a compound yearly growth rate (CAGR) of 5.23% from 2016 to 2022 (Ahmad 2019).

There are still some challenges in the recovery of xylose or XOS from lignocellulosic biomass, either in the extraction, isolation, or purification step for producing good quality of xylose and XOS which is economically feasible. However, a new paradigm shift in consumer preferences for heather and natural products as well as consumer health consciousness have significantly increased the sales and attraction for xylose and XOS, especially in the thriving food and beverage industry. In this chapter highlights will be on the xylan sources, structure, and its derivatives; extraction and isolation of xylose and XOS, purification of xylose and XOS, and potential applications of xylose and XOS in food, energy, and pharmaceutical industries.

2 Sources and Structure of Xylans, XOS, and Xylose

Hemicellulose is the base feedstock for xylose and other xylan-based sugars, particularly xylooligosaccharides (XOS), which are supposed to be utilized in the packaging, food, pharmaceutical, biomedical, cosmetic, textile, and papermaking industries (Yohana et al. 2022; Lu et al. 2021; Zhang et al. 2017a). Hemicellulose is a complex heteropolysaccharide that consists of D-xylose, L-arabinose, D-glucose, L-galactose, D-mannose, D-glucuronic acid, and D-galacturonic acid. Common composition of these sugars in hemicellulose is shown in Fig. 2. Hemicelluloses are found in several polymeric configurations, including xylan, arabinoxylan, xyloglucan, and glucuronoxylan, and develop in 26% of the dry weight of hardwoods, 22% of softwoods, and approximately 25% of agricultural waste (Zhao et al. 2020; Narisetty et al. 2022).

A dominant component in hemicellulose of lignocellulosic biomass is usually xylan. Xylan is also the second major component in lignocellulosic biomass after glucan. The content of glucan and xylan in several lignocellulosic biomass is presented in Table 1. Terrestrial plants xylans are heteropolymers having a β-(1 → 4)-d-xylopyranose backbone and branches formed of short carbohydrate chains, such as D-glucuronic acid or its 4-O-methyl ether, L-arabinose and/or different oligosaccharides, made up of D-xylose, L-arabinose, D- or L-galactose and D-glucose (Ebringerová et al. 2005; Scheller and Ulvskov 2010). Based on its primary structure in plant tissue, xylan is generally divided into homoxylans and heteroxylans, including glucuronoxylan (GX), arabinoglucuronoxylan (AGX), glucurono-arabinoxylan (GAX), arabinoxylan (AX), and complex heteroxylan (CHX) (Ebringerová et al. 2005), with structures shown in Fig. 3a–d. When hydrolyzed, the xylans are converted to XOS and xylose, depending on the severity of the process. Production of xylose and xylooligosaccharides (XOS) from hemicellulose fraction of lignocellulosic biomass would be beneficial to support the commercialization of cellulosic ethanol.

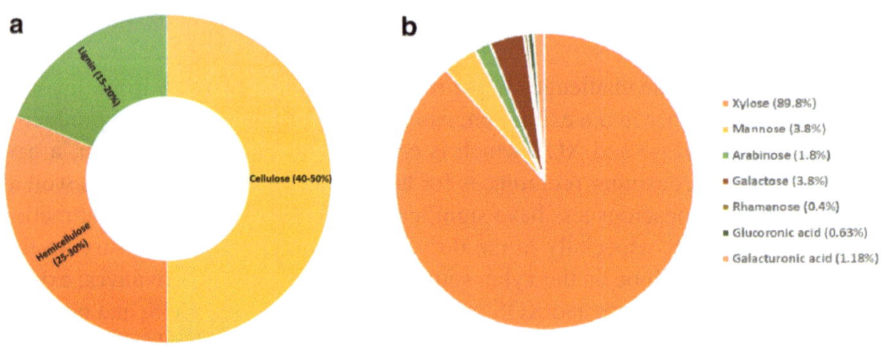

Fig. 2 General composition of lignocellulosic biomass (**a**), and neutral and acid sugars in hemicellulose (**b**) (Narisetty et al. 2022)

Table 1 Glucan and xylan content of lignocellulosic biomass

Biomass	Glucan (%)	Xylan (%)	References
Sugarcane straw	38.9	23.9	Brenelli et al. (2022)
Sugarcane bagasse	43.49	22.68	Zhang et al. (2020a)
Corn stover	34.4	22.8	Kumar et al. (2009)
Corn cob	37.9	27.83	Lee and Jeffries (2011)
Sweet sorghum bagasse	37.1	18.2	Wen et al. (2018)
Oil palm empty fruit bunch	36.8	21.5	Ho et al. (2014)
Oil palm frond	40.56	17.40	Goh et al. 2010)
Rice husk	36.83	19.66	Wu et al. (2018)
Rice straw	38.66	22.93	Wu et al. 2018)
Wheat straw	36.3	19.0	Kootstra et al. (2009)
Sweetgum	40.2	15.7	Geng et al. 2019)
Switchgrass	35.5	22.4	Geng et al. (2019)
Maple	43.2	13.0	Geng et al. (2019)
Eucalyptus	44.4	14.3	Geng et al. (2019)
Poplar	43.8	14.8	Kumar et al. (2009)
Aspen	52.4	14.9	Jun et al. (2012)
Loblolly pine	36	7.5	Rana et al. (2012)

XOS ($C_{5n}H_{8n} + {}_2O_{4n+1}$) are mixture of oligosaccharides composed of xylose molecules linked through β-1, 4-xylosidic bonds with a number of xylose residues vary between 2 and 10 (Aachary and Prapulla 2008), some have branching structures due to the presence of many side groups (Chen et al. 2021; Yan et al. 2022). As we know besides xylose, xylan usually contains α-D-glucopyranosyl uronic acid or its 4-O-methyl derivative, acetyl groups, or arabinofuranosyl residues. These structures result in some branched XOS containing these side groups which have various biological properties (Aachary and Prapulla 2011). Figure 3e shows the chemical structure of XOS. The structural properties of XOS are affected not only by the origin of the xylan-rich hemicelluloses but also by the manufacturing process (Akpinar et al. 2010; Bian et al. 2013). These variables influence the degree of polymerization (DP), monomeric units, and linkage types (Aachary and Prapulla 2011).

Xylose or D-xylopyranose (D-Xylp) is an aldopentose sugar, having chemical formula $C_5H_{10}O_5$ and molecular weight of 150.13, with a relative density of 1.525 and melting point 145–150 °C. soluble in water, hot ethanol, and pyridine, but not soluble in ether. It appears as colorless or white crystal or powder, and it has a sweet taste. Xylose has an aldehyde functional group so that it is classified as reducing sugar. The chemical structure of xylose can be seen in Fig. 3f.

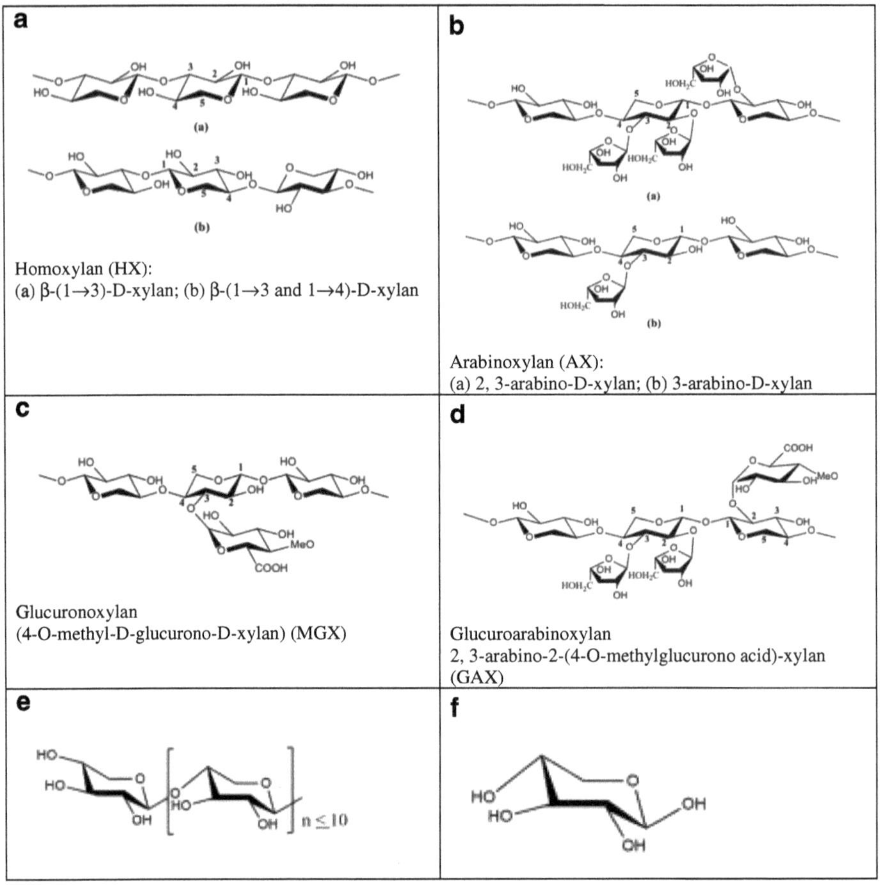

Fig. 3 Structure of xylans. **a** Homoxylan. **b** Arabinoxylan; **c** Glucuronoxylan; **d** Glucuroarabinoxylan), **e** XOS and **f** xylose (Fu et al. 2019; Pinales-Márquez et al. 2021)

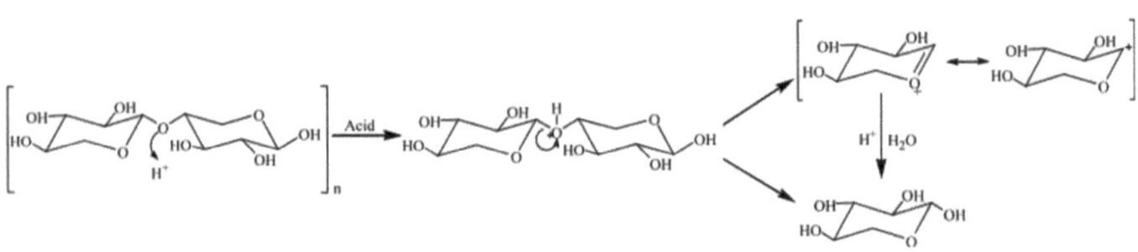

Fig. 4 Acid hydrolysis of xylan to xylose (Lu et al. 2021)

3 Extraction and Isolation of Xylose and Xylooligosaccharides

There are different extraction methods or fractionation processes used to extract xylose and XOS syrup from an abundance of lignocellulosic biomass feedstocks. In order to produce xylose and XOS efficiently, finding the most suitable, effective, and affordable method to first fractionate biomass' major chemical components and achieve the practical applications of biomass is critical. According to Fernandez (2019), many existing methods for isolating xylose start with biomass hydrolyzates or sugar in solution derived from a renewable resource and involve energy-intensive processes such as evaporation and phase transitions. This is distinct from typical sugar production, which may be quickly harvested from sugar cane or sugar beets.

3.1 Acid Hydrolysis Method

A popular method for hydrolyzing hemicellulose with a significant conversion efficiency is using diluted acid hydrolysis. Hemicellulose breakdown in the solid phase can be catalyzed by dilute acid and heat to dissolve various oligosaccharides ranging from xylose to xylotriose and above in the liquid phase. Inorganic acids, including sulfuric, hydrochloric, and nitric acids, as well as organic acids, such as oxalic, phosphoric, acetic, maleic, succinic, and citric acids can be used in the conventional diluted acid hydrolysis. Thus, by introducing acids from outside, the hemicellulose depolymerization process can be accelerated (Huang et al. 2021; Krishania et al. 2018; Paiva et al. 2009; Rahman et al. 2007; Rocha et al. 2014). In addition, Chen et al. (2020) described that the activation energy of organic acid degradation is usually higher than that of inorganic acids, but the degree of glucose and xylose dissolved in the hydrolysate with organic acids is less than that of inorganic acids, such as sulfuric acid.

The primary limitation of the acid hydrolysis approach is the production of unfavorable reactions at the same time. The acid hydrolysates usually contain three groups of undesired chemical components: (a) compounds derived from the hemicellulose structure (organic acids); (b) sugar breakdown derivatives (furfural and hydroxymethylfurfural (HMF)); and (c) excessive lignin derivatives (aromatic and polyaromatic compounds). To increase the xylose yield and decrease the creation of undesired compounds, the precise processing parameters, such as solid-to-liquid ratio, acid concentration, temperature, and duration of heating, must be followed (Delgado Arcaño et al. 2020; Harahap 2020; Shahbazi and Zhang 2010).

The two forms of acid employed in the acid-based process are concentrated acid and diluted acid. Concentrated acids not only dissolve cellulose crystals but also break down cellulose and hemicellulose into sugars and stimulate the cleavage of glycosidic bonds. In the United States, sulfuric acid converts hemicellulose and cellulose into sugars at a rate of 80–90%. However, there are risks related to the use of concentrated acids, as well as difficulties in acid recovery, which limit the acceptance of the technology. Concentrated acid can be produced efficiently at moderate temperatures, requiring less heating energy. Shahbazi and Zhang (2010) described that hemicellulose can be efficiently extracted and recovered as dissolved sugars at medium temperatures (140–190 °C), and the extraction process can be made more effective by adding an acid solution. On the contrary, dilute acid hydrolysis often requires greater temperatures with less chemicals. As a result, organic acid hydrolysis has steadily gained prominence and the issue caused by equipment corrosion can be prevented (Harahap 2020; Cheng et al. 2018; Inamuddin 2021). At temperatures below 160 °C, acid hydrolysis caused the hemicellulose to become nonhomogeneous and dissolve into soluble oligomers, such as XOS and sugar monomers, such as xylose. Furthermore, the oligomers formed during the hydrolysis of hemicellulose by a random acid attack have different degrees of polymerization (Lee et al. 1999). According to studies on the effects of conventional heating with organic and inorganic acids on the sugar monomers produced in the hydrolysate and from the saccharification and fermentation of the solid corncob residues, organic acids such as maleic acid and oxalic acid could produce more monomer sugars in the hydrolysates and more ethanol from the solid residuals (Lee and Jeffries 2011).

Some benefits of utilizing dilute acid include a faster reaction rate, lower acid consumption, lower cost than alkaline pretreatment, and no need for recycling. It is interesting to note that under particular process circumstances, diluted acid hydrolysis preferentially extracted the hemicellulose component (Harahap 2020; Cheng et al. 2018; Inamuddin 2021). The acid hydrolysis approach is a conventional method for dissolving hemicellulose to xylose or XOS. Unfortunately, numerous factors, such as chemical stability, side reactions, toxic pollution, and proton selectivity, must be addressed in the near future (Yan et al. 2022; Huang et al. 2021). The summary of some research regarding the xylose yield after acid hydrolysis method in various biomass is shown in Table 2.

According to earlier research by Zhang et al. (2017b), concentrated acetic acid was discovered to perform better than inorganic acid for hydrolysis. Hydronium ions disrupt the connections in the interiors and/or exteriors of hemicelluloses during acetic acid hydrolysis, which causes XOS to become soluble. The solid phase of

Table 2 The xylose yield after acid hydrolysis method in various biomass

Solvent (concentration)	Biomass	Xylose yield (%)	References
Oxalic acid (150 mmol/L)	Corncob	85	Cheng et al. (2018)
Oxalic acid (1.2%)	Corncob	96.1	Jin et al. (2018)
Hydrochloric acid (1%)	Corn stover	88.8	Feher et al. (2017)
Sulfuric acid (3.1%)	Sugarcane bagasse	96	Paiva et al. (2009)
Sulfuric acid (2%)	Oil palm empty fruit bunch	91.27	Rahman et al. (2007)
Sulfuric acid (6%)	Kenaf core fiber	86.50	Judiawan et al. (2019)

XOS has more exposed cellulose that is ready for cellulolytic hydrolysis due to its solubility. The majority of the severe inorganic acids have the ability to catalyze and dissolve the glycosidic linkages that connect xylose units in the skeleton of the xylan polymer, but their action is random. Consequently, much xylose rather than XOS is generated (Zhang et al. 2017b). Meanwhile, Yohana et al. (2022) discovered that the maximum XOS content from sugarcane trash and corncob was 4.74 g/L after hydrolysis with 2% oxalic acid for 45 min and 1.31 g/L after hydrolysis with 2% acetic acid for 60 min. Furthermore, according to Zhang et al. (2017a), acetic acid in corncob hydrolysate generated XOS yields of roughly 45.91%. The XOS yields from hydrochloric acid and sulfuric acid are 22.54% and 9.38%, respectively. It is also worth noting that the XOS derived from acetic acid prehydrolyzate was mostly made up of xylobiose, xylotriose, and xylotetrose, with substantially less xylopentaose and xylohexaose. This was attributed to the fact that acetic acid pretreatment resulted in a reduced degree of polymerization (DP) enrichment. Han et al. (2020) discovered that gluconic acid treated in corncob was effective to produce XOS with a yield 56.2%. Gluconic acid is a mono-carboxylic acid that could release H^+ to depolymerize hemicelluloses. It is adopted as a green and environmentally friendly solvent for the extraction of sugars.

3.2 *Alkaline Hydrolysis Method*

In complement to acid hydrolysis method, the alkaline method is frequently used to isolate hemicellulose from lignocellulosic biomass by causing cellulose to swell, rupturing the links between hemicellulose and lignin, and diluting the hemicellulose. It has been discovered that the yield of hemicellulose is affected by the type and concentration of alkali, extraction time, and temperature (Huang et al. 2021). Figure 5 illustrates the general mechanism for alkaline hydrolysis, in which the ester link between lignin's ferulic acid and the sugar residue of hemicellulose is cleaved (Lu et al. 2021).

The alkaline hydrolysis was the subject of groundbreaking work by Adams and Castagne (1951) as stated by Huang et al. (2021). The research was regarding the various hemicellulose fractions obtained from the holocellulose of wheat straw. The process also produced the D-xylose and monomethoxyl galacturonic acid complex, which was resistant to acid. Inorganic alkali solutions are extensively utilized in

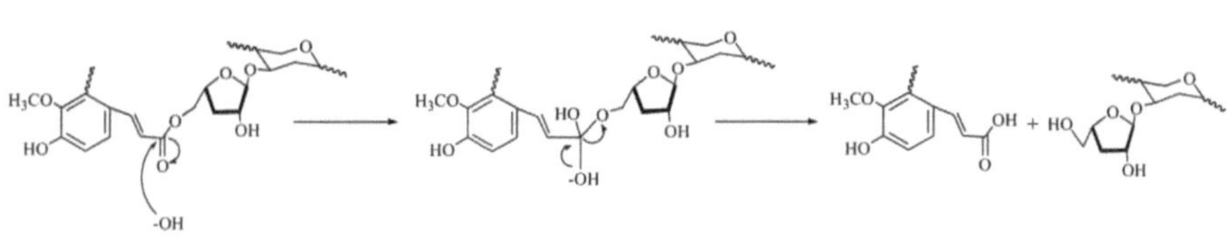

Fig. 5 Alkaline hydrolysis of xylan (Lu et al. 2021)

alkaline hydrolysis, and sodium hydroxide is one of the most prevalent. In addition, hydrogen peroxide extraction is a frequent alkaline procedure for isolating hemicellulose from biomass. High extraction rates, purity, and DP are features of hemicelluloses extracted through alkaline hydrolysis. As a result, alkaline-extracted hemicellulose is predicted to be used in a variety of industries. The standard alkaline procedure, on the other hand, involves significant pollution and costs (Lu et al. 2021).

Samanta et al. (2012) reported that sodium hydroxide was superior to potassium hydroxide in terms of xylan recovery from corncobs. Furthermore, the study discovered that increased alkali concentrations led to higher hemicellulose degradation. Besides that, a study by Jiang et al. (2019) using sorghum steam discovered that alkaline hydrolysis using 1.4% NaOH at 37.5 °C and 9.9 h resulted in a maximum xylose yield (57.7%). The study also stated that increasing alkali loading, reaction time, and temperature enhanced xylose yield. By using only alkali extraction, the yield of xylan is relatively low. Combination with other procedures, such as steam and ultrasound, might be rising the xylan yield. For instance, ultrasound-assisted alkali extraction could significantly increase XOS yield in corncobs, with a release level of 174.81 mg/g matrix (Yan et al. 2022; Kawee-ai et al. 2016).

3.3 Autohydrolysis Method

The autohydrolysis process requires specialized equipment at specific temperatures and pressures. The principle of autohydrolysis is that the hydronium ions are formed as a result of water autoionization, which results in the catalytic depolymerization of hemicelluloses to XOS and xylose. The addition of the hydronium ion to the mixture, which causes cleavage of acetyl groups in acetic acid, allows for the synthesis of XOS. When the processing severity factor is increased, the DP decreases, and the degradation of XOS into xylose increases. The downside of this procedure is the need for specialized equipment as well as the production of undesirable compounds in the XOS mixture (Samanta et al. 2015; Surek and Buyukkileci 2017).

Nabarlatz et al. (2007a) have used a variety of biomass, consisting of wheat straw, barley straw, rice husks, corncobs, olive stones, and almond shells, for XOS production using autohydrolysis. This study discovered that the XOS yield varied with xylan content and accessibility and was inversely related to the acetyl concentration in the biomass feedstock. According to Surek and Buyukkileci (2017), autohydrolysis of hazelnut shells produced XOS with the highest percentage of low-DP XOS at higher severity values. To achieve the appropriate DP, temperature and holding time must be precisely regulated. Additionally, the results indicated that the severity factor was effective in determining the patterns of oligomers, monomers, and by-products produced after autohydrolysis.

3.4 Combination Hydrolysis (Physical–Chemical Hydrolysis) Method

Physical hydrolysis includes processes using hydrothermal, steam explosion, ultrasonic, and microwave (Yan et al. 2022). A hydrothermal is an instant conversion of biomass in hot water with controlled high pressure and/or temperature, which involves the self-expansion of a material. In this process, the moisture in the biomass is vaporized under the impact of the rapid pressure transfer to a vacuum, which causes the material to expand and become more texturized. Sometimes the hydrothermal combines with alkaline hydrolysis or acid hydrolysis to separate hemicellulose to improve the yield of xylose or XOS (Lu et al. 2021; Yan et al. 2022). The research conducted by de Sá et al. (2020) found that the hydrothermal-assisted acid hydrolysis in sugarcane bagasse resulted in 80.13 mmol/L xylose. The XOS produced (35 g/L) by hydrothermal pretreatment of sugarcane bagasse with sulfuric acid was totally hydrolyzed by the acid post-hydrolysis, which happened in less than 20 min (Nakasu et al. 2017).

Harahap (2020) and Dulie et al. (2021) stated that the oligomeric form of hemicellulose predominates in the solution as a result of steam explosion circumstances. Hemicellulose oligomers need to be post-treated in order to be broken down. To maximize the performance of the steam explosion, pre-treatment is also required before the operation. The steam explosion pretreatment method has drawbacks, such as the destruction of a portion of the pentosan, despite being a cheap pretreatment method. A xylose yield of 27.58 g/g dry corn cob was obtained by Zhang et al. (2014) from the pilot-scale production of xylose using a screw-steam explosion extruder and prior sulfuric acid impregnation. On the other hand, Duangwang et al. (2016) discovered that a combination between sulfuric acid hydrolysis and superheated steam explosion in a pilot scale gave the highest yield of xylose, up to 87.58 g/kg of dry oil palm empty fruit bunches. The yield of XOS was 37%, obtained from acetic acid hydrolysis of wheat straw with steam explosion (Cao et al. 2021).

In the case of alkaline extraction of hemicellulose, the use of ultrasonic has been shown to be successful. The breaking of the polysaccharide glycoside bond brought on by ultrasonic pretreatment considerably increases the yield of the extracted hemicellulose at low temperature and short time. Additionally, hemicellulose extracted with the use of an ultrasonic method has a higher molecular weight, greater thermal stability, fewer acidic groups, fewer linked lignin molecules, and a lower branching degree (Lu et al. 2021). Sun and Tomkinson (2002) determined that the yield of hemicellulose from ultrasonic method in wheat straw hydrolysate was 1.8% higher than that of conventional alkaline hydrolysis.

Compared to conventional approaches, microwave extraction of hemicellulose has some benefits, such as energy savings and low toxicity. Additionally, the extraction of branched hemicellulose with microwave assistance is more successful, resulting in reduced yield and molecular weight of the hemicellulose recovered through microwave extraction. It has been challenging to implement in large-scale industrialization thus far (Lu et al. 2021). A study conducted by Lin et al. (2017) stated that a practical method for developing XOS production from various hemicellulose was by utilizing organic acids as catalysts and induced with microwave irradiation. Using this process, the XOS yield from beechwood xylan, corncob, and waste liquor pulp is 39.42, 27.46, and 30.89%, respectively. Another research by Hermiati et al. (2020) demonstrated that the use of maleic acid in combination with microwave assistance could result in a 24.3% yield of xylose. Meanwhile, microwave-assisted acid hydrolysis of rice husk could obtain 32.96% of xylose yield (Zhang et al. 2020b).

3.5 Solvent Hydrolysis Method

Organic solvents, such as pure organic solvent and complex organic solvent systems, can be used to separate hemicellulose. Lu et al. (2021) divided the solvent hydrolysis method into organosolv extraction, ionic liquid extraction, deep-eutectic solvent extraction, and high-pressure CO_2/H_2O technology. Organosolv extraction has the benefit of immediate hemicellulose extraction without delignification. The acetyl functional groups found in lignocellulosic biomass' cell walls are not transformed to acetic acid and then expelled. Nevertheless, organic solvents are poisonous, combustible, volatile, and difficult to decompose, which causes environmental problems (Lu et al. 2021). Liu et al. (2018) demonstrated that xylose and XOS concentration from ethanol-based auto-catalyzed organosolv method in bamboo can reach up to 15 g/L and 18 g/L, respectively.

The merits of ionic liquid (IL) extraction include excellent high-recovery effectiveness and toxic-free solvent recycling. Even though ILs are not always costly, the preparation process is frequently complex and more expensive than using conventional media. Furthermore, hemicellulose recovered by ILs frequently contains certain contaminants and is only mildly decomposed, which causes some side chains to partially break. As a result, ILs' continued commercial application is still challenging (Lu et al. 2021; Chen et al. 2014). The extraction of xylose by IL is affected by temperature and time. When temperature raised, the amount of xylan recovered in the solid fraction of corn stover is reduced, meaning that more xylan was dissolved into the hydrolysate (Xu et al. 2012). Xylose yield was also increased decreased when the temperature was increased from 50 to 110 °C, reaching maximum (75.9%) at 110 °C, but then decreased at higher temperature, implying that at higher temperature some of the dissolved xylan was degraded to lower molecular weight compound, such as furfural (Xu et al. 2012). At low temperature (70 °C), the xylose yield is increased when the duration of corn stover extraction using ionic liquid was increased from 3 to 24 h (Xu et al. 2012).

Deep-eutectic solvent (DES) extraction was an alternative solvent with 100% atomic economy compared to the previous common solvent for xylose and XOS extraction. The other advantages of DES are non-toxic, simple preparation, recyclable, low purity, and low extraction selectivity of hemicellulose (Lu et al. 2021). Research by Morais et al. (2018) shows that a yield of 14.81% was obtained when the optimized aqueous DES solutions were used to extract the hemicelluloses from

Eucalyptus globulus wood. These results are higher than the yields resulting from water or alkaline hydrolysis.

A growing number of studies have focused on high-pressure fluids as potential solvent substitutes. In light of green chemistry principles, CO_2 and H_2O are the most promising high-pressure fluids since they are renewable and inflammable and give high xylose yield. As much as 100 mol% of xylose conversion was obtained using high-pressure CO_2 in H_2O/tetrahydrofuran system (Morais and Bogel-Lukasik 2016). Furthermore, the high-pressure CO_2/H_2O method does not require an additional catalyst, in contrast to conventional hydrothermal processes. Despite its similarity comparable to the pretreatment catalyzed by a mild acid, the depressurization process for removal of CO_2 ensures that the medium acidity does not provide a risk to the environment. Unfortunately, under conditions of high pressure and temperature, a significant number of by-products are often produced (Lu et al. 2021; Morais and Bogel-Lukasik 2016).

3.6 Inorganic Salts Hydrolysis Method

Inorganic salts, such as NaCl, KCl, $CaCl_2$, $MgCl_2$, $FeSO_4$, $FeCl_3$, and other inorganic salts, do not include C–H bonds and contain environmentally favorable substances. The inorganic salt solution is mostly used to dissolve lignocellulosic biomass. The inclusion of salts can help accelerate lignocellulose hydrolysis and remove more hemicellulose. As a result, this can lower the reagent distribution cost. The researcher discovered that the addition of NaCl allowed the reaction to proceed more quickly and produce the highest amount of xylose (90%) with few secondary reaction products (Harahap 2020; Jiang et al. 2018).

3.7 Enzymatic Hydrolysis Method

Recent studies have shown that the potential of enzymatic hydrolysis is greater than that of chemical hydrolysis because of its many benefits. Since managing enzymes is simpler than handling acids, enzymatic hydrolysis is an eco-friendly process. Furthermore, it is not necessary to use expensive corrosive equipment. Enzymes create just one product from substrates because they are highly specific and selective (Inamuddin 2021).

In order to reduce xylose formation in the hydrolyzates when using enzymatic hydrolysis for XOS production, it is notable that the enzyme preparations should have decreased exoxylanase activity. The essential enzyme for generating XOS product from xylan are β-xylosidase, glycosynthases and endo- xylanases. In addition, the most popular technique for extracting XOS using xylanase is enzymatic hydrolysis because of the favorable circumstances and superior results. The extraction of XOS from xylan varies depending on the xylan source, the activity of the enzyme, and the incubation conditions. The compact lignin-hemicellulose structure may have a significant impact on its extraction from lignocellulose materials. In order to reveal more hemicelluloses and demonstrate the extraction yield, it is crucial to disrupt the compact structure. Hence, simultaneous enzymatic hydrolysis in the hemicellulose could be only used on non-recalcitrant biomass (Chen et al. 2021; Yan et al. 2022; Samanta et al. 2015).

Currently, *Aspergillus* xylanases are the most common ones utilized in XOS manufacturing. Immobilizing xylanase can increase its efficiency and reduce its cost, making it a viable option for subsequent production. Because of its beneficial impact, selectivity, and specificity, adjustable degree of hydrolysis, high yield and purity of XOS, and fewer by-products, enzymatic hydrolysis currently becomes the preferred method for producing XOS in industrial applications (Yan et al. 2022).

4 Purification of Xylose and Xylooligosaccharides

4.1 Purification of Xylooligosaccharides

Unwanted substances and oligosaccharides from a variety of DP are produced after the manufacture of XOS (Gullon et al. 2008). This unwanted substances like glucose and xylose raise the caloric value and alter the sweetening capacity of XOS and also impurities, such as furfural and HMF, have very little toxicity or adverse effects on human health in XOS with DP ranges up to 12 (Vazquez et al. 2000). Downstream processing is necessary to eliminate undesirable components and achieve the desired product (XOS of DP between 2 and 6) (Aachary and Prapulla 2011). Therefore, the next primary step in the creation of food-grade XOS is to purify the principal product by isolating the advantageous high-molecular-weight oligosaccharides from the unfavorable low-molecular-weight sugars and unwanted compounds. The crude liquors have been refined using a variety of methods to get rid of unwanted compounds and concentrate XOS with a yield from 75 to 95% as much as possible to achieve the required DP (Aachary and Prapulla 2011; Vazquez et al. 2000; Moure et al. 2006). The purification procedures used and the number of necessary stages have a direct impact on production costs and, thus, on the Minimum Product Selling Price (MPSP). For instance, purifying galacto-oligosaccharide (GOS) using activated carbon and celite for protein and color removal, ion exchange, and simulated moving bed (SMB) chromatography, known as AO-CH process, increases the MPSP linearly from 2 to 9 USD/kg as the final purity increases from 40% (unpurified) to up to 90% (Illanes et al. 2016). Understanding the above, numerous purification approaches, such as solvent extraction, adsorption using surface-active compounds, as well as chromatographic and membrane separation, have been investigated in recent decades.

4.1.1 Solvent Extraction Method in XOS Purification

Solvent extraction is often used to extract hemicellulose-based compounds from pretreatment and treatment steps in the production of XOS (Qing et al. 2013). In the first stage, vacuum evaporation is used to eliminate volatile components, concentrate XOS solutions, and removal of possible explosives (Vazquez et al. 2000; Qing et al. 2013). The solvent extraction method is useful for removing non-saccharide parts, resulting in a particularly purified liquid phase and also a solvent component containing phenolics as well as extractive-derived residues (Moure et al. 2006). The yield and degree of purification, however, are dependent on the solvent used for extraction and the kind of lignocellulosic biomass used. Organic solvent precipitation, on the other hand, allows for the extraction of XOS and/or xylose by eliminating phenolics and extractive-derived chemicals. The most often used solvents for this process are acetone, ethanol, and 2-propanol (Qing et al. 2013). Other procedures of varying nature have frequently been used, such as the work of Vegas et al. (2006) can reach 90.7% XOS purity, by applying a combination of nanofiltration, solvent extraction, and double ion-exchange chromatography.

4.1.2 Adsorption Separation Method in XOS Purification

Adsorption is a technique that has been used to separate XOS or to eliminate undesirable chemicals. The most often utilized adsorbents, such as diatomaceous earth, acid clay, titanium, bentonite, activated charcoal, aluminum oxide, silica, and other synthetic materials, are commonly used in combination with other refining procedures (Vazquez et al. 2000; Xu et al. 2019). Activated charcoal was found to be the most widely used adsorbent and also has been demonstrated to be a feasible solution for the elimination of substances derived from extractives, lignin, and carbohydrate-degradation present in XOS mixtures (Nabarlatz et al. 2007b). The XOS were initially held by activated charcoal in the study by Pellerin et al. (1991), and then were eluted with various concentrations of ethanol. A similar technique was used by Reddy and Krishnan (2014) to extract XOS generated by *B. subtilis* utilizing activated charcoal. Activated charcoal (10% w/w) was added to the culture extract, and it was retained by vacuum filtering. After being adsorbed onto charcoal, XOS was subsequently eluted with serial percentage of ethanol, resulting in XOS separation depending on molecular weight. Ion-exchange resins are used in conjunction with other purification methods to eliminate heavy metal ions, salts, negatively or positively charged organic molecules, and pigments from XOS solutions (Vazquez et al. 2000; Chen et al. 2016). Chen et al. (2016) successfully produce high XOS recovery (91.3%) and also show that the oligomer composition is not altered by the ion-exchange resin treatment.

4.1.3 Chromatographic Separation

This method utilizes high-performance liquid chromatography (HPLC), ion-exchange, affinity, and size-exclusion chromatographic separation techniques to obtain highly purified XOS at the analytical level (Geetha and Gunasekaran 2017). In the purification of XOS, chromatographic approaches such as adsorption in activated charcoal column chromatography and gel filtration chromatography (GFC) have been studied (Ho et al. 2014). Ho et al. (2014) have purified XOS that produced by agro-residue autohydrolysis using GFC, which was demonstrated to be effective in the elimination of oligosaccharides with high DP as well as undesirable small compounds, such as monosaccharides, acetic acid, and degradation chemicals (furfural, HMF and phenolics). This method also indicates a competitive advantage over membrane purification (Moniz et al. 2014). Despite the fact that the reported final XOS purities were greater than 70%, the product exhibited a wide DP range (5–40 and 3–23, respectively). As a result, if XOS having DP 2–6 is considered as the desired product, the purity was greatly overestimated. Chapla et al. (2012), on the other hand, had employed activated charcoal column chromatography for the purification of XOS generated from enzymatic saccharification of pre-extracted corncob xylan. The refined XOS preparation consists mostly of xylobiose and xylotriose, and the purity level was not specified. Study by Yang et al. (2007) had successfully isolated XOS from xylan enzymatic hydrolysis using a packed charcoal column followed by ethanol elution. The authors used this process to produce XOS mixture of 71.4% purity, mostly constituted of xylobiose, xylotriose, and xylotetrose, with a 95% XOS recovery.

SMB is a method of chromatographic separation that improves productivity of the separation and purity of products. Furthermore, this approach lowered solvent consumption, provided simple operational controls, and increased the performance of the separation in systems with limited selectivity and resolution (Li et al. 2020). In a continuous-separation mode, Choi et al. (2016) was able to recover 92.3% xylobiose from XOS with a purity of 99.5%, which will be beneficial in allowing large-scale, cost-effective manufacture of high-purity xylobiose.

4.1.4 Purification Through Membrane Filtration

Prebiotic purification to remove proteins and polysaccharides followed by ion exchange chromatography was employed in the work by Broekaert et al. (2011), but this process is not economical for industrial application. Membrane technology is currently regarded as the best downstream approach for industrially producing high-purity XOS. Membrane technology is a viable technique for concentrating and refining XOS since it has a high recovery rate, uses little energy, and is a straightforward process that does not need other compounds as a solvent.

Membrane technology techniques, such as ultrafiltration (UF) and nanofiltration (NF) seem to be the favored technologies for refining XOS due to their excellent recovery rate with no solvent (Nabarlatz et al. 2007b). The XOS generated by autohydrolysis of almond shells was successfully purified by employing commercial thin-film polymeric membranes (Nabarlatz et al. 2007b). The findings show that lignin-related low molar mass compounds are not rejected by low MWCO (1 kDa) polymeric membranes, enabling continuous diafiltration to remove impurities and recognize as an appropriate technique to purify XOS solutions produced by the autohydrolysis of lignocellulosic biomass. However, in the case of XOS, UF appears to be a pre-treatment alternative rather than a method for creating a highly pure product containing short-chain oligosaccharides (Cordova et al. 2019).

A combination of UF and NF was employed to purify XOS produced from the sequential autohydrolysis and enzymatic saccharification of almond shells to increase the elimination of tiny pollutants (Singh et al. 2019a, 2021). In the NF step, the XOS mixture was concentrated and monosaccharides and acetic acids were removed, while in the UF step components having high molecular weight, such as enzymes, were removed, with results similar to those obtained by Akpinar et al. (2007) and Kumar and Satyanarayana (2015).

Membrane that separates higher-molecular-weight molecules from oligosaccharides requires less energy and is simple to operate and scale-up (Czermak et al. 2004). However, membrane technology alone cannot perform the best technique to purify XOS, because membrane has also a drawback due to its poor performance when small molecules such as monosaccharides should be separated. The hydrolysis of previously extracted hemicellulose and autohydrolysis from lignocellulosic biomass produce XOS, which is then purified using UF membrane.

Monosaccharides have the most major impurities that have less difference in molecular weight (MW) such as glucose (MW = 180 g/mol), which difficult to separate between xylose (MW = 150 g/mol) and xylobiose (MW = 282 g/mol) (Mah et al. 2019). The membrane technology mostly does not have a very small cut-off molecular weight difference. For that reason, the study by Wijaya et al. (Wijaya et al. 2020), has performed a strategy of elimination glucose by the enzymatic reaction that does not convert cellulose to glucose before the process enters membrane separation unit. This process is a combination of alkali pretreatment and enzymatic reaction by xylanase free from cellulase (Wijaya et al. 2020). However, the research for a suitable way to purify XOS from hemicellulosic hydrolysate has not yet been resolved. Table 3 presents different techniques for recovering XOS using recent purification methods that have been published.

4.2 *Purification of Xylose*

The procedures used to convert lignocellulose resources into ethanol include polysaccharide hydrolysis to xylose and glucose, fermentation, and ethanol purification. Among these steps, acid hydrolysis is the most important in generating inhibitors, such as acetic acid, which restrict the fermentation processes in ethanol production (Mussatto and Roberto 2004). More particular, the existence of acetate can

Table 3 Techniques for recovering xylooligosaccharides (XOS) using recent purification methods have been published

Purifcation techniques	Substrate	Product	Temperature and pH of XOS production	Recovery % (DP*)	References
Anion-exchange and size exclusion chromatography	Birchwood xylan	Acidic XOS	50 °C	85 (NA)	Christakopoulos et al. (2003)
Activated carbon and ion exchange	*Miscanthus giganteus*	X2, X3, X4, X5 and X6	Autohydrolysis	91.3 (>2)	Chen et al. (2016)
Activated charcoal column chromatography	Corncob	X2 and X3	45 °C, pH 5.3	80 (NA)	Chapla et al. (2012)
Activated charcoal column chromatography	Bagasse, corncob, wheat bran, and peanut shell	XOS	60 °C, pH 7.0	95 (NA)	Yang et al. (2007)
Gel filtration chromatography	Oil palm empty fruit bunch	XOS	NA	83–85 (5–40)	Ho et al. (2014)
Gel filtration chromatography	Rice straw	XOS	Autohydrolysis	80–90 (3–54)	Moniz et al. (2014)
High-performance anion exchange	Chemical pulp	Xylohexose, Xylobiose and xylotriose	50 °C, pH 5	47 & 90.5 (NA)	Wang et al. (2018)
High-performance liquid chromatography	Corncob	XOS	50 °C, pH 8	32.5 (≤4)	Lin et al. (2011)
Nanofiltration, solvent extraction, and double ion-exchange chromatography	Rice Husks	XOS	Autohydrolysis	90.7 (NA)	Vegas et al. (2006)
Nanofiltration	Corncob meal	Xylobiose and xylotriose	55 °C, pH 5.5	74.5 (<5)	Yuan et al. (2004)
Nanofiltration	Oil palm empty fruit bunch	Xylobiose	50 °C, pH 5	90.1 (NA)	Wijaya et al. (2020)
Nanofiltration	Leaves and green tops of sugarcane plant	XOS	190 °C, pH 7	46.0, purity of 20.9 (NA)	Oliveira et al. (2022)

(continued)

Table 3 (continued)

Purifcation techniques	Substrate	Product	Temperature and pH of XOS production	Recovery % (DP*)	References
Simulated moving bed	Powder XOS	Xylobiose and XOS	NA	>92, purity of >99 of X2 form XOS (NA)	Choi et al. (2016)
Ultrafiltration	Almond shells	XOS	179 °C	58.3 (NA)	Nabarlatz et al. (2007b)
Ultrafiltration	Wheat bran	XOS	60 °C, pH 6.5	44.4 (2–5)	Geetha and Gunasekaran (2017)
Ultrafiltration, nanofiltration and ion-exchange resin	Almond shells	Xylobiose and xylotriose	50 °C, pH5.5	69.6 (<5)	Singh et al. (2021)

DP* Degree of polymerization, NA Not available

inhibit yeast fermentation development and metabolism by lowering intracellular pH (Lohmeier-vogel et al. 1998). There are several ways for removing acetic acid from hydrolysate. The detoxification procedures have been thoroughly evaluated in some studies (Mussatto and Roberto 2004; Parajo et al. 1998; Huang et al. 2008). Some of the treatments investigated for removing acetic acid include extraction, neutralization, over-liming, vacuum evaporation, steam stripping, charcoal adsorption, and ion exchange resins adsorption. There have been many membrane processes developed recently, including adsorptive membrane and membrane extraction (Han et al. 2006; Grzenia et al. 2008).

The next step, which is the separation of xylose from glucose, can be challenging. Desal-5 DK NF membrane with molecular weight cut-off 150–300 Da was used to remove acetic acid from xylose (Weng et al. 2009). Separation between xylose and glucose by tailored thin-film composite (TFC) nanofiltration membrane has been studied as well (Mah et al. 2019). Other strategy involves an enzymatic reaction as studied by Morthensen et al. (2015). They had done an enzymatic method for converting glucose to gluconic acid, followed by nanofiltration to separate xylose from gluconic acid. The method for separation of xylose from glucose using concentrated monosaccharide solutions also needs consideration (Sjoman et al. 2007).

Older methods, such as liquid chromatography (LC) by ionic liquid-modified silica can also be considered (Bi et al. 2010). The chromatographic separation method can use a cation exchange resin (Chen et al. 2018). In the future, the purification of xylose can use a combined strategy to achieve a product either ethanol or xylose as purified sugar.

5 Potential Applications of Xylose and Xylooligosaccharides

5.1 Potential Applications of Xylose

Xylose can be used in a variety of applications, either as is or as a feedstock for further chemical or biological conversion to a variety of products such as xylitol, furfural, furfuryl alcohol, levulinic acid, levulinic ester, ethanol, butanol, and hydrogen. Table 4 shows value-added products that could be created from xylose through chemical and biological catalytic processes and various applications of the products generated in food, healthcare, feed, personal care, cosmetics, pharmaceuticals, medical, materials, etc. Some important products will be discussed in more detail in the following paragraphs. The conversion of xylose through chemical process produces lower yields, and the employment of acidic catalysts and reaction operation at higher temperatures and pressures renders the process ecologically unfriendly (Narisetty et al. 2022). Due to the high cost of manufacturing and environmental incompatibility, the long-term viability of many commercial chemical processes, including xylitol production, is questionable (Narisetty et al. 2022).

5.1.1 Food and Healthcare

Xylose is usually used as a sweetener that has a high market value (Fernandez 2019). The relative sweetness of xylose is about 70% of sucrose sweetness. This natural sweetener has a low calorie, is easily absorbed into small intestine, and is not metabolized by the liver (Fernandez 2019). Therefore, it is beneficial for people with obesity, diabetes or other illnesses that should limit sugar consumption. Other health benefits of xylose are its ability to activate and promote the growth of beneficial bacteria in the human gut, such as *Bifidobacterium* and *Lactobacillus*, to prevent dental caries, to serve as dietary fiber that decreases blood fat, lowers cholesterol, and prevent colon cancer, and to be compatible with food (Hongzhang et al. 2012).

Xylose is chemically relatively inert and stable, which contributes to the strong demand for food items, particularly confectionaries, as it is easy to combine into other food products without altering the recipe. Xylose is a good flavor enhancer by inducing Maillard reaction compounds, such as amino acids and peptides in meats, fish, seafood, milk, beans, or peas. The Maillard reaction is a significant nonenzymatic process that occurs when amino compounds, such as amino acids or peptides, combine with carbonyl molecules, typically a reducing sugar like xylose or glucose (Newton et al. 2012; Sun et al. 2019; Martins et al. 2001). Maillard reaction products (MRPs) containing a variety of volatile compounds have a significant impact on food flavor and quality. D-xylose is better than glucose or sucrose in remedying the color, flavor, and fragrant of the food. The addition of 0.05–2% of D-xylose is enough to enhance the flavor of the food. The usage of xylose provides advantages

Table 4 Chemicals from xylose and their applications (Narisetty et al. 2022)

Chemicals	Applications
Chemical process	
Xylaric acid	Sequestering agents
Xylitol	Health, food pharmaceutical
Furfural	Resins, fuels, adhesives, polymers
Furfuryl alcohol	Resins, rocker propellant, adhesives, polymers
2-Methylfuran	Gasoline additive, drugs manufacturing
Levulinic acid	Precursor for fuels and chemicals
Levulinic ester	Fragrance, fuels, fuel additives
Biological process	
Ethanol	Solvent, automotive gasoline, alcohol beverages, distilled spirits, hand sanitizers, and medical antiseptics
Acetic acid	Polymeric monomers, paints, adhesives, inks, coatings, and food additives
Acetone	Plastics, cosmetics, and solvents
Lactic acid	Food, beverages, polyesters, textiles, and pharmaceuticals
Glycerol	Pharmaceuticals, food, polymers, humectant, solvents, lubricants, personal care, and household products
2,3-Butanediol	Polymers, solvents, fine chemicals, lactones, fuel additives
Succinic acid	Pharmaceutical products, surfactants, detergents, plastics, and food grade ingredients
Butanol	Lubricants, brake fluids, synthetic rubber, polymers, and automotive fuels
Isobutanol	Coatings, chemical derivatives, paints, fuel additive, and solvents
Xylitol	Confectionary, chewing gums, syrups, and odontological and pharmaceutical products
Polyhydroxybutyrate (PHB)	Biodegradable plastics

to the flavor industry that uses the Maillard reaction for the creation of thermally produced flavorings (Gaspar et al. 2012).

5.1.2 Pharmaceutical and Medical

Xylose has been used in an absorption test to evaluate small intestinal absorption since 1937 because the method is simple and reproducible. The method involves an oral dosage of xylose and its serum and urine determination after 5 h collection period. Upon ingestion of xylose, the body is expected to use 75% of its content, and the 25% remaining is excreted through the urine within 5 h (Gaspar et al. 2012). Due to the fact that xylose is absorbed unmodified by the duodenum and jejunum, its inadequate absorption makes it suitable for use as a malabsorption test (Craig and Atkinson 1988). The study confirms that the intake of 25 g D-xylose followed by a 5 h urine collection and a 1 h serum analysis is a very sensitive and specific diagnostic for malabsorption.

Earlier study by Cook (1975) shows that in healthy Africans, gamma-globulin serum levels were significantly inversely linked to D-xylose urine excretion. This indicates that when the D-xylose content in the bloodstream increases or less D-xylose is excreted in the urine, the concentration of gamma-globulin serum increases dramatically, hence enhancing the immunological response. Besides that, many other studies also show that xylose has anti-inflammatory, antiviral, antiglycemic, and anticancer properties for lung cancer (Cheudjeu 2020). These induce the idea of using xylose as an alternative therapeutic regimen for a severe case of Covid-19 (Cheudjeu 2020). Nevertheless, this hypothesis should be proved by pre-clinical and clinical trials. Xylose has a unique and useful texture-enhancing properties, which are needed in some products, such as cosmetics and personal care products.

5.1.3 Feedstock for Production of Fuels and Chemicals

The major purpose of xylose as an industrial raw material is for producing xylitol. Xylitol is a platform chemical, having five-carbon sugar alcohol that are widely used in food, confectionary, personal care, and pharmaceutical industry. The sweetness of xylitol is the same as that of sucrose, but it has lower calorific value and glycemic index (2.4 cal/g; 7%) than sucrose (4 cal/g; 60–70%). The metabolism of xylitol is insulin-independent. Xylitol has a high endothermic heat, thus as it dissolves in the mouth, a pleasant feeling of cooling and freshness is immediately noticed. These characteristics mask the unpleasant flavor of numerous pharmacological active components and excipients. Asia–Pacific, Europe, and the United States account for the majority of xylitol production. The process of producing xylitol is expensive, which prevents its production from expanding to other continents.

Currently, the method of producing xylitol from pure D-xylose on an industrial scale is a chemical process at high temperature and pressure (180 °C, 50 atm) using aluminum or nickel catalyst for hydrogenation of the xylose (Rafiqul and Sakinah 2013; Carneiro et al. 2019). This process consumes high energy and generates catalyst residues which causes environmental problem. The amount of xylitol recovered is around 50–60% of the xylan or 8–15% of the initial feedstock. The following procedures to obtain pure xylitol involve extensive separation and purification processes which are also time-consuming, energy-intensive, and costly, causing high price of the product (Saha 2003). Alternative process using bioprocess has been studied and reported. The bioprocess which usually involves xylose-fermenting yeasts to convert xylose into xylitol has gained attention, because it is low cost, operates at moderate reaction conditions, is more sustainable, and environmentally friendlier (Oktaviani et al. 2021). The yeast could be directly added to the hemicellulose hydrolysate that contains xylose, followed by the separation of xylitol from the fermentation broth using different methods, and purification of the xylitol that has been separated.

Lactic acid, also known as 2-hydroxyproponoic acid, is an optically active compound that comes in L and D forms. Lactic acid, as a platform chemical, has numerous industrial applications in food, cosmetics, polymers, and packaging. The most valuable application of lactic acid as a monomer is the production of poly-lactic acid (PLA), a commercial petrochemical polymer substitute (Narisetty et al. 2022). *Lactobacillus pentosus* and *L. brevis* naturally can ferment xylose via the pentose phosphate and phosphoketolase routes, creating lactic acid and a combination of acetic and lactic acids, respectively (Qiu et al. 2018). Other microbes that have been reported could produce lactic acid from xylose are *Pediococcus acidilactici* (Narisetty et al. 2022; Qiu et al. 2018, 2017), engineered *Escherichia coli*, *Lactobacillus delbrueckii*, *Lactobacillus lactis*, *Lactobacillus plantarum*, *Bacillus* sp., *Bacillus coagulans*, *Corynebacteria glutamicum*, and *Rhizopus oryzae* (Abedi et al. 2020).

Succinic acid is an aliphatic dicarboxylic acid with four carbon atoms that is used extensively as a precursor in the pharmaceutical, polymer, and chemical industries. Similar to lactic acid, succinic acid is a platform chemical that, due to the presence of two carboxyl acid groups, can be transformed into a range of compounds including succinic anhydride, succinic esters, 2-pyrrolidine, and polyesters for the synthesis of biodegradable plastics (Dai et al. 2020).

Ethanol/ethyl-alcohol/bioethanol is the most frequently used biofuel in the transportation industry and offers a number of benefits, including higher octane number, greater combustion efficiency, and better heat of vaporization. Bioethanol is less hazardous, rapidly biodegradable, and emits fewer airborne pollutants than petroleum fuel and other viable fuel options. Due to its hygroscopic nature, however, ethanol cannot completely replace gasoline because water vapor might damage the engine (Narisetty et al. 2022). Currently, ethanol is mixed with gasoline at varying percentages (5–20%) over the world. The combined fuels reduce emissions of hydrocarbons and greenhouse gases significantly. The generation of ethanol from xylose follows the production of X5P via the pentose phosphate route and continues via the EMP pathway (Narisetty et al. 2022). The end product of the EMP route, pyruvate, is transformed to ethanol via the intermediate acetaldehyde. Unfortunately, the common ethanol-fermenting yeast, *Saccharomyces cerevisiae*, cannot assimilate the xylose. There are some other yeasts, such as *Pachysolen tannophylus* and *Pichia stipitis*, that can use glucose and xylose to produce ethanol. However, their activities are restricted due to substrate and product-mediated inhibition. Genetic engineering of *S. cerevisiae* is needed to make the yeast assimilate both glucose and xylose (Kwak and Jin 2017). Even though xylose could be converted to ethanol, it might be more advantageous if it is converted to other chemicals that have better values than ethanol.

n-Butanol is a four-carbon alcohol with a straight chain and is regarded as a superior biofuel to ethanol due to its higher heating value, ignition issues, lower volatility, high octan number, low miscibility with water, and higher viscosity (Vivek et al. 2019). n-Butanol could be produced via chemical or biological route. The chemical process of producing n-butanol is through aldol condensation (oxo process), which consists of hydroformylation and hydrogenation of propylene. The biological process for n-butanol production is a part of acetone–butanol–ethanol (ABE) fermentation using microorganisms, such as *Clostridium* spp., well-known cell factories with ABE fermentation (Chacon et al. 2020; Gottumukkala et al. 2013). However, the manufacture of bio-butanol is hindered by low concentration and yield as well as product-mediated inhibition. Optimization of solid loading has been performed to increase the ABE fermentation and reveals that the ideal biomass loading of 10% combined with the liquid hot water (LHW) pretreatment resulted in a sustainable route for a successful sugarcane straw ABE fermentation (Pratto et al. 2020). In addition, the integrated process of hydrolysis and fermentation eliminates problems of substrate inhibition. Despite the fact that the yield of n-butanol is 20% lower than that of ethanol, the energy produced by n-butanol is 32% greater than that of ethanol (Baral and Shah 2016). With the current concentration and yield, the cost of producing biobutanol is approximately USD 1.8/L. However, further optimization of the biocatalysts and process conditions could reduce the cost to USD 0.6/L, which is equivalent to that of gasoline and other fossil fuels (Liu et al. 2016).

Hydrogen is often regarded as one of the most potentially useful alternative energies because when combined with oxygen, it produces only water and has a 2.75 times higher energy yield (122 kJ/g) than hydrocarbons (Liang et al. 2012). Fermentation of sugars to produce hydrogen is usually conducted using thermophilic anaerobic bacteria, thus, it is usually called dark fermentation. Hydrogen-producing microbes can utilize both hexoses and pentoses, while some others prefer xylose over glucose or vice versa. *Caldicellulosiruptor* members have adapted to xylose and disaccharides (cellobiose) as principal substrates rather than glucose, which may provide them a competitive advantage in thermophilic lignocellulosic breakdown (Vongkampang et al. 2021). *C. saccharolyticus*, *C. owensensis*, and *C. kristjanssonii* consume xylose faster than glucose (Zeidan and Niel 2009). In comparison to glucose, *C. kronotskyensis* prefers xylose and cellobiose the most (Vongkampang et al. 2021). Co-culture of microbes has been shown could improve hydrogen production. The hydrogen yield from the co-culture of *C. saccharolyticus* and *C. kristjanssonii* came close to the theoretical maximum stoichiometry and was much higher than that of the monoculture of either organism, suggesting the two bacteria synergistic effects on hydrogen generation (Zeidan and Niel 2009). Genetically engineered microbes could also increase hydrogen production during fermentation. Higher hydrogen yield and productivity were observed in the *Thermoanaerobacterium aotearoense* SCUT27/nfnAB mutant relative to the original strain (Li et al. 2019a).

Besides microbes, enzyme can also be used in the hydrogen production from xylose. First, the cellulose and hemicellulose of the biomass were converted to glucose and xylose, which serve as substrates for phosphorylation and hydrogen generation using enzyme mixtures. The process is continued with separation and purification of the hydrogen. The enzymatic production of hydrogen from biomass hydrolysate in aqueous solution has some benefits over the production of other biofuels because the process can use lower concentration of sugars and the product separation is not expensive (Lastname et al. 2015). Hydrogen gas could be easily separated from an aqueous sugar solution regardless of its concentration. In the case of ethanol, for energy-efficient distillation of ethanol, its concentration in the fermentation broth should be 4% minimum, which means about 9% sugars in the hydrolysates.

Furfural is an important bio-based platform chemical derived from C5 sugars, including xylose. Currently, it is industrially produced from biomass, such as corn cob or sugarcane bagasse through acid hydrolysis of hemicellulose, especially xylan, in the biomass, and followed with dehydration of the sugar monomers, especially xylose (Zhang et al. 2017a, b). The dehydration of xylose to furfural is usually conducted through chemical processes, without or with different kinds of catalysts, which is explained in quite detail and comprehensively by Delbecq et al. (2018). Furfural could be applied in wide variety of applications, such as mentioned in Table 3. Based on its physical or chemical characteristics, furfural is categorized as a hazardous substance and listed under numerous globally determined classifications of hazardous materials (Delbecq et al. 2018).

Furfuryl alcohol ($C_4H_3OCH_2OH$, 2-furylmethanol, 2 furancarbinol) is produced through hydrogenation of furfural, which is previously derived from xylose through dehydration. Both steps usually utilize chemical catalysts. Chemical production of furfuryl alcohol can give high yields and good selectivity, however, it consumes high-energy and creates environmental issues that might limit its application. The conversion of furfural to furfuryl alcohol has been conducted through biological methods by using some bacteria, such as *Bacillus coagulans*, *Bacillus cereus*, and *Escherichia coli* that have been genetically engineered. This method has gained attraction due to its energy-saving and environmentally friendly performance (Liang et al. 2022). Perez et al. (2017) introduced a one-step conversion of xylose to furfuryl alcohol via a sulfated zirconia-supported Pt catalyst with balanced acid and metal sites. In a one-step conversion of xylose to furfuryl alcohol, the reaction takes place on vicinal acid-metal pare sites formed on single multifunctional catalysts, whereas in a two-step process furfuryl alcohol is generated through consecutive dehydration hydrogenation reactions on independent acid and metal sites on dual catalytic systems. Furfuryl alcohol is widely used in the production of synthetic fiber, rubber resin, furan resin, vitamin C, lubricant, lysine, and hypergolic rocket fuel (Liang et al. 2022; Millán and Sixta 2020).

Levulinic acid, sometimes referred to as 4-oxopentanoic acid or gamma ketovaleric acid, is a C-5 fatty acid with the chemical formula $C_5H_8O_3$ and one of the key platform chemicals (Morone et al. 2015). Levulinic acid has a ketone carbonyl group (C=O) and an acidic carboxyl group (COOH), which endows it with the capacity to react with various functional groups to generate a vast array of derivatives, making it a good platform chemical (Morone et al. 2015; Bozell et al. 2000). Due to the presence of highly reactive carbonyl and carboxyl groups, levulinic acid can be converted into a variety of high-value-added chemicals and other bioproducts (Elumalai et al. 2016; Galletti et al. 2012; Rackemann et al. 2011). Most studies focus on the generation of levulinic acid from starch or cellulose, or through C6 sugars, and fewer studies are using C5 sugars, such as xylose. Furfural, which was produced from C5 sugars, such as xylose, could be transformed into furfuryl alcohol (FA) through a transfer-hydrogenation (TH), and subsequently, into levulinic acid through a hydrolytic ring opening reaction (Lange et al. 2009; Chamnankid et al. 2014). While levulinic acid has been successfully produced in the lab, its commercialization has been hampered by a number of factors. These include the high cost of raw materials and equipment, the low yield of levulinic acid due to undesirable side reactions, the difficulties in efficient product recovery, the high cost of recovering the catalyst, the inefficiency of the process economy, and the high cost of energy (Morone et al. 2015).

Polyhydroxybutyrate (PHB), is a biodegradable polymer that is made up of 3-hydroxybutyrate and can reach a degree of polymerization of 2000. PHB is a member of the polyhydroxyalkanoate group, that can be used as an alternative to plastics that are made from petrochemicals (Li et al. 2019b; Lee et al. 2021). The physical and mechanical properties of PHB such as optical purity, a high melting point, crystallinity, and desirable water and gas barrier properties, are similar and comparable to those of commercial polymers or plastics derived from fossils (Saratale et al. 2019). Unfortunately, the cost of producing PHB (USD 2.25–2.75/lb) is three to four times that of conventional plastics (USD 0.60–0.87/lb), limiting its use in the industrial and commercial sectors (Saratale et al. 2019; Singh et al. 2019b). The high price of PHB is due to the low productivity, content, and yield as well as the high price of carbon substrate (Singh et al. 2019c). PHB is produced by numerous microbes, including *Ralstonia*, *Halomonas*, and *Bacillus* species, that consume sugar monomers, including glucose, fructose, xylose, arabinose, etc. However, there are only a small number of strains capable of producing PHB from xylose, the second abundant and inexpensive carbon source in lignocellulosic biomass (Lee et al. 2021). *R. eutropha*, the strain with the highest PHB-producing capacity, cannot readily absorb and metabolize xylose (Lee et al. 2021; Bhatia et al. 2018; Lopes et al. 2009). Some efforts to increase the yield of PHB have been reported, for example by screening for xylose-utilizing microbes, co-culturing of the microbes (Lee et al. 2021), or optimizing sugar compositions during fermentation (Li et al. 2019b). Some other solutions to produce more economical PHB are co-production with other value-added products, design of an appropriate bioreactor system, and production of PHB using synthetic biology (Singh et al. 2019c).

5.2 *Potential Applications of XOS*

XOS has been reported used as antioxidant, prebiotic, plant growth regulator, cosmetics, gelling agent, and for the treatment of diabetes, arteriosclerosis, and colon cancer (Ahmad 2019). XOS is commercially appealing for use as animal feed, food, beverage, and pharmaceutical ingredients (Pinales-Márquez et al. 2021). Furthermore, XOS is utilized in the manufacture of micro or nanoparticles and hydrogels for medication administration and therapies, particularly for the prevention of gastrointestinal problems (Shimoda et al. 2011; Gupta et al. 2016). The relative sweetness of xylobiose and XOS is 0.34 and 0.26, respectively (Park et al. 2017).

5.2.1 Antioxidant

Antioxidant is a compound that can delay or prevent oxidation of a substrate by scavenging or neutralizing free radicals. The free radical scavenging capability of XOS is a result of the efficient release of total phenolic compounds and the transfer of the hydrogen atom from the phenolic compounds (Huang et al. 2005; Gowdhaman and Ponnusami 2015). The precise mechanism of action of XOS in reducing the negative effects of oxidative stress has yet to be established by scientific evidence (Samanta et al. 2015). The scavenging capacity of an antioxidant is determined using various methods, for example, 2,2-diphenyl-1-picrylhydrazyl (DPPH) assay, beta carotene bleaching assay, 2,2A-azinobis(3-ethylbenzthiazoline)-6-sulfonic acid (ABTS) assay, and ferric reducing antioxidant power (FRAP) assay (Antolovich et al. 2002).

The antioxidant activity of XOS depends on several factors, such as types of feedstocks or sources of XOS, methods of XOS extraction, concentration and degree of polymerization of XOS, and total phenolic contents in XOS (Bian et al. 2013; Samanta et al. 2015; Gowdhaman and Ponnusami 2015). A DPPH assay demonstrates that the antioxidant activity of XOS derived from sugarcane bagasse is dose-dependent (Bian et al. 2013). The antioxidant activity of the XOS-rich fraction recovered from wood using ethyl acetate extraction ($EC50 = 0.39$ g/l) was equivalent to 61% of that presented by butylated hydroxy anisole (BHA) and 12 times higher than that of butylated hydroxy toluene (BHT), and the fraction contained up to 0.43 g of gallic acid equivalents per 100 g of dry wood (González et al. 2004). The EC50 (0.23 g/l) was similar to that of BHA when the acid hydrolysis was performed under circumstances that led to extracts with maximum antioxidant-specific activity (González et al. 2004). The varied concentrations of XOS from corn cob were shown to have a range of scavenging abilities between 9.7 and 74.2% (Gowdhaman and Ponnusami 2015). The antioxidant activity of XOS was shown to have an IC50 value of 1 mg/ml, which is comparable to earlier results on the antioxidant activity of XOS extracted from maize and sugarcane bagasse (Aachary and Prapulla 2008; Gowdhaman and Ponnusami 2015). Study by Rashad et al. (2016) reveals that the XOS derived from a variety of agricultural wastes demonstrated concentration-dependent antioxidant activity. This finding is consistent with prior findings regarding XOS derived from wheat, millet brans, and sugarcane bagasse (Bian et al. 2013; Veenashri and Muralikrishna 2011; Rivas et al. 2013). The overall antioxidant capacity of XOS mixes is heavily influenced by both the total phenolic acid concentration and the type of those phenolic acids (Veenashri and Muralikrishna 2011). The existence of a higher quantity of total phenolic acid contents in XOS combinations formed from orange peels and mango peels compared to other XOS mixtures may account for their considerably stronger antioxidant activity (Rashad et al. 2016).

5.2.2 Prebiotics

A prebiotic is "a nondigestible food element that promotes the health of the host by selectively encouraging the growth and/or activity of one or a restricted number of bacteria in the colon" (Gibson and Roberfroid 1995). A more precise definition of prebiotic is a fermented element that induces certain changes in the composition and/or activity of the microbiota of the gastrointestinal tract and offers health benefits (Aachary and Prapulla 2011; Gibson et al. 2004). The current daily dose of XOS that is recommended to generate a prebiotic effect is low, which allows them to compete favorably in terms of pricing. In order to achieve an essential regulatory status and boost their commercial value, however, additional in vivo experimental data is required (Amorim et al. 2019).

Several researchers have documented the prebiotic effects of XOS. The XOS streams utilized as carbon sources for the in vitro fermentability evaluation with human fecal inoculum exhibited prebiotic effects by stimulating the formation of lactate, formate, and SCFA (acetate, propionate, and butyrate), as well as alterations in the selected bacterial populations (*Bifidobacterium* genus, *Lactobacillus*, *Enterococcus* group and *Bacteroides*, *Prevotella* group) (Buruiana et al. 2017). It appears that XOS promotes gut health by selectively stimulating the growth of bifidobacteria and lactobacilli. Additionally, XOS reduces the number of potentially pathogenic organisms (Chen et al. 2021). In vitro fermentation of the putative probiotic *L. brevis* strain validated prebiotic property of XOS-mixture derived from wheat (Faryar et al. 2015).

5.2.3 Food Processing

According to Deutschmann and Dekker (2012), XOS is resistance to high temperature up to 100 °C and has a good adaptability in large range of pH (2.5–8). XOS has a better resistance to low pH (<4) and high temperature (>90 °C) than other oligosaccharides and inulin, which is beneficial in food processing point of view and makes the XOS can be used in carbonated drinks, low-pH juices, and acidic foods (Aachary and Prapulla 2011). Current study also shows that XOS is stable under high pressure processing (100–600 MPa) combined with heat treatment (100 °C) (Silva et al. 2022). In food industry, XOS can also be used as a low-calorie sweetener or feedstock for producing low-calorie sweeteners, such as xylitol. XOS has been reported as flavor enhancer in formulating a beverage (Gupta et al. 2016). The addition of XOS in a non-alcoholic carbonated drink has a positive impact with other sweeteners, such as mixture of ace sulfame K and aspartame (Gupta et al. 2016). The addition of XOS in beverages significantly enhances full body character of the beverages without any drawback of off-flavor perception or mouth feel (Gupta et al. 2016, 2012).

5.2.4 Other Biological Functions

XOS has a wide range of biological activities, including anti-inflammatory, antioxidant, antitumor, and antimicrobial properties (Chen et al. 2021). Experiments in rats show that XOS reduces concentration of sugars and lipids in blood of diabetic rats, improves calcium absorption in rats, enhances phagocytic activity of neutrophils in mice, and improves gastrointestinal health of rats (Liang et al. 2012). XOS is non-cariogenic, preserve pancreatic insulin production, and promote intestinal mineral absorption (Moure et al. 2006). In human health, XOS has been reported to reduce risk of colon cancer and have a cytotoxic effect on human leukimia cells (Liang et al. 2012).

6 Concluding Remarks

Lignocellulosic biomass is a potential feedstock because it is widely available, and relatively inexpensive. It is very beneficial for producing high-value-added products, including fuels and chemicals, that can be used in various fields and applications. The production of xylose and XOS from lignocellulosic biomass has become interesting as a part of lignocellulosic biorefinery, which could promote economic viability of the biorefinery. Despite significant progress in xylan-based bioproduction over the last few decades, many challenges remain to be overcome before xylan can be used as a feedstock at the industrial level. The first challenge comes from the feedstock itself, which usually varies or not uniform, the scattered existence, and sometimes has problem with continuous supply. In the process and technology sector, it is important to select the appropriate pretreatment or fractionation of biomass which could be applied in a wide variety of biomass, and subsequently facilitate the extraction, isolation, and purification of xylose and XOS from the biomass using different kinds of techniques available. The choice of extraction, isolation, and purification should be feasible technologically and economically. For example, the process should be able to recover more xylose or XOS, minimize the production of undesirable products, and retain more cellulose in solid residue. Xylose and XOS could be used as they are, or to be further transformed to different kinds of chemicals that could be used for fuels and various applications in food, health care, pharmaceutical, cosmetics, and personal care. The challenge in this transformation is to find a robust, practical, inexpensive, efficient, and environmentally friendly process, to generate products of economic value. Standardized methods of process and analysis of the products are also important to obtain consistent quality and standardized bioproducts.

References

Aachary AA, Prapulla SG (2008) Corncob-Induced endo-1, 4-D-Xylanase of Aspergillus oryzae MTCC 5154: production and characterization of Xylobiose from Glucuronoxylan. J Agric Food Chem 56:3981–3988. https://doi.org/10.1021/jf073430i

Aachary AA, Prapulla SG (2011) Xylooligosaccharides (XOS) as an emerging prebiotic: microbial synthesis, utilization, structural characterization, bioactive properties, and applications. Compr Rev Food Sci Food Saf 10:2–16. https://doi.org/10.1111/j.1541-4337.2010.00135.x

Abedi E, Mohammad S, Hashemi B (2020) Heliyon Lactic acid production—Producing microorganisms and substrates sources-state of art. Heliyon 6:e04974. https://doi.org/10.1016/j.heliyon.2020.e04974

Adams G, Castagne A (1951) Hemicelluloses of wheat straw. Can J Chem 29:109–122

Ahmad R (2019) Better health with plant oligosaccharides: trends and future. Second Int Conf Food Agric 601–607

Akpinar O, Ak O, Kavas A et al (2007) Enzymatic production of xylooligosaccharides from cotton stalks. J Agric Food Chem 55:5544–5551. https://doi.org/10.1021/jf063580d

Akpinar O, Gunay K, Yilmaz Y et al (2010) Enzymatic processing and antioxidant activity of agricultural waste autohydrolysis liquors. BioResources 5:699–711

Amorim C, Silvério SC, Gonçalves RFS et al (2019) Downscale fermentation for xylooligosaccharides production by recombinant Bacillus subtilis 3610. Carbohydr Polym 205:176–183. https://doi.org/10.1016/j.carbpol.2018.09.088

Antolovich M, Prenzler PD, Patsalides E et al (2002) Methods for testing antioxidant activity. Analyst 127:183–198. https://doi.org/10.1039/b009171p

Baral NR, Shah A (2016) Techno-economic analysis of cellulosic butanol production from corn stover through acetone-butanol-ethanol (ABE) fermentation techno-economic analysis of cellulosic butanol production from corn stover through. Energy Fuels 30:5779–5790. https://doi.org/10.1021/acs.energyfuels.6b00819

Bhatia SK, Gurav R, Choi T et al (2018) Bioconversion of plant biomass hydrolysate into bioplastic (polyhydroxyalkanoates) using Ralstonia eutropha 5119. Bioresour Technol 271:306–315. https://doi.org/10.1016/j.biortech.2018.09.122

Bi W, Zhou J, Row KH (2010) Analytica chimica acta separation of xylose and glucose on different silica-confined ionic liquid stationary phases. Anal Chim Acta 677:162–168. https://doi.org/10.1016/j.aca.2010.08.004

Bian J, Peng F, Peng XP et al (2013) Structural features and antioxidant activity of xylooligosaccharides enzymatically produced from sugarcane bagasse. Bioresour Technol 127:236–241. https://doi.org/10.1016/j.biortech.2012.09.112

Bozell JJ, Moens L, Elliott DC et al (2000) Production of levulinic acid and use as a platform chemical for derived products. Resour Conserv Recycl 28:227–239

Brenelli LB, Bhatia R, Djajadi DT et al (2022) Xylo-oligosaccharides, fermentable sugars, and bioenergy production from sugarcane straw using steam explosion pretreatment at pilot-scale. Bioresour Technol 357:127093. https://doi.org/10.1016/j.biortech.2022.127093

Broekaert W, Courtin C, Delcour J (2011) (12) Patent Application Publication (10) Pub. No.: US 2011/0020498A1. 1

Buruiana CT, Gómez B, Vizireanu C, Garrote G (2017) Manufacture and evaluation of xylooligosaccharides from corn stover as emerging prebiotic candidates for human health. Lwt 77:449–459. https://doi.org/10.1016/j.lwt.2016.11.083

Cao R, Liu X, Guo J, Xu Y (2021) Comparison of various organic acids for xylo-oligosaccharide productions in terms of pKa values and combined severity. Biotechnol Biofuels 14:1–8. https://doi.org/10.1186/s13068-021-01919-9

Carneiro CVGC, de Silva FCP, Almeida JRM (2019) Xylitol production: identification and comparison of new producing yeasts. Microorganisms 7:1–15. https://doi.org/10.3390/microorganisms7110484

Chacon SJ, Matias G, Ferreira C et al (2020) Enabling butanol production from crude sugarcane bagasse hemicellulose hydrolysate by batch-feeding it into molasses fermentation. Ind Crop Prod 155:112837. https://doi.org/10.1016/j.indcrop.2020.112837

Chamnankid B, Ratanatawanate C, Faungnawakij K (2014) Conversion of xylose to levulinic acid over modified acid functions of alkaline-treated zeolite Y in hot-compressed water. Chem Eng J 258:341–347. https://doi.org/10.1016/j.cej.2014.07.036

Chapla D, Pandit P, Shah A (2012) Bioresource technology production of xylooligosaccharides from corncob xylan by fungal xylanase and their utilization by probiotics. Bioresour Technol 115:215–221. https://doi.org/10.1016/j.biortech.2011.10.083

Chen L, Sharifzadeh M, Mac Dowell N et al (2014) Inexpensive ionic liquids: [HSO4] based solvent production at bulk scale. Green Chem 16:3098–3106. https://doi.org/10.1039/c4gc00016a

Chen MH, Bowman MJ, Cotta MA et al (2016) Miscanthus × giganteus xylooligosaccharides: purification and fermentation. Carbohydr Polym 140:96–103. https://doi.org/10.1016/j.carbpol.2015.12.052

Chen X, Li H, Sun S et al (2018) Co-production of oligosaccharides and fermentable sugar from wheat straw by hydrothermal pretreatment combined with alkaline ethanol extraction. Ind Crops Prod 111:78–85. https://doi.org/10.1016/j.indcrop.2017.10.014

Chen Y, Yan Z, Liang L et al (2020) Comparative evaluation of organic acid pretreatment of eucalyptus for kraft dissolving pulp production. Materials (basel) 13. https://doi.org/10.3390/ma13020361

Chen Y, Xie Y, Ajuwon KM et al (2021) Xylo-Oligosaccharides, preparation and application to human and animal health: a review. Front Nutr 8:1–10. https://doi.org/10.3389/fnut.2021.731930

Cheng B, Zhang X, Lin Q et al (2018) A new approach to recycle oxalic acid during lignocellulose pretreatment for xylose production. Biotechnol Biofuels 11:1–9. https://doi.org/10.1186/s13068-018-1325-3

Cheudjeu A (2020) Since January 2020 Elsevier has created a COVID-19 resource centre with free information in English and Mandarin on the novel coronavirus COVID-19. The COVID-19 resource centre is hosted on Elsevier Connect, the company's public news and information

Choi J, Park H, Park C et al (2016) Highly efficient recovery of xylobiose from xylooligosaccharides using a simulated moving bed method. J Chromatogr A 1465:143–154. https://doi.org/10.1016/j.chroma.2016.08.063

Christakopoulos P, Katapodis P, Kalogeris E et al (2003) Antimicrobial activity of acidic xylo-oligosaccharides produced by family 10 and 11 endoxylanases 31:171–175

Cook G (1975) Relation between malaria, serum gamma globulinconcentration and the D-xylose absorption test. Trans R Soc Trop Med Hyg 69:143–145

Cordova A, Astudillo C, Illanes A (2019) Membrane technology for the purification of enzymatically produced oligosaccharides. In: Galankis CM (ed) Separation of functional molecules in food by membrane technology, 1st edn. Elsevier Inc., pp 113–153

Craig RM, Atkinson AJ (1988) D-xylose testing: a review. Gastroenterology 95:223–231. https://doi.org/10.1016/0016-5085(88)90318-6

Czermak P, Ebrahimi M, Grau K et al (2004) Membrane-assisted enzymatic production of galactosyl-oligosaccharides from lactose in a continuous process, 232:85–91. https://doi.org/10.1016/j.memsci.2003.11.024

Dai Z, Guo F, Zhang S, Key S (2020) Bio-based succinic acid: an overview of strain development, substrate utilization, and downstream purification. Biofuels, Bioprod Biorefining 14:965–985. https://doi.org/10.1002/bbb.2063

de Oliveira MG, Forte MBS, Franco TT (2022) A serial membrane-based process for fractionation of xylooligosaccharides from sugarcane straw hydrolysate. Sep Purif Technol 278:119285. https://doi.org/10.1016/j.seppur.2021.119285

de Sá LRV, Faber M de O, da Silva ASA et al (2020) Biohydrogen production using xylose or xylooligosaccharides derived from sugarcane bagasse obtained by hydrothermal and acid pretreatments. Renew Energy 146:2408–2415. https://doi.org/10.1016/j.renene.2019.08.089

De Paiva JE, Maldonade IR, Scamparini ARP (2009) Xylose production from sugarcane bagasse by surface response methodology. Rev Bras Eng Agric e Ambient 13:75–80. https://doi.org/10.1590/S1415-43662009000100011

Delbecq F, Wang Y, Muralidhara A et al (2018) Hydrolysis of hemicellulose and derivatives-a review of recent advances in the production of furfural. Front Chem 6. https://doi.org/10.3389/fchem.2018.00146

Delgado Arcaño Y, Valmaña García OD, Mandelli D et al (2020) Xylitol: a review on the progress and challenges of its production by chemical route. Catal Today 344:2–14. https://doi.org/10.1016/j.cattod.2018.07.060

Deutschmann R, Dekker RFH (2012) From plant biomass to bio-based chemicals: latest developments in xylan research. Biotechnol Adv 30:1627–1640. https://doi.org/10.1016/j.biotechadv.2012.07.001

Duangwang S, Ruengpeerakul T, Cheirsilp B et al (2016) Pilot-scale steam explosion for xylose production from oil palm empty fruit bunches and the use of xylose for ethanol production. Bioresour Technol 203:252–258. https://doi.org/10.1016/j.biortech.2015.12.065

Dulie NW, Woldeyes B, Demsash HD, Jabasingh AS (2021) An insight into the valorization of hemicellulose fraction of biomass into furfural: catalytic conversion and product separation. Waste Biomass Valoriz 12:531–552. https://doi.org/10.1007/s12649-020-00946-1

Ebringerová A, Hromadkova Z, Heinze T (2005) Hemicellulose. Adv Polym Sci 1–67. https://doi.org/10.1007/b136816

Elumalai S, Agarwal B, Sangwan RS (2016) Thermo-chemical pretreatment of rice straw for further processing for levulinic acid production. Bioresour Technol 218:232–246. https://doi.org/10.1016/j.biortech.2016.06.037

Faryar R, Linares-Pastén JA, Immerzeel P et al (2015) Production of prebiotic xylooligosaccharides from alkaline extracted wheat straw using the K80R-variant of a thermostable alkali-tolerant xylanase. Food Bioprod Process 93:1–10. https://doi.org/10.1016/j.fbp.2014.11.004

Feher A, Feher C, Rozbach M, Barta Z (2017) Combined approaches to Xylose production from corn Stover by dilute acid hydrolysis. Chem Biochem Eng Q 31:77–87. https://doi.org/10.15255/CABEQ.2016.913

Fernandez L (2019) Purification of biomass hydrolyzates for the promotion of xylose purification of biomass hydrolyzates for the promotion of xylose diester formation diester formation

Fu G-Q, Hu Y-J, Bian J et al (2019) Isolation, purification, and potential applications of Xylans

Galletti AMR, Antonetti C, De Luise V et al (2012) Levulinic acid production from waste biomass. BioResources 7:1824–1835

Gaspar EM, Nunes IS, Lopes JF (2012) Xylose as a dietary sugar: sources and analysis. In: Xu P, Luo M (eds) Xylose: production, consumption, and health benefits. Nova Science Publishers Inc., New York, pp 43–68

Geetha K, Gunasekaran P (2017) Purification of endoxylanase from bacillus pumilus B20 for production of prebiotic xylooligosaccharide syrup; an in vitro study. Iran J Biotechnol 15:232–240. https://doi.org/10.15171/ijb.1494

Geng W, Narron R, Jiang X et al (2019) The influence of lignin content and structure on hemicellulose alkaline extraction for non-wood and hardwood lignocellulosic biomass. Cellulose 26:3219–3230. https://doi.org/10.1007/s10570-019-02261-y

Gibson GR, Roberfroid MB (1995) Dietary modulation of the human colonie microbiota: introducing the concept of prebiotics. J Nutr 125:1401–1412

Gibson GR, Probert HM, Van LJ et al (2004) Dietary modulation of the human colonic microbiota: updating the concept of prebiotics. Nutr Res Rev 17:259–275. https://doi.org/10.1079/NRR200479

Goh CS, Tan HT, Lee KT, Mohamed AR (2010) Optimizing ethanolic hot compressed water (EHCW) cooking as a pretreatment to glucose recovery for the production of fuel ethanol from oil palm frond (OPF). Fuel Process Technol 91:1146–1151. https://doi.org/10.1016/j.fuproc.2010.03.029

González J, Cruz JM, Domínguez H, Parajó JC (2004) Production of antioxidants from Eucalyptus globulus wood by solvent extraction of hemicellulose hydrolysates. Food Chem 84:243–251. https://doi.org/10.1016/S0308-8146(03)00208-5

Gottumukkala LD, Parameswaran B, Valappil SK et al (2013) Biobutanol production from rice straw by a non acetone producing Clostridium sporogenes BE01. Bioresour Technol 1–6. https://doi.org/10.1016/j.biortech.2013.01.046

Gowdhaman D, Ponnusami V (2015) Production and optimization of xylooligosaccharides from corncob by Bacillus aerophilus KGJ2 xylanase and its antioxidant potential. Int J Biol Macromol 79:595–600. https://doi.org/10.1016/j.ijbiomac.2015.05.046

Grzenia DL, Schell DJ, Wickramasinghe SR (2008) Membrane extraction for removal of acetic acid from biomass hydrolysates 322:189–195. https://doi.org/10.1016/j.memsci.2008.05.030

Gullon P, Moura P, Esteves MP et al (2008) Assessment on the fermentability of xylooligosaccharides from rice husks by probiotic bacteria. J Agric Food Chem 56:7482–7487. https://doi.org/10.1021/jf800715b

Gupta P, Agarwal P, Hegde P (2012) A review on xylooligosaccharides. Int J Res Pharm 3:71–74

Gupta P, Agrawal P, Hegde P et al (2016) Xylooligosaccharide-A valuable material from waste to taste: a review. J Environ Res Dev 10

Han B, Carvalho W, Canilha L, Silverio S (2006) Adsorptive membranes versus resins for acetic acid removal from biomass hydrolysates 193:361–366. https://doi.org/10.1016/j.desal.2005.07.052

Han J, Cao R, Zhou X, Xu Y (2020) An integrated biorefinery process for adding values to corncob in co-production of xylooligosaccharides and glucose starting from pretreatment with gluconic acid. Bioresour Technol 307:123200. https://doi.org/10.1016/j.biortech.2020.123200

Harahap BM (2020) Degradation techniques of hemicellulose fraction from biomass feedstock for optimum xylose production: a review. J Keteknikan Pertan Trop dan Biosist 8:107–124. https://doi.org/10.21776/ub.jkptb.2020.008.02.01

Hermiati E, Oktaviani M, Ermawar RA et al (2020) Optimization of xylose production from sugarcane trash by microwave-maleic acid hydrolysis. Reaktor 20:81–88. https://doi.org/10.14710/reaktor.20.2.81-88

Ho AL, Carvalheiro F, Duarte LC et al (2014) Production and purification of xylooligosaccharides from oil palm empty fruit bunch fibre by a non-isothermal process. Bioresour Technol 152:526–529. https://doi.org/10.1016/j.biortech.2013.10.114

Hongzhang C, Qin H (2012) Biomass refinery—Xylose: production, consumption and health benefits. In: Xu P, Luo M (eds) Xylose: production, consumption, and health benefits. Nova Science Publishers Inc., New York, pp 95–122

Huang D, Ou B, Prior RL (2005) The chemistry behind antioxidant capacity assays 53:1841–1856. https://doi.org/10.1021/jf030723c

Huang H, Ramaswamy S, Tschirner UW, Ramarao BV (2008) A review of separation technologies in current and future biorefineries. Sep Purif Technol 62:1–21. https://doi.org/10.1016/j.seppur.2007.12.011

Huang LZ, Ma MG, Ji XX et al (2021) Recent developments and applications of hemicellulose from wheat straw: a review. Front Bioeng Biotechnol 9:1–14. https://doi.org/10.3389/fbioe.2021.690773

Illanes A, Guerrero C, Vera C et al (2016) Lactose-derived prebiotics: a process perspective. Elsevier Inc., Amsterdam

Inamuddin, Khan A (2021) Sustainable bioconversion of waste to value added products

Jiang Z, Budarin VL, Fan J et al (2018) Sodium chloride-assisted depolymerization of xylo-oligomers to xylose. ACS Sustain Chem Eng 6:4098–4104. https://doi.org/10.1021/acssuschemeng.7b04463

Jiang K, Ding S, Tang B (2019) Optimization of dilute NaOH pretreatment at mild temperatures for monomeric sugar release from sorghum pith using response surface methodology. BioResources 14:3411–3431. https://doi.org/10.15376/biores.14.2.3411-3431

Jin LQ, Zhao N, Liu ZQ et al (2018) Enhanced production of xylose from corncob hydrolysis with oxalic acid as catalyst. Bioprocess Biosyst Eng 41:57–64. https://doi.org/10.1007/s00449-017-1843-6

Judiawan W, Sudiyani Y, Nurnasari E (2019) Conversion of hemicellulose from kenaf core fiber to xylose through dilute sulfuric acid hydrolysis. J Kim Terap Indones 21:14–22. https://doi.org/10.14203/jkti.v21i1.412

Jun A, Tschirner UW, Tauer Z (2012) Hemicellulose extraction from aspen chips prior to kraft pulping utilizing kraft white liquor. Biomass Bioenerg 37:229–236. https://doi.org/10.1016/j.biombioe.2011.12.008

Kawee-ai A, Srisuwun A, Tantiwa N, et al (2016) Influence of pretreatment with sonocatalytic–synergistic Fenton reaction and its antioxidant potentials. Ultrason Sonochem 184–192. https://doi.org/10.1016/j.ultsonch.2015.12.018.

Kootstra AMJ, Beeftink HH, Scott EL, Sanders JPM (2009) Comparison of dilute mineral and organic acid pretreatment for enzymatic hydrolysis of wheat straw. Biochem Eng J 46:126–131. https://doi.org/10.1016/j.bej.2009.04.020

Krishania M, Kumar V, Sangwan RS (2018) Integrated approach for extraction of xylose, cellulose, lignin and silica from rice straw. Bioresour Technol Reports 1:89–93. https://doi.org/10.1016/j.biteb.2018.01.001

Kumar R, Mago G, Balan V, Wyman CE (2009) Physical and chemical characterizations of corn stover and poplar solids resulting from leading pretreatment technologies. Bioresour Technol 100:3948–3962. https://doi.org/10.1016/j.biortech.2009.01.075

Kumar V, Satyanarayana T (2015) Generation of xylooligosaccharides from microwave irradiated agroresidues using recombinant thermo-alkali-stable endoxylanase of the polyextremophilic bacterium Bacillus halodurans expressed in Pichia pastoris. Bioresour Technol 179:382–389. https://doi.org/10.1016/j.biortech.2014.12.049

Kwak S, Jin YS (2017) Production of fuels and chemicals from xylose by engineered Saccharomyces cerevisiae: a review and perspective. Microb Cell Fact 16:1–15. https://doi.org/10.1186/s12934-017-0694-9

Lange J, Van De GWD, Haan RJ (2009) Conversion of furfuryl alcohol into ethyl levulinate using solid acid catalysts. Chemsuschem 2:437–441. https://doi.org/10.1002/cssc.200800216

Lee YY, Iyer P, Torget R (1999) Dilute-acid hydrolysis of lignocellulosic biomass. Adv Biochem Eng Biotechnol 65

Lee JW, Jeffries TW (2011) Efficiencies of acid catalysts in the hydrolysis of lignocellulosic biomass over a range of combined severity factors. Bioresour Technol 102:5884–5890. https://doi.org/10.1016/j.biortech.2011.02.048

Lee SM, Lee HJ, Kim SH et al (2021) Screening of the strictly xylose-utilizing Bacillus sp. SM01 for polyhydroxybutyrate and its co-culture with Cupriavidus necator NCIMB 11599 for enhanced production of PHB. Int J Biol Macromol 181:410–417. https://doi.org/10.1016/j.ijbiomac.2021.03.149

Li Y, Hu J, Qu C et al (2019a) Engineered Thermoanaerobacterium aotearoense with nfnAB knockout for improved hydrogen production from lignocellulose hydrolysates. Biotechnol Biofuels 12:1–14. https://doi.org/10.1186/s13068-019-1559-8

Li M, Eskridge KM, Wilkins MR (2019b) Optimization of polyhydroxybutyrate production by experimental design of combined ternary mixture (glucose, xylose and arabinose) and process variables (sugar concentration, molar C: N ratio). Bioprocess Biosyst Eng 42:1495–1506. https://doi.org/10.1007/s00449-019-02146-1

Li Y, Xu J, Yu W, Ray AK (2020) Multi-objective optimization of sequential simulated moving bed for the purification of xylo-oligosaccharides. Chem Eng Sci 211:115279. https://doi.org/10.1016/j.ces.2019.115279

Liang Y (2012) Utilization of xylose for producing bio-based fuels. In: Xu P, Luo M (eds) Xylose: production, consumption, and health benefits. Nova Science Publishers Inc., New York, pp 69–94

Liang J, Ji L, He J et al (2022) Chemoenzymatic conversion of biomass-derived D-xylose to furfuryl alcoholwith corn stalk-based solid acid catalyst and reductase biocatalyst in aDeep eutectic solvent-water system. Processes 10. https://doi.org/10.3390/pr10010113

Lin Y, Tseng M, Lee W (2011) Production of xylooligosaccharides using immobilized endo-xylanase of Bacillus halodurans. Process Biochem 46:2117–2121. https://doi.org/10.1016/j.procbio.2011.08.008

Lin Q, Li H, Ren J et al (2017) Production of xylooligosaccharides by microwave-induced, organic acid-catalyzed hydrolysis of different xylan-type hemicelluloses: optimization by response surface methodology. Carbohydr Polym 157:214–225. https://doi.org/10.1016/j.carbpol.2016.09.091

Liu J, Guo T, Wang D, Shen X (2016) Enhanced butanol production by increasing NADH and ATP levels in Clostridium beijerinckii NCIMB 8052 by insertional inactivation of Cbei_4110. Appl Microbiol Biotechnol 100:4985–4996. https://doi.org/10.1007/s00253-016-7299-9

Liu J, Gong Z, Yang G et al (2018) Novel kinetic models of xylan dissolution and degradation during ethanol based auto-catalyzed organosolv pretreatment of bamboo. Polymers (Basel) 10. https://doi.org/10.3390/polym10101149

Lohmeier-vogel EM, Sopher CR, Lee H (1998) Intracellular acidification as a mechanism for the inhibition by acid hydrolysis-derived inhibitors of xylose fermentation by yeasts 75–81

Lopes M, Rocha R, Zanotto S et al (2009) Screening of bacteria to produce polyhydroxyalkanoates from xylose. World J Microbiol Biotechnol 25:1751–1756. https://doi.org/10.1007/s11274-009-0072-9

Lu Y, He Q, Fan G et al (2021) Extraction and modification of hemicellulose from lignocellulosic biomass: a review. Green Process Synth 10:779–804. https://doi.org/10.1515/gps-2021-0065

Mah KH, Yussof HW, Seman MNA, Mohammad AW (2019) Separation and purification technology optimisation of interfacial polymerization factors in thin film composite (TFC) polyester nano filtration (NF) membrane for separation of xylose from glucose. Sep Purif Technol 209:211–222. https://doi.org/10.1016/j.seppur.2018.07.041

Martins S, Jongen W, Boekel V, Martinus A (2001) A review of Maillard reaction in food and implications to kinetic modelling. Trends Food Sci Technol 11:364–373

Millán GG, Sixta H (2020) Towards the green synthesis of furfuryl alcohol in a one-pot system from xylose: a review. Catalysts 10:1–23. https://doi.org/10.3390/catal10101101

Moniz P, Pereira H, Duarte LC, Carvalheiro F (2014) Hydrothermal production and gel filtration purification of xylo-oligosaccharides from rice straw. Ind Crop Prod 62:460–465. https://doi.org/10.1016/j.indcrop.2014.09.020

Morais ARC, Bogel-Lukasik R (2016) Highly efficient and selective CO2-adjunctive dehydration of xylose to furfural in aqueous media with THF. Green Chem 18:2331–2334. https://doi.org/10.1039/c5gc02863a

Morais ES, Mendonça PV, Coelho JFJ et al (2018) Deep eutectic solvent aqueous solutions as efficient media for the solubilization of hardwood xylans. Chemsuschem 11:753–762. https://doi.org/10.1002/cssc.201702007

Morone A, Apte M, Pandey RA (2015) Levulinic acid production from renewable waste resources: bottlenecks, potential remedies, advancements and applications. Renew Sustain Energy Rev 51:548–565. https://doi.org/10.1016/j.rser.2015.06.032

Morthensen T, Luo J, Meyer AS et al (2015) High performance separation of xylose and glucose by enzyme assisted nano filtration 492:107–115. https://doi.org/10.1016/j.memsci.2015.05.025

Moure A, Gullón P, Domínguez H, Parajó JC (2006) Advances in the manufacture, purification and applications of xylo-oligosaccharides as food additives and nutraceuticals. Process Biochem 41:1913–1923. https://doi.org/10.1016/j.procbio.2006.05.011

Mussatto SI, Roberto IC (2004) Alternatives for detoxification of diluted-acid lignocellulosic hydrolyzates for use in fermentative processes: a review. Bioresour Technol 93:1–10. https://doi.org/10.1016/j.biortech.2003.10.005

Nabarlatz D, Torras C, Garcia-Valls R, Montané D (2007a) Purification of xylo-oligosaccharides from almond shells by ultrafiltration. Sep Purif Technol 53:235–243. https://doi.org/10.1016/j.seppur.2006.07.006

Nabarlatz D, Ebringerová A, Montané D (2007b) Autohydrolysis of agricultural by-products for the production of xylo-oligosaccharides. Carbohydr Polym 69:20–28. https://doi.org/10.1016/j.carbpol.2006.08.020

Nakasu PYS, Chagas MF, Costa AC, Rabelo SC (2017) Kinetic study of the acid post-hydrolysis of xylooligosaccharides from hydrothermal pretreatment. Bioenergy Res 10:1045–1056. https://doi.org/10.1007/s12155-017-9864-1

Narisetty V, Cox R, Bommareddy R et al (2022) Valorisation of xylose to renewable fuels and chemicals, an essential step in augmenting the commercial viability of lignocellulosic biorefineries. Sustain Energy Fuels 6:29–65. https://doi.org/10.1039/D1SE00927C

Newton AE, Fairbanks AJ, Golding M et al (2012) The role of the Maillard reaction in the formation of flavour compounds in dairy products—Not only a deleterious reaction but also a rich source of flavour compounds. Food Funct 3:1231–1241. https://doi.org/10.1039/c2fo30089c

Oktaviani M, Mangunwardoyo W, Hermiati E (2021) Characteristics of adapted and non-adapted Candida tropicalis InaCC Y799 during fermentation of detoxified and undetoxified hemicellulosic hydrolysate from sugarcane trash for xylitol production. Biomass Convers Biorefinery. https://doi.org/10.1007/s13399-021-02087-4

Parajo JC, Dominguez H, Dominguez JM (1998) Biotechnological production of xylitol. Part 3: operation in culture media made from lignocellulose hydrolysate. Bioresour Technol 66:25–40

Park H, Seo S, Yoo S (2017) Relative sweetness and sweetness quality of xylobiose. Food Sci Biotechnol 26:689–696. https://doi.org/10.1007/s10068-017-0109-z

Pellerin P, Gosselin M, Lepoutre JP et al (1991) Enzymic production of oligosaccharides from corncob xylan. Enzyme Microb Technol 13:617–621. https://doi.org/10.1016/0141-0229(91)90074-K

Perez RF, Canhaci SJ, Borges LEP, Fraga MA (2017) One-step conversion of xylose to furfuryl alcohol on sulfated zirconia-supported Pt catalyst—Balance between acid and metal sites. Catal Today 289:273–279. https://doi.org/10.1016/j.cattod.2016.09.003

Pinales-Márquez CD, Rodríguez-Jasso RM, Araújo RG et al (2021) Circular bioeconomy and integrated biorefinery in the production of xylooligosaccharides from lignocellulosic biomass: a review. Ind Crops Prod 162. https://doi.org/10.1016/j.indcrop.2021.113274

Pratto B, Chandgude V, De SR et al (2020) Biobutanol production from sugarcane straw: defining optimal biomass loading for improved ABE fermentation. Ind Crop Prod 148:112265. https://doi.org/10.1016/j.indcrop.2020.112265

Qing Q, Li H, Kumar R, Wyman CE (2013) Xylooligosaccharides production, quantification, and characterization in context of lignocellulosic biomass pretreatment. Aqueous Pretreat Plant Biomass Biol Chem Convers to Fuels Chem 391–415. https://doi.org/10.1002/9780470975831.ch19

Qiu Z, Gao Q, Bao J (2017) Constructing xylose-assimilating pathways in Pediococcus acidilactici for high titer D-lactic acid fermentation from corn stover feedstock. Bioresour Technol 245:1369–1376. https://doi.org/10.1016/j.biortech.2017.05.128

Qiu Z, Gao Q, Bao J (2018) Engineering Pediococcus acidilactici with xylose assimilation pathway for high titer cellulosic L-lactic acid fermentation. Bioresour Technol 249:9–15. https://doi.org/10.1016/j.biortech.2017.09.117

Rackemann DW, Doherty WOS, Crops T (2011) The conversion of lignocellulosics to levulinic acid. Biofuels Bioprod Biorefining 5:198–214. https://doi.org/10.1002/bbb

Rafiqul ISM, Sakinah AMM (2013) Processes for the production of xylitol-a review. Food Rev Int 29:127–156. https://doi.org/10.1080/87559129.2012.714434

Rahman SHA, Choudhury JP, Ahmad AL, Kamaruddin AH (2007) Optimization studies on acid hydrolysis of oil palm empty fruit bunch fiber for production of xylose 98:554–559. https://doi.org/10.1016/j.biortech.2006.02.016

Rana D, Rana V, Ahring BK (2012) Producing high sugar concentrations from loblolly pine using wet explosion pretreatment. Bioresour Technol 121:61–67. https://doi.org/10.1016/j.biortech.2012.06.062

Rashad MM, Mahmoud AE, Nooman MU et al (2016) Production of antioxidant xylooligosaccharides from lignocellulosic materials using Bacillus amyloliquifaciens NRRL B-14393 xylanase. J Appl Pharm Sci 6:30–36. https://doi.org/10.7324/JAPS.2016.60606

Reddy SS, Krishnan C (2014) Production of xylooligosaccharides in SSF by Bacillus subtilis KCX006 producing β-xylosidase-free endo-xylanase and multiple xylan debranching enzymes. Prep Biochem Biotechnol 37–41. https://doi.org/10.1080/10826068.2014.970694

Rivas S, Conde E, Moure A et al (2013) Characterization, refining and antioxidant activity of saccharides derived from hemicelluloses of wood and rice husks. Food Chem. https://doi.org/10.1016/j.foodchem.2013.03.008

Rocha MVP, Rodrigues THS, de Albuquerque TL et al (2014) Evaluation of dilute acid pretreatment on cashew apple bagasse for ethanol and xylitol production. Chem Eng J 243:234–243. https://doi.org/10.1016/j.cej.2013.12.099

Rollin et al (2015) High-yield hydrogen production from biomass by in vitro metabolic engineering—Mixed sugars coutilization and kinetic modelling.pdf

Saha BC (2003) Hemicellulose bioconversion. J Ind Microbiol Biotechnol 30:279–291. https://doi.org/10.1007/s10295-003-0049-x

Samanta AK, Senani S, Kolte AP et al (2012) Production and in vitro evaluation of xylooligosaccharides generated from corn cobs. Food Bioprod Process 90:466–474. https://doi.org/10.1016/j.fbp.2011.11.001

Samanta AK, Jayapal N, Jayaram C et al (2015) Xylooligosaccharides as prebiotics from agricultural by-products: Production and applications. Bioact Carbohydrates Diet Fibre 5:62–71. https://doi.org/10.1016/j.bcdf.2014.12.003

Santibáñez L, Henríquez C, Corro-Tejeda R, et al (2021) Xylooligosaccharides from lignocellulosic biomass: a comprehensive review. Carbohydr Polym 251. https://doi.org/10.1016/j.carbpol.2020.117118

Saratale RG, Saratale GD, Cho SK et al (2019) Pretreatment of kenaf (Hibiscus cannabinus L.) biomass feedstock for polyhydroxybutyrate (PHB) production and characterization. Bioresour Technol 282:75–80. https://doi.org/10.1016/j.biortech.2019.02.083

Scheller HV, Ulvskov P (2010) Hemicelluloses. Annu Rev Plant Biol 61:263–289. https://doi.org/10.1146/annurev-arplant-042809-112315

Shahbazi A, Zhang B (2010) Dilute and concentrated acid hydrolysis of lignocellulosic biomass. Bioalcohol Prod Biochem Convers Lignocellul Biomass 143–158. https://doi.org/10.1533/9781845699611.2.143

Shimoda K, Hamada H, Hamada H (2011) Synthesis of xylooligosaccharides of daidzein and their anti-oxidant and anti-allergic activities. Int J Mol Sci 12:5616–5625. https://doi.org/10.3390/ijms12095616

Silva EK, Arruda HS, Mekala S et al (2022) Xylooligosaccharides and their chemical stability under high-pressure processing combined with heat treatment. Food Hydrocoll 124. https://doi.org/10.1016/j.foodhyd.2021.107167

Singh RD, Banerjee J, Sasmal S et al (2018) High xylan recovery using two stage alkali pretreatment process from high lignin biomass and its valorisation to xylooligosaccharides of low degree of polymerisation. Bioresour Technol 256:110–117. https://doi.org/10.1016/j.biortech.2018.02.009

Singh RD, Gracy C, Muir J, Arora A (2019a) Green and clean process to obtain low degree of polymerisation xylooligosaccharides from almond shell. J Clean Prod 241:118237. https://doi.org/10.1016/j.jclepro.2019.118237

Singh MK, Rai PK, Rai A, Singh S (2019b) Poly-β-hydroxybutyrate production by the cyanobacterium Scytonema Geitleri Bharadwaja under varying environmental conditions. Biomolecules 9:198

Singh AK, Srivastava JK, Chandel AK et al (2019c) Biomedical applications of microbially engineered polyhydroxyalkanoates: an insight into recent advances, bottlenecks, and solutions. Appl Microbiol Biotechnol 103:2007–2032. https://doi.org/10.1007/s00253-018-09604-y
Singh RD, Muir J, Arora A (2021) Concentration of xylooligosaccharides with a low degree of polymerization using membranes and their effect on. Biofuels Bioprod Biorefining 61–73. https://doi.org/10.1002/bbb.2145
Sjoman E, Manttari M, Niystrom M et al (2007) Separation of xylose from glucose by nanofiltration from concentrated monosaccharide solutions. J Memb Sci 292:106–115. https://doi.org/10.1016/j.memsci.2007.01.019
Sun RC, Tomkinson J (2002) Characterization of hemicelluloses obtained by classical and ultrasonically assisted extractions from wheat straw. Carbohydr Polym 50:263–271. https://doi.org/10.1016/S0144-8617(02)00037-1
Sun F, Cui H, Zhan H et al (2019) Aqueous preparation of Maillard reaction intermediate from glutathione and xylose and its volatile formation during thermal treatment. J Food Sci 84:3584–3593. https://doi.org/10.1111/1750-3841.14911
Surek E, Buyukkileci AO (2017) Production of xylooligosaccharides by autohydrolysis of hazelnut (Corylus avellana L.) shell. Carbohydr Polym 174:565–571. https://doi.org/10.1016/j.carbpol.2017.06.109
Vazquez M, Alonso JL, Dominguez H, Parajo J (2000) Xylooligosaccharides: manufacture and applications. Trends Food Sci Technol 11:387–393
Veenashri BR, Muralikrishna G (2011) In vitro anti-oxidant activity of xylo-oligosaccharides derived from cereal and millet brans—A comparative study. Food Chem 126:1475–1481. https://doi.org/10.1016/j.foodchem.2010.11.163
Vegas R, Luque S, Alvares JR et al (2006) Membrane-assisted processing of xylooligosaccharide-containing liquors. J Agric Food Chem 54:5430–5436. https://doi.org/10.1021/jf060525w
Vivek N, Nair LM, Mohan B, Chandrasekharan S (2019) Bio-butanol production from rice straw—Recent trends, possibilities, and challenges. Bioresour Technol Reports 7:100224. https://doi.org/10.1016/j.biteb.2019.100224
Vongkampang T, Sreenivas K, Engvall J et al (2021) Characterization of simultaneous uptake of xylose and glucose in Caldicellulosiruptor kronotskyensis for optimal hydrogen production. Biotechnol Biofuels 14:1–11. https://doi.org/10.1186/s13068-021-01938-6
Wang Y, Cao X, Zhang R et al (2018) Evaluation of xylooligosaccharide production from residual hemicelluloses of dissolving pulp by acid and enzymatic hydrolysis. RSC Adv 8:35211–35217. https://doi.org/10.1039/c8ra07140c
Wen H, Chen H, Cai D et al (2018) Integrated in situ gas stripping-salting-out process for high-titer acetone-butanol-ethanol production from sweet sorghum bagasse. Biotechnol Biofuels 11:1–12. https://doi.org/10.1186/s13068-018-1137-5
Weng YH, Wei HJ, Tsai TY, Chen WH, Wei TY, Hwang WS,Huang CP (2009) Separation of acetic acid from xylose by nanofiltration. Sep Purif Technol 67(1):95–102. https://doi.org/10.1016/j.seppur.2009.03.030
Wijaya H, Sasaki K, Kahar P et al (2020) High enzymatic recovery and purification of xylooligosaccharides from empty fruit bunch via nanofiltration. Processes 8
Wu J, Elliston A, Le Gall G et al (2018) Optimising conditions for bioethanol production from rice husk and rice straw: Effects of pre-treatment on liquor composition and fermentation inhibitors. Biotechnol Biofuels 11:1–13. https://doi.org/10.1186/s13068-018-1062-7
Xu F, Shi YC, Wang D (2012) Enhanced production of glucose and xylose with partial dissolution of corn stover in ionic liquid, 1-Ethyl-3-methylimidazolium acetate. Bioresour Technol 114:720–724. https://doi.org/10.1016/j.biortech.2012.03.023
Xu F, Chen J, Yang G et al (2019) Combined treatments consisting of calcium hydroxide and activate carbon for purification of xylo-oligosaccharides of pre-hydrolysis liquor. Polymers (Basel) 11. https://doi.org/10.3390/polym11101558

Yan F, Tian S, Chen H et al (2022) Advances in xylooligosaccharides from grain byproducts: extraction and prebiotic effects. Grain Oil Sci Technol 5:98–106. https://doi.org/10.1016/j.gaost.2022.02.002

Yang C-H, Yang S-F, Liu W-H (2007) Production of xylooligosaccharides from xylans by extracellular xylanases from thermobifida fusca. J Agric Food Chem 55:3955–3959

Yohana T, Fahrezi MZM, Lianawati A et al (2022) Preliminary study on the utilization of sugarcane trash and corncob for xylo-oligosaccharides and xylose production through dilute acid hydrolysis. Springer Nature Singapore

Yuan QP, Zhang H, Qian ZM, Yang XJ (2004) Pilot-plant production of xylo-oligosaccharides from corncob by steaming, enzymatic hydrolysis and nanofiltration 1079:1073–1079. https://doi.org/10.1002/jctb.1071

Zeidan AA, Van Niel EWJ (2009) Developing a thermophilic hydrogen-producing co-culture for efficient utilization of mixed sugars. Int J Hydrog Energy 34:4524–4528. https://doi.org/10.1016/j.ijhydene.2008.07.092

Zhang Y, Liu W, Liu C et al (2014) Retrogradation behaviour of high-amylose rice starch prepared by improved extrusion cooking technology. Food Chem 158:255–261. https://doi.org/10.1016/j.foodchem.2014.02.072

Zhang H, Xu Y, Yu S (2017a) Co-production of functional xylooligosaccharides and fermentable sugars from corncob with effective acetic acid prehydrolysis. Bioresour Technol 234:343–349. https://doi.org/10.1016/j.biortech.2017.02.094

Zhang H, Zhou X, Xu Y, Yu S (2017b) Production of xylooligosaccharides from Waste Xylan, obtained from viscose fiber processing, by selective hydrolysis using concentrated acetic acid. J Wood Chem Technol 37:1–9. https://doi.org/10.1080/02773813.2016.1214154

Zhang W, Zhang X, Lei F, Jiang J (2020a) Co-production bioethanol and xylooligosaccharides from sugarcane bagasse via autohydrolysis pretreatment. Renew Energy 162:2297–2305. https://doi.org/10.1016/j.renene.2020.10.034

Zhang M, Wang Y, Hui KS et al (2020b) Microwave-assisted acid-catalyzed hydrolysis of hemicelluloses in rice husk into xylose microwave-assisted acid-catalyzed hydrolysis of hemicelluloses in rice husk into xylose. https://doi.org/10.1088/1755-1315/513/1/012016

Zhao Y, SUn H, Yang B, Weng Y (2020) Hemicellulose-based film: potential green films for food packaging. Polymers (Basel) 12:1–14. https://doi.org/10.3390/polym12081775

Carotene Production from Biomass Waste

Made Tri Ari Penia Kresnowati and Dianika Lestari

Abstract Carotenoids, a group of substances that belongs to the terpenoid group, are widely used for food colouring to give yellow, orange, or red colour in food products. In particular, β-carotene can be produced by extraction from biological substance, such as vegetables, fruits, oils, or microbial fermentation; or produced synthetically via chemical reactions. This article presents a review of β-carotene production method, in particular from biomass waste. As a case study, the potential of β-carotene production from Oil Palm Empty Fruit Bunches (OPEFB), the major biomass waste from the palm oil industry, is evaluated.

Keywords β-carotene · Extraction · Oil palm empty fruit bunches · Production

1 Introduction

Food colorants are the additive that is widely used by the food industry to improve the appearance of food products. Carotene is one of the colorants that is widely used in food industry to give yellow, orange or red color. Carotene also has antioxidant properties and beta carotene, in particular, is the precursor of vitamin A. Thereby it is a food additive that also has a nutritional value.

Carotene can be produced synthetically via chemical reactions or obtained from natural sources. Some of the potential sources for carotene are vegetables such as carrots, tomatoes, and spinach; vegetable oils such as palm oil and buriti oil, or microorganisms such as algae, bacteria, and fungi/yeast (Riberio et al. 2011).

Palm oil is one of Indonesian top plantation commodities. Data shows that in 2021 the Indonesian oil palm plantation area was spread at 14.6 million hectares (Indonesian Statistic Bureau, www.bps.go.id) producing about 46.9 million tonnes of crude palm oil or CPO (Palm Oil Association, https://gapki.id). Besides producing CPO, palm oil industries also coproduce a wide array of biomass waste such as oil

M. T. A. P. Kresnowati (✉) · D. Lestari
Department of Food Engineering, Faculty of Industrial Technology, Insitut Teknologi Bandung, Bandung, Indonesia
e-mail: kresnowati@cheitb.id

M. A. R. Lubis et al. (eds.), *Biomass Conversion and Sustainable Biorefinery*, Green Energy and Technology, https://doi.org/10.1007/978-981-99-7769-7_12

palm empty fruit bunches, oil palm fibers, and shells. The oil palm empty fruit bunches (OPEFB) are the leftover biomass after the pressing or extraction of palm oil from the fresh fruit bunches. Thereby OPEFB still contains traces of palm oil and carotene. Palm oil can be one of the potential raw materials for the beta-carotene industry. Producing carotene from palm oil and its biomass waste can be an additional value to the palm oil industry.

This paper presents a review of β-carotene production method. Detail evaluation is presented for the potential of β-carotene production from OPEFB, the major biomass waste from the palm oil industry.

2 Carotene

Carotenes are hydrophobic organic molecules containing 40 carbon atoms that belong to the terpenoid group. Carotenes contain 8 isoprene structure and does not contain any oxygen atom. Carotenes absorb ultraviolet, violet, and blue light, scatter orange, red and yellow light. The term ‘carotene’ itself was pinned from carrot, from which it was first extracted. In carrots, carotene is used as the molecule to capture (sun)light energy for photosynthesis.

The two primary carotene isomers are α-carotene and β-carotene. These molecules have terminal β-ionone rings on both ends and differ only in the positions of double bonds (Fig. 1a, b). Another carotene isomer is lycophene, which lacks of β-ionone ring (Fig. 1c). Another class of molecule sharing the 8-isoprenes-terpenoid group is called xanthophyll. Some examples of xanthophyll are luthein, fucoxanthin and astaxanthin. The difference between carotene and xanthophyll is that the first group does not contain any oxygen atom and thus purely hydrocarbon whereas the second one contains some oxygen atoms. Both carotene and xanthophyll are grouped as the carotenoids.

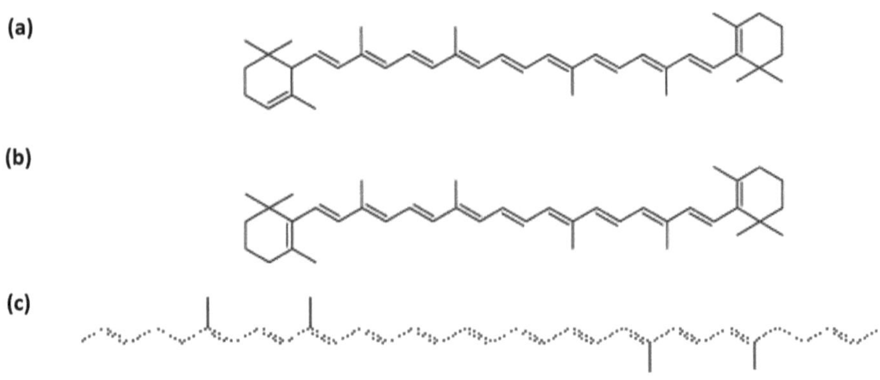

Fig. 1 Molecular structure of α-carotene (**a**), β-carotene (**b**), lycophene (**c**)

The presence of β-ring in carotene contributes to the pro-vitamin A activity of this molecule, which is the main nutritional function of carotene. β-carotene can be cleaved by a particular enzyme called β-carotene-15,15′-dioxygenase to produce retinal, which is vitamin A (Grune et al. 2010). On the other hand, the presence of double bonds in carotene molecules allows the binding of this molecule to singlet oxygen, providing an antioxidant property. Overall, these valuable properties of carotene make this compound a desirable food additive and widely applied as red, orange colorants in food products. Commercially carotenes are applied as food colorants, nutraceutical supplement, feed additives, and also in cosmetics and pharmaceutical (Bhosale and Bernstein 2005; Schmidt-Dannert 2000; Mortensen 2009).

β-carotene is sold as capsules, powders, or concentrates either in water-dispersible form or in oil-based form. A quick survey indicates the price of β-carotene ranges from 14 to 600 US$/kg, depending on its purity. The global β-carotene market size was reported to be 432.2 million USD in 2015 and is expected to reach 621.0 million USD in 2024 (Grand View Research 2024).

3 Commercial Production of β-Carotene

3.1 Synthetic Route of β-Carotene Production

Most of commercial β-carotene is produced via chemical synthetic route. The chemical synthetic route uses β-ionone as the starting chemical precursor. This molecule can be obtained from natural resources, such as turpentine or lemon grass oil, or produced from acetone, acetylene, or butadiene (Riberio et al. 2011; Tedeschi 2003; Vani et al. 2001). The reactions involve the combining of two or more smaller molecules to give the required carbon skeleton of β-carotene, that is 40 carbon atoms, either by the Grignard coupling, elimination, and partial hydrogenation reaction or by the Wittig condensation reaction. The reaction was pioneered by Roche, and now the two main β-carotene producers are DSM and BASF. These companies have been producing β-carotene from the early fifties and now supply up to 85% of world β-carotene demands.

3.2 Natural β-Carotene

B-carotene is naturally present in plants and microorganisms. Some potential sources of β-carotene are listed in Table 1. Commonly, these organisms contain not only β-carotene but also other carotenoids, despite in lower quantity. Thereby natural β-carotene provides more health benefits and can be consumed in larger quantities

(Riberio et al. 2011). The growing health awareness of the usage of chemical additives in food encourages the growing demand of natural β-carotene.

Palm oil and buriti oil are reported as potential sources for plants-derived carotene. Their carotene content is much higher than carrots, from where carotene was first extracted. The processes applied to obtain carotene from vegetable/fruit commonly include extraction and refinery (i.e. concentration/crystallization) (Fig. 2). A more complex process is required to obtain carotene from vegetable oil, considering that carotene is a hydrophobic compound that well dissolves in oil. A review on various carotene extraction methods from CPO is provided by Othman et al. (2010) and Sudibyo and Sardjono (2015). The commercial process includes transesterification process followed by molecular distillation (Ooi et al. 1994).

Wide range of microorganisms can produce carotenoids. Information on strategies for optimizing the carotenoid content such that it can be applied at commercial scale, however, is still limited. Nevertheless, using microorganisms as the platform to produce natural carotene is always an interesting alternative considering the fast doubling time or fast rate of growth of microbial cells.

An example of commercial microbial source of carotene is *Dunaliella salina*, the halotolerant unicellular microalgae. Applying various kinds of stresses during microalgae cultivation has been reported to increase the carotenoegenesis in this species such as high salt concentration, light stress, suppression of Nitrogen and Phosphorous sources, or addition of organic solvent (Riberio et al. 2011). Other reported commercial microbial source of carotene is fungus *Blakeslea trispora*. Similarly,

Table 1 Potential sources of β-carotene

Resources		Carotene yield (μg/g)	References
Vegetables/plants	Carrots (*Daucus carota*)	85–174	Rodriguez-Amaya (2001)
	Palm oil (*Elaeis guineensis*)	470–700	Rodriguez-Amaya (2001)
	Sweet potato (*Ipomoea batatas*)	160–226	Rodriguez-Amaya (2001)
	Buriti oil (*Mauritia vinifera*)	1150–3380	Rodriguez-Amaya (2001)
Microorganisms/algae	*Dunaliella salina*	46600	García-González et al. (2005)
Microorganisms/fungi	*Blakeslea trispora*	67000	Choudhari et al. (2008)
	Phycomyces blakesleeanus	32800	Mehta et al (1997)
	Neurospora sp.	1190	Gmoser et al. (2018), Priatni (2016)
Microorganism/yeast	*Rhodotorula sp.*	800–900	Buzzini et al. (2005), Wang et al. (2008)

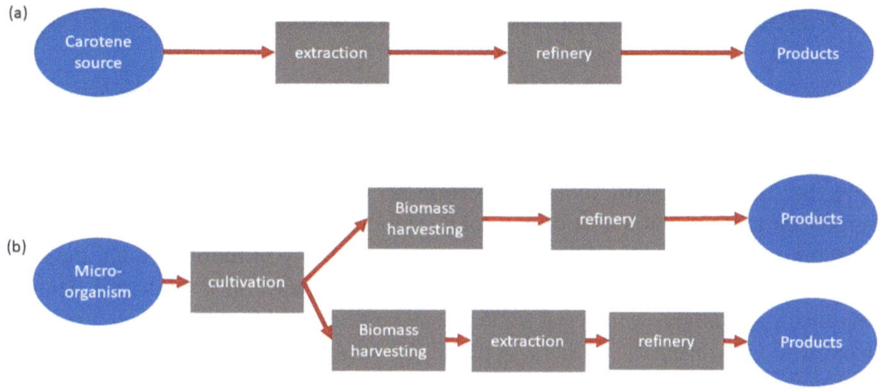

Fig. 2 Block flow diagram of natural carotene production from natural resources: **a** From oil, fruit, or vegetable, **b** from microorganism

some stimuli or inducers should be added to the cultivation/fermentation to accumulate carotene. Some reported inducers are lighting or addition of n-dodecane or β-ionone to the cultivation (Riberio et al. 2011). Wild type of bacterial species/strains are rarely reported to produce significant amount of carotene. Some studies explore the use of genetically modified bacteria using *Escherichia coli* or *Serratia marcensens* as the bacterial platform (Bogasz-Radomska and Harasym 2018). Recently some studies have been focused on *Neurospora sp.* for producing carotene (Gmoser et al. 2018; Priatni 2016). This is of particular interest as this fungal species is commonly used in preparing traditional fermented food in Indonesia such as oncom.

The general scheme of microbial-based carotene production includes the cultivation of the microorganisms (microalgae/fungi/yeast/bacteria) followed by the downstream separation process (Fig. 2). The first step of the downstream separation process is biomass separation, for example by centrifugation. The products can be sold as dried biomass containing carotenoids or raw carotene. In this case, the biomass is dried and pressed. The products can also be sold as concentrate or purified carotene. In this case, the biomass is disrupted, before carotene is extracted and concentrated. The concentrated carotene can further be purified by crystallization.

Overall, the extraction process is the critical step in natural carotene production. Reviews on various carotene extraction methods are given for example by Saini and Keum (2018) and Rammuni et al. (2019).

Major natural β-carotene producer is Vitatene (Spain) which produces β-carotene from fungi *Blakeslea trispora*, and Aquacarotene Limited (Australia) which produces β-carotene from algae *Dunaliella salina*. Besides producing β-carotene synthetically, BASF also has microalgae plants to produce natural β-carotene in Australia. Vitatene is now part of DSM, another major producer of synthetic β-carotene.

4 Carotene Production from Biomass Waste

Despite the wide distribution of natural carotene in plants and microorganism, they only comprise of a small portion of the cell which leads to the high cost of natural β-carotene. The use of low cost substrate, such as biomass waste, as the carotene source or as the substrate for microbial cultivation for carotene production was suggested to reduce the production cost of natural β-carotene (Igreja et al. 2021).

Palm oil is the most important plantation commodity in Indonesia. On the other hands Table 1 shows that the carotenoid content of palm oil is high that palm oil becomes a potential raw material for carotene production. Most of carotenoid content in palm oil lost during the refinery, in particular bleaching and deodorization, in order to produce the light colour oil that is demanded by market. These processes employ high temperature and pressure such that carotenoid molecules are degraded into smaller compounds (Ooi et al. 1994). For producing carotene from palm oil, it is then suggested to use crude palm oil (CPO) as the raw material.

The palm oil industry also coproduces a lot of biomass waste. In the plantation, there are oil palm fronds and oil palm tree trunks, that need to be regenerated periodically. Within the boundary of the palm oil processing plants there are OPEFB, oil palm shells, and oil palm fibers. The oil palm fibers are produced as biomass waste after the oil palm extraction process. The oil palm shells are produced as biomass waste after the palm kernel extraction process. The empty fruit bunches are produced as biomass waste after the threshing of oil palm fruits from the fresh fruit bunches. These biomass waste still contains palm oil residue and therefore it can be expected to contain materials that dissolve in it.

Some literatures reported the presence of carotene in palm oil residue. Masni (2004) evaluated the use of palm oil mill wastes, in particular the oil palm pressed fibers, as the source of carotenoid. The employed processes were extraction followed by chromatography absorption for purification. The concentration of the produced carotene concentrate was up to 11580 ppm. Beside the oil palm fiber, Kupan et al. (2016) also evaluated the use of OPEFB as the source of carotene. The employed processes were extraction followed by absorption and gave carotene concentrate up to 1414 and 702 ppm succeedingly for the pressed fibers and empty fruit bunches. These literatures show that more carotene can be extracted from the fibers, but fibers are produced much less than the empty fruit bunches in the palm oil mills (Prasertsan and Prasertsan 1996). Moreover, the oil palm fibers are usually burned to fuel the boiler in the palm oil mills.

4.1 Empty Fruit Bunch as an Alternative Raw Material for Carotene Production

As has been discussed previously, empty fruit bunches are produced as biomass waste after the threshing of oil palm fresh fruit bunches from the fresh fruit bunches and thus still contain palm oil residue and all materials that dissolve in it. Research have been conducted to develop the carotene production process from OPEFB further.

Direct extraction of OPEFB gave low concentration of carotene extract, in the order of 1.94–2.54 ppm (Kresnowati et al. 2020) which indicated the need of further refinery of the carotene extract. Anshori et al. (2022) reported that the type of solvent used, different parts of OPEFB that was extracted, as well as the cut size of OPEFB affect the carotene recovery during the extraction. Kresnowati et al. (2020) also reported the freshness of the OPEFB, or in another word how long the OPEFB has been dumped nearby the palm oil mill plant, affected the maximum carotene that can be extracted. Fresh OPEFB still contains a lot of palm oil residue, thus considerable amount of carotene can be extracted. After just a few days, yellow orange colour fungi would spontaneously grow on OPEFB, and consistently a significant increase in the carotene content can be expected. After the growth of yellow orange colour fungi, the fungal growth would be dominated by the white colour fungi untill the OPEFB became rotten. Followingly, the carotene content decrease. Although the results are still limited, the potential is quite promising, the yield of carotene from empty fruit bunches was estimated up to 180 μg/g which was comparable to that of carrots and sweet potatoes. Overall, the natural sequential fungal growth during the storage of OPEFB in the dumpsite nearby the oil palm mill plants indicated that the presence of fungi increased the carotenoid content of the OPEFB, it also indicated the potential OPEFB as the substrate for fungal fermentation for carotene production. In this case, cheap biomass waste materials are used to supply important precursors that are needed for carotene production, thereby no expensive chemical precursors need to be added. In addition, the empty fruit bunches themselves also contain carotene which can increase the overall yield of carotene from the process. In total, the carotene can be obtained from both the empty fruit bunches and the fungi. The yellow orange colour fungi that is naturally grown on OPEFB was identified as *Neurospora sp.*

OPEFB is lignocellulosic material that is rich in Carbon, but limited Nitrogen content. Indeed, preliminary research on nutrient addition on the solid state fermentation of *Neurospora sp.* by using OPEFB as the substrate, such as urea as the Nitrogen source, and $MgSO_4$ as both the Sulphur source and Magnesium source, affected fungal growth and carotene production (Tommy and Purba 2020). Other fermentation parameters that affected the fungal growth and carotene production are the water activity, lighting, temperature, and fermentation duration (Anshori 2020; Tambunan and Rucita 2021). Overall, reengineering of the solid state fermentation of *Neurospora sp.* on OPEFB substrate may improve carotene production.

The idea of using palm oil wastes as the source of carotene may first sound tedious, in particular when compared to using the crude palm oil. However, Manurung et al. (2017) showed that the extraction of carotene or in general the reextraction of palm oil from oil palm waste is indeed reducing the greenhouse gas emission of the palm oil industry. Considering that empty fruit bunches are cheap biomass waste that is available in abundance, the idea of utilizing empty fruit bunches as the source of carotene offers great potential. The proposed process for carotene production from OPEFB as well as factors that affecting the process is presented in Fig. 3.

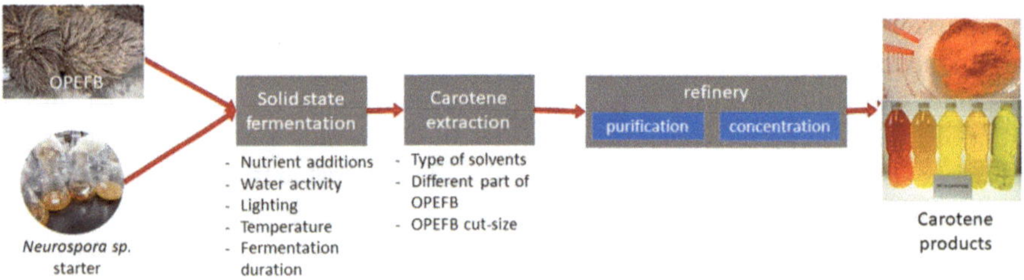

Fig. 3 Carotene production from OPEFB

Using data of Prasertsan and Prasertsan (1996) that about 1 ton of OPEFB are coproduce during the production of 1 ton of CPO, and potential carotene yield from empty fruit bunch of 180 μg/g, we estimate 5600 ton of empty fruit bunches will be required to produce 1 ton of carotene. This means that only 2% of the available empty fruit bunches will be required to supply the demanded carotene as the provitamin A for Indonesian population.

5 Conclusion

The abundance of oil palm empty fruit bunches (OPEFB), as the waste from palm oil mills, offers an interesting alternative to utilizing them as the raw material for carotene. The proposed process for carotene production from empty fruit bunches consists of fungal fermentation followed by extraction and purification.

Acknowledgements This research was funded by Indonesian Minister of Research Technology and Higher Education through Penelitian Dasar Unggulan Perguruan Tinggi scheme.

References

Anshori M (2020) Production of β-carotene food coloring from empty fruit bunches (EFB) utilizes neurospora sp.: optimization of fermentation conditions and scale-up production, graduate research thesis, Chemical Engineering Department, Institut Teknologi Bandung

Anshori M, Jafar RM, Lestari D, Kresnowati MTAP (2022) Production of carotenoids from oil palm empty fruit bunches: selection of extraction methods. J. Eng. Technol. Sci. 54(3):220303. https://doi.org/10.5614/j.eng.technol.sci.2022.54.3.3

Bhosale P, Bernstein PS (2005) Microbial xanthophylls. Appl Microbiol Biotechnol 68(4):445–455. https://doi.org/10.1007/s00253-005-0032-8

Bogasz-Radomska L, Harasym J (2018) β-carotene—Propertoes and production methods. Food Quality Safety 2:69–74. https://doi.org/10.1093/fqsafe/fyy004

Buzzini P, Martini A, Gaetani M, Turchetti B, Pagnoni UM, Davoli P (2005) Optimization of carotenoid production by Rhodotorula graminis DBVPG 7021 as a function of trace element

concentration by means of response surface analysis. Enzyme Microb Technol 36:687–692. https://doi.org/10.1016/j.enzmictec.2004.12.028
Choudhari SM, Ananthanarayan L, Singhal RS (2008) Use of metabolic stimulators and inhibitors for enhanced production of β-carotene and lycopene by Blakeslea trispora NRRL 2895 and 2896. Biores Technol 99(8):3166–3173. https://doi.org/10.1016/j.biortech.2007.05.051
García-González M, Moreno J, Manzano JC, Florencio FJ, Guerrero MG (2005) Production of Dunalliela salina biomass rich in 9-cis-β-carotene and lutein in a closed tubular photobioreactor. J Biotechnol 115(1):81–90. https://doi.org/10.1016/j.jbiotec.2004.07.010
Gmoser R, Ferreira JA, Lundin M, Raherzadeh MJ, Lennartsson PR (2018) Pigment production by the edible filamentous fungus neurospora intermedia. Fermentation 4(1):11. https://doi.org/10.3390/fermentation4010011
Grand View Research (2024) Beta-carotene market analysis by source (Algae, fruits & vegetables, & synthetic), by application (Food & beverages, dietary supplements, cosmetics, & animal feed) and segment forecasts to 2024. https://www.grandviewresearch.com/industry-analysis/beta-carotene-market
Grune T, Lietz G, Palou A, Ross C, Stahl A, Tang G, Thurnham D, Yin S, Biesalski HK (2010) β-carotene is an important vitamin A source for humans. J Nutr 140(12):2268S-2285S. https://doi.org/10.3945/jn.109.119024
Igreja WS, Maia FA, Lopes AS, Chisté RC (2021) Biotechnological production of carotenoids using low cost-substrates is influenced by cultivation parameters: a review. Int J Mol Sci 22(16):8819. https://doi.org/10.3390/ijms22168819
Kresnowati MTAP, Lestari D, Anshori M, Jafar RM (2020) Production of carotenoids from oil palm empty fruit bunches. IOP Conf Ser: Earth Environ Sci 460 012025. https://doi.org/10.1088/1755-1315/460/1/012025
Kupan S, Hamid H, Kulkarni A, Yusoff M (2016) Extraction and analysis of beta carotene recovery in CPO and oil palm waste by using HPLC. ARPN J Eng Appl Sci 11(4):2184–2188
Manurung H, Silalahi J, Siahaan D, Elisa J (2017) The re-extraction of oil from oil palm empty fruit brunch residues and oil palm mesocarp fibers and measures in reducing greenhouse gas emission. Asian J Agric Biol 5(4):337–345
Masni (2004) Study on utilization of palm oil mill wastes as a source of carotenoid, dissertation IPB University
Mehta BJ, Salgado LM, Bejarano ER, Cerdá-Olmedo E (1997) New mutants of Phycomyces blaskeleeanus for β-carotene production. Appl Environ Microbiol 63(9):3657–3661. https://doi.org/10.1128/aem.63.9.3657-3661.1997
Mortensen A (2009) Supplements. In: Britton G, Liaaen-Jensen F, Pfander H (eds) Carotenoids: nutrition and health, vol 5. Ch.4. Birkhauser Verlag Basel, pp 67–82. ISBN 978-3-7643-7500-3
Ooi C, Choo Y, Yap S, Basiron Y, Ong A (1994) Recovery of carotenoids from palm oil. J Am Oil Chem Soc 71(4):423–426. https://doi.org/10.1007/BF02540524
Othman N, Manan ZA, Wan Alwi SR, Sarmidi MR (2010) A review of extraction technology for carotenoids and vitamin E recovery from palm oil. J Appl Sci 10:1187–1191. https://doi.org/10.3923/jas.2010.1187.1191
Prasertsan S, Prasertsan P (1996) Biomass residues from palm oil mills in Thailand: an overview on quantity and potential usage. Biomass Bioenerg 11(5):387–395. https://doi.org/10.1016/S0961-9534(96)00034-7
Priatni SRI (2016) Review: potential production of carotenoids from Neurospora. Nusantara Biosci 6:63–68
Rammuni MN, Ariyadasa TU, Nimarshana PHV, Attalage RA (2019) Comparative assessment on the extraction of carotenoids from microalgal sources: Astaxanthin from H. pluvialis and β-carotene from D. salina. Food Chem 277:128–134. https://doi.org/10.1016/j.foodchem.2018.10.066
Riberio BD, Barreto DW, Coelho MAZ (2011) Technological aspects of β-carotene production. Food Bioprocess Technol 4:693–770

Rodriguez-Amaya DB (2001) A guide to carotenoid analysis in foods. ILSI Human Nutrition Institute, One Thomas Circle, NW, Washington DC, 20005-5802, p 64

Saini RK, Keum Y (2018) Carotenoid extraction methods: a review of recent developments. Food Chem 240:90–103. https://doi.org/10.1016/j.foodchem.2017.07.099

Schmidt-Dannert C (2000) Engineering novel carotenoids in microorganisms. Curr Opin Biotechnol 11(3):255–261. https://doi.org/10.1016/s0958-1669(00)00093-8

Sudibyo A, Sardjono (2015) The extraction and its separation technology for carotenoids recovery from palm oil: a review. J Riset Teknol Ind 9(1)

Tedeschi RJ (2003) Acetylene. In: Meyers RA (ed) Encyclopedia of physical science and technology, 3rd edn. Academic Press, pp 55–89. https://doi.org/10.1016/B0-12-227410-5/00004-1

Tambunan NA and Rucita JB (2021) Production of carotene from oil palm empty fruit bunch (OPEFB) through fermentation, undergraduate research thesis, Chemical Engineering Department, Institut Teknologi Bandung

Tommy ED, Purba MS (2020) β-carotene production from oil palm empty fruit bunches (OPEFB) which were grown by Neurospora sp., undergraduate research thesis, Food Engineering Department, Institut Teknologi Bandung

Vani PVSN, Chida AS, Srinivasan R, Chandrasekharam M, Singh AK (2001) Synthesis of β-ionone. Synth Commun 31(2):219–224. https://doi.org/10.1081/SCC-100000202

Wang S-L, Chen D-J, Deng B-W, Wu X-Z (2008) Effects of high hydrostatic pressure on the growth and β-carotene production of Rhodotorula Glutinis. Yeast 25:251–257

Effect of Single Clove Black Garlic Extract on Lipid Accumulation During Adipocyte Differentiation Using 3T3-L1 Cell Line

Efri Mardawati, Erin Nur Fitriani, Desi Ayu Sundari, Tenny Putri Wikayani, Nurul Qomarilla, Nova Rachmadona, Eko Fuji Ariyanto, Roni Kastaman, Totok Pujianto, and Devi Maulida Rahmah

Abstract Obesity, defined as an excessive adipose tissue mass, is a major factor in increasing the risk of serious diseases such as heart disease, hypertension, and diabetes. Obesity is associated with the expansion of adipose tissue by excessive dietary fat intake, which results in adipocyte hyperplasia and hypertrophy. Thus, inhibiting adipocyte differentiation and accumulation of lipids are important targets for preventing obesity. As the mechanism of single clove black garlic (SCBG) extract affecting lipid metabolism in adipocytes remains unclear, this study aimed to evaluate the effect of SCBG extract on lipid metabolism in mature 3T3-L1 adipocytes. The analysis revealed that SCBG extract contained 23.15 mg/g of polyphenol and 9.75 mg/g of flavonoid compounds. SCBG extract had stronger capacities to scavenge α,α-diphenyl-β-picrylhydrazyl (DPPH) than fresh single clove garlic (FSCG) with half-maximal inhibitory concentration (IC50) values of 0.602 mg/mL. The treatment of SCBG extracts at a concentration of 2.5–7.5 mg/mL had a cytotoxic effect that reduced cell viability. However, there was no significant difference between the

E. Mardawati (✉) · E. N. Fitriani · D. A. Sundari · R. Kastaman · T. Pujianto · D. M. Rahmah
Department of Agroindustrial Technology, Faculty of Agroindustrial Technology, Universitas Padjadjaran, Jatinangor 45363, Indonesia
e-mail: efri.mardawati@unpad.ac.id

E. Mardawati · N. Rachmadona · T. Pujianto · D. M. Rahmah
Research Collaboration Center for Biomass and Biorefinery Between BRIN and Universitas Padjadjaran, Jatinangor 45363, Indonesia

T. P. Wikayani · N. Qomarilla
Cell Culture Laboratory, Faculty of Medicine, Universitas Padjadjaran, Jl. Eckyman No. 38, Bandung 40161, Indonesia

N. Rachmadona
Department of Chemistry, Faculty of Mathematics and Natural Sciences, Universitas Padjadjaran, Jatinangor 45363, Indonesia

E. F. Ariyanto
Division of Biochemistry and Molecular Biology, Department of Biomedical Sciences, Faculty of Medicine, Universitas Padjadjaran, Jatinangor 45363, Indonesia

M. A. R. Lubis et al. (eds.), *Biomass Conversion and Sustainable Biorefinery*, Green Energy and Technology, https://doi.org/10.1007/978-981-99-7769-7_13

concentration of extract to the cell viability of adipocytes ($p < 0.05$). Furthermore, SCBG extract at 2.5 and 5 significantly reduced lipid accumulation ($p < 0.05$), and 7.5 mg/mL significantly reduced lipid accumulation ($p < 0.01$) compared to cell control indicating potential in anti-obesity effect.

Keywords Black garlic · Phenolic · Flavonoid · Obesity · Lipid accumulation · 3T3-L1 C

1 Introduction

Obesity is a chronic, systemic disease defined as a pathologically increased fat mass associated with an increased health risk (Bischoff and Schweinlin 2020). About 35 percent of the adult population worldwide is overweight and obese, according to a World Health Organization (WHO) scientific survey (WHO 2018). Obesity might increase the risk of developing chronic diseases such as diabetes, cardiovascular, musculoskeletal disorders, and behavioral disorders (Chooi et al. 2019). Obesity is an abnormal or excessive fat accumulation in adipose tissue (Torres-Villarreal et al. 2019). Fat accumulation in adipose tissue involves two metabolic processes: an increase in lipid size in adipocytes (hypertrophy) and an increase in the number of adipose cells (hyperplasia). The process of differentiation of preadipocytes into mature adipocytes is also called the adipogenesis process (Bahmad et al. 2020). 3T3-L1 cells are a cell line used widely as a cell model in learning to control molecular adipogenesis and are associated with obesity (Romao and Guan 2015).

The 3T3-L1 cell line is beneficial in identifying molecular markers, transcription factors (TFs), and various interactions that occur in adipogenesis (Zebisch et al. 2012). The amount of differentiation on the 3T3-L1 cell can be seen from the accumulation of lipids after cells are given several pro-differentiating agents. These agents include insulin, dexamethasone (DEX), and 1-methyl-3-isobutyl xanthine (IBMX), which can increase intracellular cAMP levels through the presence of fetal bovine serum (FBS) (Morrison and McGee 2015). 3T3-L1 cells have been used extensively to evaluate the effects of compounds or nutrients on adipogenesis, to establish the molecular mechanisms underlying adipogenesis, and to evaluate the potential applications of various compounds and nutrients in the treatment of obesity (Kang et al. 2016). In particular, compounds such as quercetin (Eseberri et al. 2015) and resveratrol (Chang et al. 2016) are known to inhibit adipogenesis in 3T3-L1 adipocytes.

Garlic *(Allium Sativum Linn)* is a plant considered to have a pharmacological effect that can be useful in various treatments, including obesity (Batiha et al. 2020). The organosulfur compounds contained in garlic have been shown to have various biological activities. The four most critical organosulfur compounds, which are considered to be the main biological agents, are diallyl sulfide (DAS), diallyl disulfide (DADS), diallyl trisulfide (DATS), and allyl-methyl trisulfide (Yang et al. 2018). A study by Li et al. (2017) evaluated the effects of a garlic compound (DATS) on pre-adipocytes

3T3-L1 and the mechanisms involved. The test results indicated that the administration of (20–75 μM) DATS could inhibit CCAAT or enhancer-binding protein (C/EBP α and β). Proliferator-activated receptor (PPAR) c mRNA that causes decreased expression of fatty acid synthase (FAS) and lipid accumulation in 3T3-L1 cells suggests that DATS compounds may inhibit the differentiation of preadipocyte cells 3T3-L1 becomes an adipocyte. Garlic is therefore considered to be useful for the prevention of obesity.

Black garlic is a derivative of garlic products obtained from the fermentation process for a while (1–2 months) at a controlled high temperature (60–90 °C) with high humidity (80–90%) (Angeles et al. 2016). Compared to fresh garlic, black garlic does not have a distinctive taste and a sharp scent that customers are less interested in due to the reduced content of allicin compounds that are unstable at high temperatures. Allicin is converted during fermentation into more stable compounds such as S-allyl cysteine (SAC), bioactive alkaloids, polyphenols, and flavonoids known to be antioxidants (Kimura et al. 2017).

Currently, research into black garlic products is still based on the form of multi-clove garlic, and research into the potential of single-clove garlic is still limited. In research by Chen et al. (2019), a comparison was made with the bioactive content of 4 varieties of garlic, including multi-clove and single-clove garlic, and the results of this study showed that the content of bioactive compounds in single-clove garlic was higher compared to multi-clove garlic. Considering the potential of single-clove garlic varieties, currently evaluating the potential of black garlic with single-clove garlic as raw materials against 3T3-L1 anti-obesity is still very limited. In this study, we examine the effect of single clove black garlic (SCBG) extract on lipid accumulation in the adipocyte differentiation process using 3T3-L1 cell lines as a reference for the potential use of SCBG in the treatment of obesity.

2 Materials and Method

2.1 Preparation of Single Clove Black Garlic (SCBG) Extract

Single-clove garlic was manufactured for 21 days. SCBG at 1 kg was homogenized and extracted with 70% methanol (MeOH; 5L) for 24 h. The mixture was then filtered with Whatman No. 3 filter paper. The total filtrate was concentrated by evaporation, and the residue of SCBG was dissolved in deionized water.

2.2 Determination of Bioactive Compound

The method described previously by Kim et al. (2013) with slight modification. Briefly, 3 g of each sample was weighed in an extraction tube, and 10 mL of 70%

methanol was added. The extract was mixed for 20 min. The extracts were centrifuged at 6000 rpm for 10 min. The supernatant was decanted in a graduated conical tube. The extraction step was repeated third times. Both extracts were pooled, and the volume was adjusted to 10 mL with 70% methanol. One milliliter of the extract was diluted with water to 50 mL.

2.2.1 Total Phenolic Compound

The total polyphenol compound was determined spectrophotometrically, using gallic acid as a standard, according to the method described by the International Organization for Standardization (ISO) 14502-1. Briefly, 1.0 mL of the diluted sample extract was transferred in triplicate to separate tubes containing 5.0 mL of a 1/10 dilution of Folin Ciocalteu's reagent in water. Then, 4.0 mL of a sodium carbonate solution (7.5%, w/v) was added. After 1 h at room temperature, the absorbance was read at $(\lambda) = 756$ nm. All values were expressed as mg gallic acid equivalents (GAE)/mg dry matter of the garlic sample.

2.2.2 Total Flavonoid Compound (TFC)

The total flavonoid compound was determined using a colorimetric method described previously by Kim et al. (2013). A portion of 0.5 mL extract was taken, and 0.5 mL of 2% ethanolic solution of $AlCl_3$ was added. After 1 h at room temperature, the absorbance was read at $(\lambda) = 420$ nm. All values were expressed as mg quercetin equivalents (QE)/mg dry matter of garlic.

2.3 *Determination of Antibacterial Activity*

Antibacterial activity analysis refers to clinical and international laboratory committee standards (Riyanti et al. 2020). There are several test procedures, including the affiliation of tools and materials, the provision of test bacteria, preparation and standardization of bacterial suspense, and analysis of antibacterial activity.

Petri dishes containing the media are to be prepared at room temperature for 10–15 min. Vortex bacterial suspense until homogeneous, then insert the swab into the suspension and apply it to the media evenly. Let stand for 15 min. Then, prepare a paper disk that has been saturated by the sample and store it on the top layer of the agar media. Then, incubate at 37 °C for 24 h. After that, observations were made using the calipers (measuring the diameter/radius of the zone of inhibition).

2.4 Determination of Antioxidant Activity

The free radical scavenging activity of SCBG was determined based on the scavenging activity of the stable DPPH (2,2-diphenyl-1-picrylhydrazyl) free radical using the method described by Choi et al. (2014) with slight modification. A sample of SCBG 5 mg dissolved in 5 mL methanol solution as a stock solution, stock solutions diluted on a range of 0.2, 0.4, 0.6, and 0.8 (mg/mL) and 0.2 mL DPPH was added dissolved in methanol solution (1 mL). After incubating the solution at room temperature in the dark for 30 min, the absorbance was measured at (λ) = 517 nm.

2.5 Cell Culture and Cell Viability Assay

3T3-L1 preadipocyte cells were cultured using 12-well plates in DMEM medium (Stigma, USA) containing 10% FBS (Stigma, USA) in a humidified atmosphere of 5% CO_2 at 37 °C. To induce adipogenesis, the cells were cultured for 2 days until confluency. This induction was in the presence of the differentiation mixture containing 0.5 mM methyl-isobutyl-xanthine (Sigma, USA), 1 μM dexamethasone (Sigma, USA), and 10 μg/ml insulin (Sigma, USA).

Cell viability assay was performed according to the method described by Torres-Villarreal et al. (2019). Four-well plates were used in this experiment. Each well was added with different concentrations of C. longa extracts (0, 2.5, 5, and 7.5 mg/mL), and cells were then incubated for 48 h.

The cells were incubated for 48 h. After 48 h, on day 3, the medium was replaced with DMEM containing 10 μg/m insulin to optimize glucose uptake into the cells and lipogenesis during the differentiation process. On day 5, the medium was then replaced with DMEM. On day 7, the culture medium was replaced again with DMEM. On day 12, the optimal adipocyte differentiation was obtained in concentration control (0 mg/mL).

2.6 Determination of Lipid Accumulation by Oil Red O Staining

Oil Red O staining was performed using Kim's protocol (Nam et al. 2018) with minor modifications. The morphology of the cells was examined under an inverted microscope, and the images were captured. The colored oil droplets were dissolved in 100% isopropanol, and the relative lipid accumulation content was measured by reading the absorbance at a wavelength (λ) = 520 nm with an ELISA reader.

2.7 Statistical Analysis

All experiments were carried out in triplicate, and data were expressed as mean ± standard deviation (SD) using SPSS 25.0 version. One-Way analysis of variance (ANOVA) and Duncan's multiple comparison tests were used to determine the significance of the difference among samples with a significance level of ($p < 0.05$).

3 Result and Discussion

3.1 Bioactive Compound of Black Garlic

3.1.1 Total Phenolic Compound

Hydroxycinnamic acid derivatives are garlic's major phenolic acid compounds (Kim et al. 2013). There is an increase in hydroxycinnamic acid and phenolic acid derivatives during the development phase of black garlic products (Kimura et al. 2017). In making black garlic, the content of phenolic compounds may be caused by the process of creating complex compounds bound by esterification and glycolysis reactions, increasing free phenolic forms (Lu et al. 2017).

The total phenolic compound in SCBG was (23.15 ± 0.15 mg GAE/g dry basis), which was 2.5 times higher compared to TPC content in FSCG (9.26 ± 0.20 mg GAE/g dry basis). These results are consistent with those obtained by Choi et al. (2014), who reported that, compared to fresh garlic (13.91 ± 1.62 mg GAE/g dry basis), the concentration of polyphenol compounds in black garlic heated for 21 days was significantly higher (58.33 ± 1.90 mg GAE/g dry basis). In another study, Jang et al. (2018) showed. an increase in total polyphenol content in samples of black garlic containing (43.01 ± 2.85 mg GAE/g dry basis), compared to fresh garlic (2.86 ± 0.14 mg GAE/g dry basis). The difference between the value of phenolic compounds may be related to the different sources of garlic varieties. The increase of TPCs in SCBG compared to fresh single-clove garlic (FSCG) is shown in Fig. 1.

The antioxidant activity possessed by phenolic compounds was further established for treating various diseases. In this case, phenolic compounds are also considered essential in stabilizing lipids against peroxidation and inhibiting different forms of oxidizing enzymes to better treat obesity (Anyanwu et al. 2020). The mitochondrial-targeted antioxidant action of phenolic compounds can be a potential mechanism to treat obese inflammation. An array of phenolic compounds has displayed AMPK-activating properties in adipocyte models (Zhang and de Mejia 2020). According to Wang et al. (2014), intake of polyphenol compounds can help prevent obesity by reducing lipogenesis, increasing lipolysis, stimulating fatty acid oxidation (FA), inhibiting adipocyte differentiation and growth, weakening the inflammatory response, and suppressing the occurrence of oxidative stress.

Fig. 1 A comparison of total phenolic compound (mg GAE/g dry basis) of FSCG and SCBG extract. Data represent mean ± SD, $n = 3$; $*P < 0.05$ significant difference between the two groups

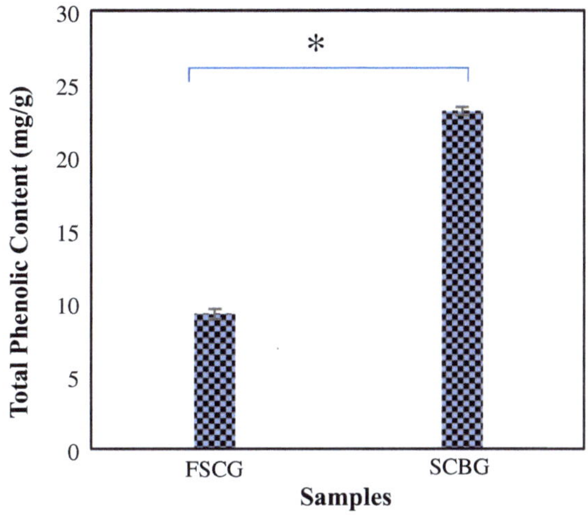

3.1.2 Total Flavonoid Compound

Several in vitro studies have shown that flavonoid class compounds affect adipocyte cells, where flavonoids can reduce cell viability and proliferation, inhibit triglyceride aggregation, promote lipolysis, and reduce inflammation. (Herranz-López et al. 2017). The anti-adipogenic effect of flavonoids is mainly via their effect on the regulation of several pathways, such as induction of apoptosis, suppression of key adipogenic transcription factors, activation of AMPK and Wnt pathways, inhibition of clonal expansion, and cell-cycle arrest (Khalilpourfarshbafi et al. 2019). Quercetin and structurally similar luteolin are ubiquitous dietary flavones found in a wide range of herbs, and their anti-obesity effects are well established; this is believed to be mediated by increasing the expression of AMPK, which subsequently reduces the differentiation and proliferation of human preadipocytes 3T3-L1 and induces apoptosis (Woon and Toh 2014).

Flavonoid compound (TFCs) levels have been expressed as Quercetin equivalents. The increase of TFCs in SCBG compared to FSCG is shown in Fig. 2. From the results, the total flavonoids in the SCBG sample were (9.75 ± 0.27 mg QE/g dry basis). This level increased by 6 times when compared to the total flavonoid content FSCG (1.35 ± 0.11 mg QE/g dry basis).

Quercetin has been shown to minimize intracellular ROS in the hypertrophic adipocyte model (Leiherer et al. 2016). Anthocyanins can suppress lipid accumulation in adipocytes due to widespread inhibition of transcription factors that control lipogenesis, such as active peroxisome receptors and binding proteins to the CCAAT conjugator (Lee et al. 2014). In studies with animal models, flavonoid compounds have also shown positive results in preventing and treating obesity. Based on research from You et al., flavonoids can increase energy expenditure or inhibit food intake through various processes that suppress oxidative stress and release gastrointestinal

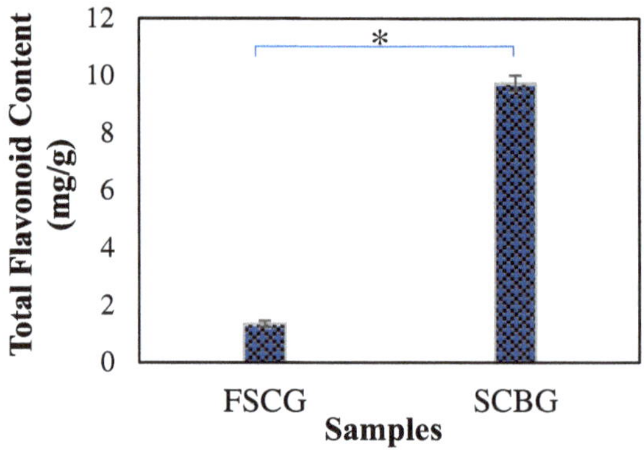

Fig. 2 A comparison of total flavonoid compound (mg QE/g dry basis) of FSCG and SCBG extract. Data represent mean ±SD, $n = 3$; $*P < 0.05$ significant difference between the two groups

Table 1 Comparison of total phenolic and total flavonoid compound

Samples	Total phenolic compound (mg/g)	Total flavonoid compound (mg/g)
FSCG	9.26 ± 0.20^a	1.35 ± 0.11^a
SCBG	23.15 ± 0.15^b	9.75 ± 0.27^b

Data represent mean ±SD, $n = 3$; values by different letters were indicated significantly different ($p < 0.05$), according to Duncan's test

Different symbols indicate significant differences in each aspect/column (consisting of: total phenolics, total flavonoids, and antioxidant activity) based on the post-hoc test (DMRT, $p < 0.05$)

hormones. The recapitulation of total phenolic and total flavonoid compounds of FSCG and SCBG is shown in Table 1.

In addition, black garlic's total polyphenol and flavonoid contents significantly increased during the aging period. It shows the potential use of black garlic in various diseases, including obesity.

3.2 *Antibacterial Activity*

The antibacterial activity of SCBG can be seen based on the size of the precise zone diameter (mm) formed around the paper disk against the bacteria tested, namely *Escherichia coli and Staphylococcus aureus.* A clear zone is an area not overgrown with bacteria and looks more apparent than the surrounding area. The greater the inhibitory zone formed, the greater the ability of antibacterial activity. Based on the analysis made on samples against the growth of *E. coli and S. aureus,* a zone of inhibition formed around the paper disk. The inhibition zone decreased for the sample with more longer heating duration.

The antibacterial activity of *S. aureus* looks better when compared to *E.coli* due to differences in the cell wall structure in the test bacteria. *E.coli* is a gram-negative

Table 2 Comparison on minimum inhibition of antibacterial activities

Samples	Inhibition Zone (mm)	
	E. coli	S. aureus
FSCG	4.5	6.58
SCBG	2	0.65

Data represent the measurement results of the clear zone average diameter

bacterium that has a cell wall, higher lipid content, and a multilayer cell wall structure consisting of lipoprotein, phospholipid outer membrane, and lipopolysaccharide, which causes gram-negative bacterial cell walls to be difficult to penetrate by antibacterial substances while gram-positive bacteria have a layer of peptidoglycan on the outside and have less role in effective defense permeability. The measurement results of the apparent zone average of FSCG and SCBG diameter are shown in Table 2.

The results showed that FSCG in this study showed higher antibacterial activity compared to black garlic. The compound that acts as an antibacterial is an organosulfur compound, which is allicin (Lawson and Hunsaker 2018). However, allicin compounds are unstable against high temperatures (Zhang et al. 2015). The heating carried out in producing black garlic with high temperatures causes the loss of and damage to the allicin compound. Compared to fresh garlic, black garlic does not cause a strong taste because of reduced allicin compounds converted into antioxidant compounds such as S-ally-cysteine (SAC), bioactive alkaloids, polyphenols, and flavonoid compounds (Botas et al. 2019).

3.3 *Antioxidant Activity*

The antioxidant activity of SCBG extracts was investigated using the DPPH scavenging assay. The DPPH assay is widely used due to the relatively short time required for the analysis. The method is based on the scavenging of DPPH by antioxidant compounds, which includes a reduction reaction that decolorizes the DPPH methanol solution (Lu et al. 2017). The DPPH radical scavenging activity indicates the ability of an antioxidant compound to donate electrons or hydrogen, thereby converting DPPH into a more stable molecule with a reduced absorbance (Wu et al. 2020). The DPPH radical scavenging activity of black garlic samples is shown in Fig. 3.

Based on the analysis of the percent inhibition value of the sample against DPPH radicals, it was found that there was an increase in the percent inhibition of SCBG compared to FSCG. There was an increase in the percent inhibition in the sample where the FSCG extract had a percent inhibition range (of 9.07–60.6%) while the SCBG extract was (41.10–95.78%). The DPPH free radical scavenging activity of SCBG within 21 days of the heating process was significantly higher by approximately 3-fold from FSCG ($p < 0.05$). The increase in the percent inhibition of SCBG extract may be due to an increase in antioxidant compounds in black garlic, such as

Fig. 3 Radical scavenging activity determined using DPPH method. The logistic model in origin was used to calculate half-inhibition (IC50). "X" represents the different FSCG and SCBG extract concentrations, and "Y" represents the radical scavenging activity rate

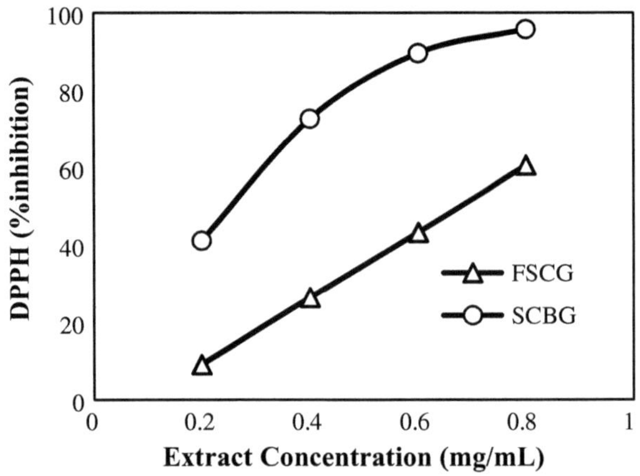

total phenolic and flavonoid compounds, where there is an increase in the content of these compounds, which are also known to have antioxidant activity (Bae et al. 2012).

The concept of the half maximal inhibitory concentration (IC_{50}) is extensively used in the pharmaceutical world to measure its effectiveness in inhibiting biological or biochemical functions. The IC_{50} value indicates the inhibitor concentration required to inhibit a given biological or biochemical function by half. In other words, large IC_{50} values denote inhibitors that interact less effectively with an enzyme than those with small IC_{50} values (Caldwell et al. 2012). The IC_{50} value is inversely related to the percentage value of inhibition or the ability of the compound to inhibit free radical activity, which is related to the concentration of an extract.

The IC_{50} value of FSCG and SCBG is shown in Table III. Based on the result, SCBG has a lower IC_{50}, indicating a more significant free radical scavenging activity (Table 3).

The abilities to scavenge DPPH radical (%) of SCBG extract (4 different concentrations) at the same time were used to calculate the IC_{50}. According to Duncan's test, values by different letters were indicated significantly differently ($p < 0.05$).

Bae et al. (2012), Choi et al. (2014), and Jang et al. (2018) have described similar results for DPPH scavenging activity and reducing power using samples of fresh garlic purchased in local Korean markets and subjected to heat treatment by the authors.

Table 3 Comparison of IC_{50} value of single fresh clove black garlic (FSCG) and single clove black garlic (SCBG)

Samples	IC_{50} (mg/mL)
FSCG	0.602 ± 0.91^{a}
SCBG	0.320 ± 0.36^{b}

Different symbols indicate significant differences in each aspect/column (consisting of: total phenolics, total flavonoids, and antioxidant activity) based on the post-hoc test (DMRT, $p < 0.05$)

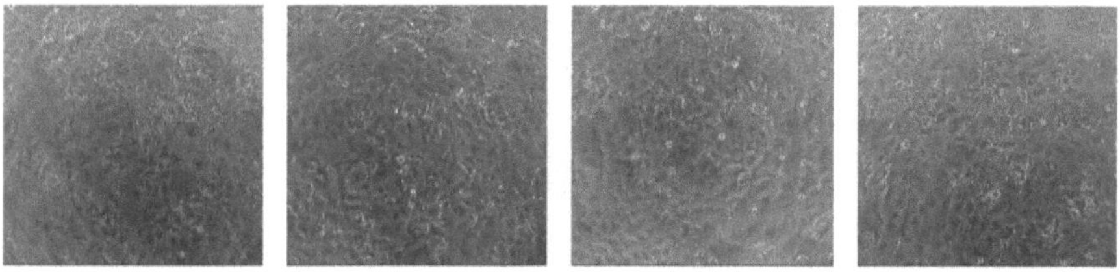

Fig. 4 Representative photographs under a 200x magnification microscope for the cell viability effect of SCBG extract to the 3T3-L1 cells line

3.4 *Cell Viability*

When a new drug, either derived from natural or synthesized material, is being developed, it is necessary to check its safety for the host cell or its cytotoxic effect on cells. This test is known as the cell viability test. Among the viability tests that depend on converting substrates to chromogenic products by living cells, the MTT test is one of the most widely used tests (Kumar et al. 2018).

To investigate the effect of SCBG extract on viability in mature 3T3-L1 adipocytes, the cell was treated with various concentrations of SCBG extract for 72 h. When viewed from the condition of the photogenic observation cells under the microscope, there was no significant damage to the 3T3-L1 cells affected by the SCBG extract. Representative photographs under a 200x magnification microscope for the cell viability effect of SCBG extract are shown in Fig. 4.

Based on cell viability measurement, in preadipocytes, SCBG extract at 2.5, 5, and 7.5 mg/mL significantly decreased cell viability, resulting in cell viability values of 84.70 ± 0.80, 79.12 ± 2.92 and $79.37 \pm 4.30\%$. A diagram of the percent cell viability can be seen in Fig. 5.

There was a decrease in the viability of cells given SCBG extract at 2.5, 5, and 7.5 mg/ mL compared to the control. However, the concentration of the extract did not significantly decrease cell viability or the number of dead cells. To ensure that SCBG extract can inhibit lipid accumulation, several tests, such as inhibiting CCAAT or enhancer-binding protein (C/EBP α and β) and proliferator-activated receptor (PPAR) c mRNA, should be performed.

3.5 *Effects of Single Clove Black Garlic on Lipid Accumulation in 3T3-L1 Preadipocytes Differentiation*

The inhibitory effect of SCBG extract on lipid accumulation was evaluated by Oil Red O (ORO) staining. ORO staining is widely used to detect lipids in cells and tissues. The increase in adipocytes is known to be closely related to the accumulation of lipid

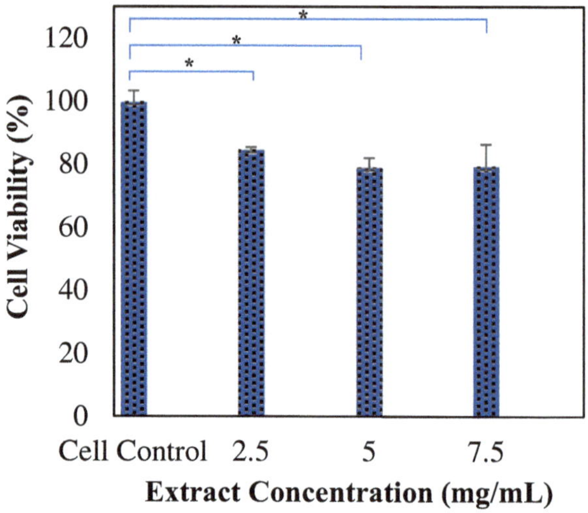

Fig. 5 Cell viability. Cell viability was determined by CCK-8 assay. 3T3-L1 preadipocytes were treated with 0, 2.5, 5, and 7.5 mg/mL SCBG extract for 72 h. Data represent mean ±SD, $n = 9$; *$P < 0.05$ significant difference between the two groups

content (Kang et al. 2021). Therefore, ORO-stained cells indicate the degree of lipid accumulation in 3T3-L1 adipocytes.

The differentiation of 3T3-L1 preadipocytes was initiated with an inducer after two days of contact inhibition (when preadipocytes exited from the growth cycle). 3T3-L1 preadipocytes differentiated with the treatment of SCBG extract at the indicated concentration for 18 days. After the differentiation of preadipocytes along with the treatment of SCBG for 18 days, ORO staining and subsequent quantification were performed to examine intracellular lipid accumulation.

As seen in Fig. 6, microscopic observations show that the treatment of black garlic extract can reduce lipid data in the 3T3-L1 cell line. The lipid droplets in differentiation media-treated cells became larger with a deeper red color; however, the treatment decreased these phenomena, which is indicated by the reduction of red cells. It could be related to the synthesis of triglycerides in mature adipocytes and the increased hydrolysis of intracellular triglycerides, which in turn prevents the accumulation of lipids in adipocytes, thereby inhibiting fat cell hypertrophy.

After Oil-Red O staining, the stained oil droplets were dissolved in 100% isopropanol, and the relative triglyceride content was measured by reading the absorbance at the wavelength (λ) = 520 nm with an ELISA reader. Based on the results of the calculation of the percentage of accumulated lipids, black garlic extract at concentrations of 2.5, 5, and 7.5 mg/mL resulted in lipid accumulation values of 76.15 ± 6.50, 71.93 ± 8.49 and 57.84 ± 4.10%, diagram of the accumulation of lipid content can be seen in Fig. 7.

The measurement results of cell lipid accumulation added with black garlic, and the value was smaller when compared to control cells ($p < 0.05$), ($p < 0.01$), the value of lipid accumulation in the sample significantly decreased with the higher concentration of the extract added.

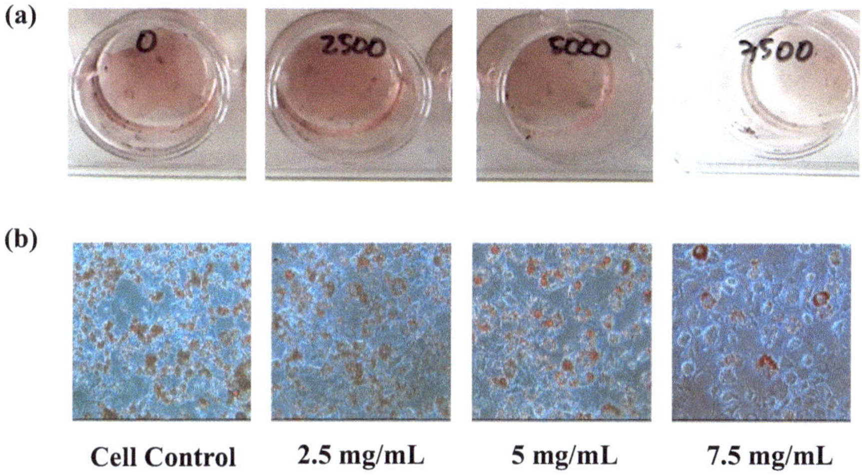

Fig. 6 Lipid accumulation. 3T3-L1 preadipocytes were treated with SCBG extract with 0, 2.5, 5, and 7.5 mg/mL. Mature 3T3-L1 adipocytes were stained with oil red O after 18 days. **a** Lipid accumulation was observed in cell culture media. **b** Microscopic pictures were taken at 200x magnification

Fig. 7 Lipid accumulation. Values were measured from isopropanol elution of the oil red O stained cells. Data represent mean ±SD, $n = 9$; $*P < 0.05$ significant difference between the two groups. $**P < 0.01$ significant difference between the two groups

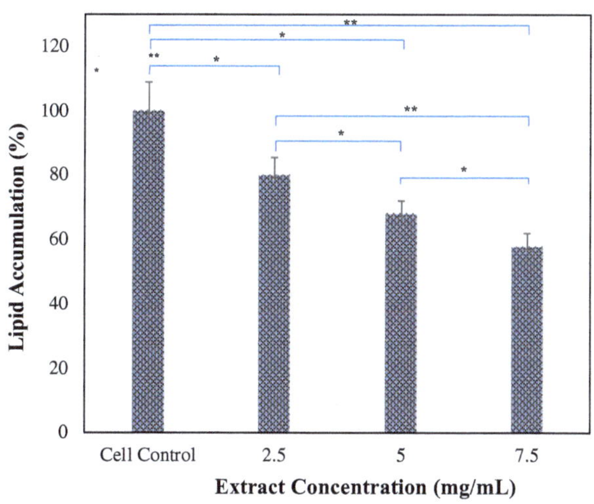

Wu et al. specifically tested melanoidin compounds in black garlic to determine their anti-obesity effect, based on the results of research on the effect of melanoidin on C57BL/6 J mice with obesity induced by a high-fat diet (HFD) that the administration of melanoidin orally, has a significant effect in reducing weight gain and white adipose tissue weight and reversing glucose tolerance, especially at high doses. Nam et al. reported that aged black garlic (ABG) extract at 2.5 and 5 mg/mL significantly reduced protein expression of proliferator-activated receptor c (PPARc) and perilipin in mature 3T3-L1 adipocytes. The hormone-sensitive lipase (HSL) and Ser563-pHSL levels were significantly reduced by treatment with 5 mg/mL of ABG extract.

4 Conclusion

Single-clove black garlic (SCBG), produced by aging single-clove garlic for 21 days, has shown higher antioxidant properties when compared to fresh single-clove garlic (FSCG). Measurement results of lipid accumulation added with SCBG. The value was lower compared to control cells ($p < 0{,}05$), ($p < 0.01$), and the value of lipid accumulation in the sample decreased significantly with a higher concentration of the added extract. These results suggest that SCBG extract may have anti-lipogenic effects on adult 3T3-L1 adipocytes that could be associated with potential. These results suggest that black garlic extract may have anti-lipogenic effects on mature 3T3-L1 adipocytes, which could be associated with the potential for black garlic in treating obesity.

Acknowledgements This research was funded by Directorate of Higher Education and Universitas Padjadjaran, Indonesian Ministry of National Education (DIKTI) under the scheme of Science and Technology Application Research. We would also like to express our gratitude to Dr. Afiat Berbudi for the kind gift of 3T3-L1 cell line and to Universitas Padjadjaran for the research fund. This study was conducted at Cell Culture Laboratory, Faculty of Medicine, Universitas Padjadjaran, Indonesia

References

Angeles TMM, Jesús PA, Rafael MR, Tania MA (2016) Evolution of some physicochemical and antioxidant properties of black garlic whole bulbs and peeled cloves. Food Chem 199:135–139
Anyanwu GO, Kolb AF, Bermano G (2020) Antiobesity functional leads and targets for drug development. In: Phytochemicals as lead compounds for new drug discovery, pp 143–160
Bae SE, Cho SY, Won YD, Lee SH, Park HJ (2012) A comparative study of the different analytical methods for analysis of S-allyl cysteine in black garlic by HPLC. LWT – Food Sci Technol 46:532–535
Bahmad HF, Daouk R, Azar J, Sapudom J, Teo JCM, Abou-Kheir W, Al-Sayegh M (2020) Modeling adipogenesis: current and future perspective. Cells 9:2326
Batiha GES, Beshbishy AM, Wasef LG, Elewa YHA, Al-Sagan AA, El-Hack MEA, Taha AE, Abd-Elhakimm YM, Devkota HP (2020) Chemical constituents and pharmacological activities of garlic (*Allium sativum* L.): a review. Nutrients 12:1–21
Bischoff SC, Schweinlin A (2020) Obesity therapy. Clin Nutr ESPEN 38:9–18
Botas J, Fernandes Â, Baros L, Alves MJ, Carvalho AM, Ferrerira ICFR (2019) A comparative study of black and white *Allium sativum* L.: nutritional composition and bioactive properties. Molecules 24:1–11
Caldwell GW, Yan Z, Lang W, Masucci JA (2012) The IC50 concept revisited. Curr Top Med Chem 12:1282–1290

Chang CC, Lin KY, Peng KY, Day YJ, Hung LM (2016) Resveratrol exerts anti-obesity effects in high-fat diet obese mice and displays differential dosage effects on cytotoxicity, differentiation, and lipolysis in 3T3-L1 cells. Endocrine J 63:169–178

Chen C, Cai J, Liu SQ, Qiu GL, Wu XG, Zhang W, Chen C, Qi WL, Wu Y, Liu ZB (2019) Comparative study on the composition of four different varieties of garlic. Peer J 2:1–17

Choi IS, Cha HS, Lee YS (2014) Physicochemical and antioxidant properties of black garlic. Molecules 19:16811–16823

Chooi YC, Ding C, Magkos F (2019) The epidemiology of obesity. Metab: Clin Exp 92:6–10

Eseberri I, Miranda J, Lasa A, Churruca I, Portillo MP (2015) Doses of quercetin in the range of serum concentrations exert delipidating effects in 3T3-L1 preadipocytes by acting on different stages of adipogenesis, but not in mature adipocytes. Oxid Med Cell Longev 2015:1–11

Herranz-López M, Borrás-Linares I, Olivares-Vicente M, Gálvez J, Segura-Carretero A, Micol V (2017) Correlation between the cellular metabolism of quercetin and its glucuronide metabolite and oxidative stress in hypertrophied 3T3-L1 adipocytes. Phytomedicine 25:25–28

Jang HJ, Lee HJ, Yoon DK, Ji DS, Kim JH, Lee CH (2018) Antioxidant and antimicrobial activities of fresh garlic and aged garlic by-products extracted with different solvents. Food Sci Biotechnol 27:219–225

Kang MC, Kang N, Ko SC, Kim YB, Jeon YJ (2016) Anti-obesity effects of seaweeds of Jeju Island on the differentiation of 3T3-L1 preadipocytes and obese mice fed a high-fat diet. Food Chem Toxicol 90:36–44

Kang YM, Kang HA, Cominguez DC, Kim SH, An HJ (2021) Papain ameliorates lipid accumulation and inflammation in high-fat diet-induced obesity mice and 3T3-L1 adipocytes via AMPK activation. Int J Mol Sci 22:1–16

Khalilpourfarshbafi M, Gholami K, Murugan DD, Sattar MZA, Abdullah NA, (2019) Differential effects of dietary flavonoids on adipogenesis. Eur J Nutr 58:5–25

Kim JS, Kang OJ, Gweon OC (2013) Comparison of phenolic acids and flavonoids in black garlic at different thermal processing steps. J Funct Foods 5:80–86

Kimura S, Tung YC, Pan MH, Su NW, Lai YJ, Cheng KC (2017) Black garlic: a critical review of its production, bioactivity, and application. J Food Drug Anal 25:62–70

Kumar P, Nagarajan A, Uchil PD (2018) Analysis of cell viability by the MTT assay. Cold Spring Harb Protoc 6:469–471

Lawson LD, Hunsaker SM (2018) Allicin bioavailability and bioequivalence from garlic supplements and garlic foods. Nutrients 10:1–49

Lee B, Lee M, Lefevre M, Kim HR (2014) Anthocyanins inhibit lipogenesis during adipocyte differentiation of 3T3-L1 preadipocytes. Plant Foods Hum Nutr 69:137–141

Leiherer A, Stoemmer K, Muendlein A, Saely CH, Kinz E, Brandtner EM, Fraunberger P, Drexel H (2016) Quercetin impacts expression of metabolism- and obesity-associated genes in SGBS adipocytes. Nutrients 8:1–12

Li Y, Rong Y, Bao L, Nie B, Ren G, Zheng C, Amin R, Arnold RD, Jeganathan RB, Huggins KW (2017) Suppression of adipocyte differentiation and lipid accumulation by stearidonic acid (SDA) in 3T3-L1 cells. Lipids Health Dis 16:1–10

Lu X, Li N, Qiai X, Qiu Z, Liu P (2017) Composition analysis and antioxidant properties of black garlic extract. J Food Drug Anal 25:340–349

Morrison S, McGee SL (2015) 3T3-L1 adipocytes display phenotypic characteristics of multiple adipocyte lineages. Adipocyte 4:295–302

Nam H, Jung H, Kim Y, Kim B, Kim KH, Par SJ, Suh JG (2018) Aged black garlic extract regulates lipid metabolism by inhibiting lipogenesis and promoting lipolysis in mature 3T3-L1 adipocytes. Food Sci Biotechnol 27:575–579

Riyanti, Balansa W, Liu Y, Sharma A, Mihajlovic S, Hartwig C, Leis B, Rieuwpassa FJ, Ijong FG, Wägele H, König GM, Schäberle TF (2020) Selection of sponge-associated bacteria with high potential for the production of antibacterial compounds. Sci Rep 10:19614

Romao JM, Guan LL (2015) Adipogenesis and obesity. In: MicroRNA in regenerative medicine, pp 539–565

Torres-Villarreal D, Camacho A, Castro H, Ortiz-Lopez R, de la Garza AL (2019) Anti-obesity effects of kaempferol by inhibiting adipogenesis and increasing lipolysis in 3T3-L1 cells. J Physiol Biochem 75:83–88
Wang S, Moustaid-Moussa N, Chen L, Mo H, Shastri A, Su R, Bapat P, Kwun IS, Shen CL (2014) Novel insights of dietary polyphenols and obesity. J Nutr Biochem 25:1–18
WHO (2018) World Health Organization — Obesity. Retrieved from WHO website: http://www.who.int/topics/obesity/en/
Woon ECY, Toh JDW (2014) Antiobesity effects of natural products from an epigenetic perspective. Stud Nat Prod Chem 41:161–193
Wu J, Liu Y, Dou Z, Wu T, Liu R, Sui W, Jin Y, Zhang M (2020) Black garlic melanoidins prevent obesity, reduce serum LPS levels and modulate the gut microbiota composition in high-fat diet-induced obese C57BL/6J mice. Food Funct 11:9585–9598
Yang C, Li L, Yang L, Lǚ H, Wang S, Sun G (2018) Anti-obesity and Hypolipidemic effects of garlic oil and onion oil in rats fed a high-fat diet. Nutr Metab 15:1–8
Zebisch K, Voigt V, Wabitsch M, Brandsch M (2012) Protocol for effective differentiation of 3T3-L1 cells to adipocytes. Anal Biochem 425:88–90
Zhang Q, de Mejia EG (2020) Protocatechuic acid attenuates adipogenesis-induced inflammation and mitochondrial dysfunction in 3T3-L1 adipocytes by regulation of AMPK pathway. J Funct Foods 69:1–12
Zhang X, Li N, Lu X, Liu P, Qiao X (2015) Effects of temperature on the quality of black garlic. J Sci Food Agric 96:2366–2372

Physical and Organoleptic Characteristic of Bread Substituted with *Spirulina Platensis*

Souvia Rahimah, Muhammad Fadly Sonjaya, Robi Andoyo, Awalina Satya, Siti Nurhasanah, and Tjandra Chrismadha

Abstract Spirulina is a microalgae that can be used as an alternative protein source to improve the physical and sensory properties of baked products. This article outlines the effects of *Spirulina platensis* substitution on the physical, i.e., specific volume, porosity, texture, and color index, and the organoleptic characteristics of the bread. This study allowed us to identify the optimal concentration of spirulina addition which ranges from 2 to 4% (dry matter). At this concentration, the bread has a reduced specific volume, crumb hardness, and lightness degree. Contrarily, there is a significant increase in bread porosity and gumminess. In terms of the organoleptic parameters, spirulina substitution is acceptable to be added within 2–4%, though the hedonic trend decreases as the concentration increases.

Keywords Spirulina platensis · Bread · Organoleptic · Protein alternative · Substitution

1 Introduction

Spirulina platensis is a microalgae that contains a protein content of 60–70% (dry weight) followed by levels of carbohydrates, fats, and water ((Danesi et al. 2004; Rosa et al. 2015). The protein is composed of essential and non-essential amino acids which are supported by other components such as γ linoleic acid, carotenoids, linoleic acid, arachidonic acid, vitamins, iron, calcium, phosphorus, RNA and DNA nucleic acids, chlorophyll, and phycocyanin (Singh et al. 2015).

Spirulina has been added to various food products and processed products (Rabelo et al. 2013; Saharan and Jood 2021). Apart from its high protein content, spirulina

S. Rahimah (✉) · M. F. Sonjaya · R. Andoyo · S. Nurhasanah
Department of Food Industrial Technology, Faculty of Agro-Industrial Technology, Universitas Padjadjaran, Jatinangor, Indonesia
e-mail: souvia@unpad.ac.id

A. Satya · T. Chrismadha
Research Centre for Limnology and Water Resources, National Research and Innovation Agency, KST Soekarno, Cibinong, Indonesia

M. A. R. Lubis et al. (eds.), *Biomass Conversion and Sustainable Biorefinery*, Green Energy and Technology, https://doi.org/10.1007/978-981-99-7769-7_14

also has a group of bioactive compounds in the form of pigments which are useful as antioxidants, anti-inflammatories, anti-microbials, and preventive effects for certain diseases (Borowitzka 2009; Lafarga et al. 2020). Substitution of spirulina into food products is one way to increase the functional value of food products.

The trend to consume functional foods is increasing globally. As a result, the food industry has gained around 30% more profits than the previous year (2019) (BPS 2020). Functional food products are categorized into several product groups, i.e. fermented products (tape, yogurt, kefir), plant-based analogue products (meat analogue, cheese analogue, and rice analogue), and fortified products (vitamin D fortified milk and vitamin E palm oil) (Stübler et al. 2020). This trend was also followed by an increase in the consumption of bakery products. Therefore, the development of functional food in bread products by utilizing spirulina has great potential for the food industry.

The development of bread as a bakery product into functional food has been carried out by many national and international researchers. The development was carried out to improve the quality of bread from the nutritional aspect, physicochemical, and organoleptic characteristics into functional food (Sadeghi et al. 2019; Pinto et al. 2014; Pan 2017).

Gluten-free bakery products are one of the most developed product by replacing wheat flour with other flours such as soy flour and sweet potato flour (Pătraşcu et al. 2017; Lafarga 2018). Gluten-free bread is intended for consumers who have certain medical conditions such as gluten intolerance, allergies, and celiac disease so they cannot consume gluten-based products (Quiñones et al. 2015; Skendi et al. 2021). Gluten-free bakery products are also currently in great demand by consumers who are concerned about environmental issues on the basis of the negative impact of land clearing for wheat fields (Recchia et al. 2019). The high market demand for gluten-free bread products has encouraged researchers to develop other functional breads, especially gluten-free bread.

Gluten-free bread dough, which uses composite or non-wheat flour has rheological characteristics that have lower elasticity and cohesive values than flour-based bread dough which has a high gluten content (Cappelli et al. 2020). Gluten-free bread also has a crumbly texture and color image that is less attractive to consumers. This deficiency can be overcome by using several strategies, one of which is to use alternative materials that can mimic the viscoelastic characteristics of the gluten network. These active ingredients include hydrocolloids, enzymes, emulsifiers, and alternative protein sources (Cappelli et al. 2020). However, until now no technology has been found that can exactly match the characteristics of wheat flour-based bread with gluten-free bread.

One method that can be used to improve the characteristics of gluten-free bread is to add alternative protein sources such as Spirulina. Spirulina has a protein content that reaches 50–70% (dry weight). According to Cauvain and Young in (Selmo and Salas-Mellado 2014), the addition of alternative protein sources can help regulate the characteristics of gluten-free dough, preventing the product from having a brittle texture so that it can maintain the volume of bread.

The addition of spirulina to bread products has been shown to significantly increase protein, mineral, and active compounds such as phenolics which have antioxidant capacity content (Matos et al. 2017). The concentration of spirulina added to the dough needs to be considered because the high protein content in spirulina will affect the final properties of a product (de Nogueira and Steel 2018; Khemiri et al. 2020). One of the characteristics that need attention is the functional characteristics of the dough which includes the rheological properties of the dough (viscosity, extensibility and elasticity), dough development time, and dough stability (Singh et al. 2015).

2 Conventional and Alternative Bread Formulations and Methods

Bread is a bakery product that is composed of wheat flour, water, and yeast as the main ingredients. In addition, salt, sugar, and emulsifiers are used which will strengthen the crumb structure and give a certain flavor (Cauvain 2012). Bread is produced by going through three stages, namely: kneading, fermentation (proofing), and heating (baking/steaming) where each stage has its own role in forming the desired structure, texture, color, and characteristics (Manano et al. 2021; Yang et al. 2020).

Not only the ingredients used, but also each stage that the dough goes through each has its own critical point that determines the final quality of the bakery product (Cappelli et al. 2020). Therefore, bread making is good experimentally and industrial scale, must know the changes in the parameters of each stage. The final quality in question includes texture, pore uniformity, volume, color, and taste which is a parameter of consumer acceptance (Ekafitri and Isworo 2014).

According to Table 1, the formulations and methods used in bread making affect the physical quality of bread products, in this case the specific volume and the resulting color index. Research conducted by Amoriello et al. (2021) is a control parameter for making normal bread dough using basic ingredients without any additions. This research uses the no time dough method where the fermentation time is only around 20–30 min. The specific volume produced is 2.83 ml/g. Unlike the research conducted by (Babajide et al. 2014), the specific volume produced ranges from 3.95 to 5.40 ml/g depending on the concentration of sugar or honey added. Sugar added in bread making is in the range of 5–20% of the total flour mass (Babajide et al. 2014). Both studies show that the difference in methods affects the specific volume of bread produced This is in line with the statement put forward (Zhou et al. 2014) that the straight dough method will produce a higher specific volume because the dough has increased water absorption capacity during the fermentation process.

Pătraşcu et al. (2017) and Manano et al. (2021) conducted a study using two different methods with specific volumes which showed the same value (1.57 ml/g). This contrasts with other studies where the method influenced the specific volume results. The figure shows that the method is not one of the factors that determine the specific volume of bread. There is interaction involvement between the ingredients

Table 1 Bread characteristics based on dough formulation and method

Method	Ingredients	Specific volume spesifik (mL/g)	Color index	References
No-time dough	Flour, yeast, salt, sugar, water	2,83		Amoriello et al. (2021)
Straight dough	Flour, yeast, oil, salt, water	3,00		Saharan and Jood (2021)
Straight dough	Composite flour (cassava: wheat), yeast, honey (10–50%), water, butter	3,95–5,40	Crust: L* = 45,25–39,18; a* = 11,58–12,25; b* = 23,99–27,88 Crumb: L* = 64, 23–50,8 a* = 10,00–14,05; b* = 18,14–33,65	Babajide et al. (2014)
No-time dough	Composite flour (rice, cornstach, potato), psyllium husk, yeast, salt, sugar, protein extract, soybean, gum guar	1,57	Crumb: L* = 76,94 a* = 3,2 b* = 27,5	Pătraşcu et al. (2017)
No-time dough	Composite flour (rice, cornstach, potato), psyllium husk, yeast, salt, sugar, protein extract, egg flour, gum guar	2,26	Warna crumb: L* = 66,6 a* = 2,57 b* = 26,61	Pătraşcu et al. (2017)
Straight dough	Composite flour (cassava 10–60%), yeast, salt, sugar, margarine	1,57–2,1	–	Pătraşcu et al. (2017)

used which will affect the physical characteristics of the bread. Pătraşcu et al. (2017) used composite flour, hydrocolloids, and protein extracts as additives to improve the characteristics of gluten-free bread. Meanwhile, Manano et al. (2021) does not use any additional ingredients other than the main ingredient. Thus, although the no-time dough method cannot produce optimal volume, if the right additional ingredients are used, the expected characteristics can be achieved.

3 Comparison of Spirulina Platensis Components and Their Interaction on Bread Characteristics

3.1 Effect of Protein on Bread Characteristics

Spirulina is a microalgae that belongs to Cyanobacteria (blue-green) group with a cylindrical trichome shape and lives naturally in lakes, rivers, swamps, and oceans with different concentrations depending on the influence of growth factors (Morais Junior et al. 2020). Spirulina has been known as a food additive and labeled as a superfood because it contains nutrients that have a positive effect on the human body when consumed in sufficient levels. Components such as protein, essential fatty acids, vitamins, minerals, chlorophyll, and phycocyanin are the components contained in spirulina which can have a positive effect on the body. In addition, spirulina is able to provide linoleic acid (ALA), linolenic acid, stearidonic acid, eicosapentaenoic acid, docosahexanoic acid, and arachidonic acid (Vo et al. 2015) (Table 2).

Spirulina type algae have the highest protein group compared to other types of algae. This is confirmed by comparing the protein levels of spirulina (46–71%) in various alternative protein sources such as soybeans (36–42%), chickpeas (17–22%), mushrooms (19–35%), and peanuts. green (23–25%) The data shows that microalgae belonging to the Spirulina group have the highest protein content and great potential to improve the quality of gluten-free bread (Christwardana et al. 2013) (Table 3).

The use of Chlorella algae with composite flour significantly reduced the brightness level of the bread and intensified the green and yellow colors of the bread (Diprat et al. 2020). This is in line with the research by Selmo and Salas-Mellado (2014) who used Spirulina where the brightness level of the mixture decreased significantly and the green color became darker (L*: 34.59). This color change is caused by pigments that are owned by algae where the Chlorella and spirulina pigments that contribute to giving the color are chlorophyll and phycocyanin, respectively (Fradinho et al. 2020; Ścieszka and Klewicka 2020).

Table 2 Nutritional content of *Spirulina platensis*

Parameter	Content (%)
Protein	56–62
Fat	4–6
Carbohidrate	17–25
Linoleic acid (Gamma)	0.8
Chlorofil	0.8
Phycoyanin	6.7–11.7
Caroten	0.43
Zeaxanthin	0.1
Water	3–6

Christwardana et al. (2013)

Table 3 Alternative protein applications on gluten-free bread

Protein source	Concentration (%)	Flour control	References
Chlorella powder (*Chlorella sorokiniana*)	2.1, 4.2	25% rice flour, 58.3% corn flour, 16.7% pea flour	Diprat et al. (2020)
Spirulina platensis powder (strain LEB18)	1–4	100% rice flour	Selmo and Salas-Mellado (2014)
Acheta domesticus flour	2, 6, 10	80% corn flour 20% potato flour	Kowalczewski et al. (2019)
Eggs white powder	5, 10, 15	Commercial gluten-free flour (potato starch, cassava starch, sorghum)	Aprodu and Banu (2021)
Eggs white powder	2, 4	100% Tepung Beras	Phongthai et al. (2016)
Soybean powder	2, 4, 6	100% rice flour	Srikanlaya et al. (2018)
Potato protein extract	2, 6, 10	80% corn flour 20% potato flour	Witczak et al. (2017)
Pea powder	5, 10	50% rice flour 50% corn flour	Pico et al. (2019)
Canola protein extract	3, 6, 9	100% wheat flour 100% rice flour	Salah et al. (2019)

Chlorella and spirulina both have a relatively high protein composition. As a result, when given chlorella algae (4.2%), the porosity of the crumb was much higher compared to the control (Diprat et al. 2020). Selmo and Salas-Mellado (2014) stated that increasing the concentration of spirulina would decrease its specific volume. The presence of high protein content in the dough causes competition between protein in algae and starch in the dough for hydration. If these two components are not optimally hydrated, it will be difficult to maintain the texture and structure of the gluten-free bread (Bhattarai et al. 2016).

Changes in bread volume also occur when adding animal protein such as egg whites and milk to the gluten-free dough. Phongthai et al. (2016) explained that there was an increase in the volume of bread up to an addition of 2%, then the volume of bread experienced a downward trend up to the addition of 4, but there was an increase in the specific volume of bread up to the addition of 10% albumin.

The application of vegetable protein to gluten-free bread dough, on the other hand, decreases the volume of the dough on bread. Vegetable proteins such as soy and pea protein isolate only affect the porosity of the crumb, where the crumb has larger pores than the control (Srikanlaya et al. 2018; Pico et al. 2019). This is because gluten-based dough has a high proline content, compared to gluten-free dough. Proline is a type of amino acid that plays a role in volume development during the proofing period. Gluten-free dough has high levels of lysine and arginine and low levels of proline (Skendi et al. 2021).

3.2 *The Effect of Spirulina on the Physical Characteristics of Bread*

The physical characteristics of bread are one of the important parameters of bread quality that can affect consumer acceptance. These characteristics include specific volume, porosity, texture profile, and color index. The following table shows the effect of spirulina on the physical characteristics of bakery products (Table 4).

Based on research by Amoriello et al. (2021), the addition of spirulina caused the dough development time to decrease from 9.1 to 1 min. According to him, the interaction between fat, fiber, and protein components is the cause of reduced dough development time due to the weakening of the gluten network so that it can increase the diffusion of water into the dough. This causes the required mixing time to be significantly reduced.

The addition of spirulina also increased the specific volume of the dough by 5% from the control. In contrast to research conducted by Selmo and Salas-Mellado (2014) where the addition of spirulina actually reduced the specific volume of the dough like the majority of other algae. The decrease in specific volume can be caused by the interaction between fiber and non-gluten protein, and the dilution of starch in wheat flour. These causes have implications for decreasing extensibility and weakening of the gluten network which of course affects the volume of bread.

Table 4 Physical characteristics of bread on the addition of *Spirulina platensis*

Product type	Flour	Spirulina concentration (%)	Results		References
			With Spirulina	Control	
Bread	Soft wheat	1, 2.5, 4	Volume: 3,02–2,93 mL/g Porosity: 38,4–56,5% –L* crumb: 54,55–37,40	–Volume: 2,83 mL/g –Porositas: 18,9% –L* crumb: 74,68	Amoriello et al. (2021)
Baguette	Hard wheat	10	Gumminess: 2,23 N Springiness: 10,56 mm Cohesiveness: 0,43 Hardness: 20 g	Gumminess: 0,84 N Springiness: 8,43 mm Cohesiveness: 0,46 Hardness: 55 g	Sanjari et al. (2018)
Bread	Rice flour	1–4	Volume: 2,93–3,10 mL/g Firmness: 436,39–316,25 L* crumb: 34,59–53,21		Selmo and Salas-Mellado (2014)

Based on several studies, the addition of algae significantly affects the color of the crumb and crust on bread. The decrease in L*, a*, and b* values was affected by the high chlorophyll content in the algae. In other words, the higher the concentration added, the lower the brightness level (L*). In spirulina, the b* value decreases with increasing spirulina concentration, due to the high phycocyanin content (Bhattarai et al. 2016; Sanjari et al. 2018; Skendi et al. 2021).

Spirulina affects the texture profile of bakery products (Saharan and Jood 2021). There was an increase in the elasticity parameter from the research that has been done. This can occur due to the interaction between the protein in spirulina and the protein in the dough. The interaction is a hydrogen bond between amidehydroxyl and hydroxyl-carbonyl groups with polar compound groups. This bond can increase the hardness and resistance of the dough.

Nevertheless, the study stated that there was a decrease in the hardness value in the sample with the addition of spirulina. According to Nikoozade et al. (2011) in Sanjari et al. (2018), this can occur due to the fiber content which prevents the release of some water into the air because it is bound by the fiber. In addition, fiber also reacts with molecules in wheat flour starch which has implications for inhibiting the retrogradation process in the final bread product.

3.3 The Effect of Spirulina on the Organoleptic Characteristics of Bread

Spirulina added to flour-based food products ranges from 1 to 8% with the best acceptance in the 2–4% concentration range. The significant change that occurs due to the addition of spirulina is the change in the intensity of the green color which is directly proportional to the concentration added. The application of spirulina also makes bakery products have a distinctive spirulina taste that can be accepted by consumers up to a concentration of 4%. If viewed from the shelf life, bread products with the addition of spirulina can last up to two days at room temperature and four days at refrigerator temperature.

Saharan and Jood (2021) conducted organoleptic testing of bread products with the addition of spirulina concentrations of 2, 4, and 6% on 10 semi-trained panelists. Panelists assessed the color, appearance, aroma, texture, taste, and overall acceptance using the hedonic method with 9 scales (9 point hedonic rating scale) on bread stored at room temperature and refrigerator temperature. Overall, in two different conditions, bread with the addition of 2% spirulina was the most preferred with a score of 8.10 (very like), then 7.60 and 7.16 (quite like) at concentrations of 4 and 6%. While the control value is at 8.00 (very favorable).

Research conducted by Saharan and Jood (2021) states that the shelf life of bread stored at room temperature in terms of its organoleptic acceptability is 2 days of shelf life. The organoleptic quality of bread stored at refrigator temperature can be maintained for up to 4 days of shelf life.

Batista et al. (2019) conducted a study on the evaluation of the organoleptic cracker added to four types of microalgae at a concentration of 2% which included Arthospira platensis, Chlorella vulgaris, Tetraselmis suecica, and Phaeodactylum tricornutum. This research was conducted by 30 untrained panelists with an age range of 19–38 years. Parameters that are then assessed are color, aroma, taste, texture, and overall acceptance. In his research, it was known that the control crackers had a higher score (>4) compared to the treatment sample. However, the samples with the addition of microalgae had a texture that resembled the control sample. Spirulina platensis in this study had advantages in color acceptance, aroma, taste, and overall acceptance (3–4) compared to other types of microalgae. Research conducted by Fradinho et al. (2020) on the addition of 2% spirulina to gluten-free pasta products was tested on 31 untrained panelists. The panelists assessed the parameters of texture, color, flavor, extentability, and overall acceptability. Fradinho et al. (2020) stated that panelists preferred control samples without the addition of microalgae, even though the quantitative differences were not that significant. However, the level of consumer preference for products with the addition of 2% spirulina is satisfactory (>3.7).

Evaluation of hedonic quality on a scale of 10 in biscuit products was carried out by Singh et al. (2015). The test involved 10 semi-trained panelists consisting of students and academic staff of Indian universities. Parameters tested include color intensity, taste, sweetness, graininess, and crispiness. Of the five parameters tested, spirulina had a significant impact only on the color intensity and taste parameters.

The color intensity in this study specifically assesses the intensity of the green color. In this study, it was stated that the increase in the concentration of spirulina platensis was directly proportional to the color intensity value. This is caused by the pigments contained in spirulina in the form of chlorophyll-a, chlorophyll b, carotenoids, and phycocyanins which contribute to giving certain colors to food products (Nuhu 2013).

The taste score on the organoleptic test for this biscuit product ranged from 7.38 to 9.02. Based on the data obtained, the addition of spirulina powder concentration lowered the taste assessment of this product. This is in line with research conducted by Saharan and Jood (2021) where the addition of spirulina concentration in bakery products actually reduces the taste value of a product. This is because spirulina has a strong enough aroma to change the original aroma of a product. One of the characteristics of spirulina itself is that it has a fairly fishy smell.

4 Conclusion

The addition of spirulina to bread products can affect the specific volume of bread, porosity, color intensity, and organoleptic properties on parameters of color, aroma, taste, and overall acceptability. The addition of spirulina to bread products is generally at a concentration of 2–4% where there is a decrease in specific volume and brightness level (L*), as well as an increase in crumb porosity. In terms of its organoleptic properties, the addition of spirulina to bakery products was acceptable to the panelists at a concentration of 2–4%.

References

Amoriello T, Mellara F, Amoriello M, Ceccarelli D, Ciccoritti R (2021) Powdered seaweeds as a valuable ingredient for functional breads. Europ Food Res Technol 247(10): 2431–2443. https://doi.org/10.1007/s00217-021-03804-z

Aprodu I, Banu I (2021) Effect of starch and dairy proteins on the gluten free bread formulation based on quinoa. J Food Meas Character 15:2264–2274. https://doi.org/10.1007/s11694-021-00826-9

Babajide J, Adeboye A, Shittu T (2014) Effect of honey substitute for sugar on rheological properties of dough and some physical properties of cassava-wheat bread performance and bone morphometry of two breeds of finishing broiler chickens reared on different housing systems view project management systems for cockerel chickens View project

BPS [Badan Pusat Statistik, Indonesia Statistic] (2020) Statistictable. BPS Jakarta, retrived from https://www.bps.go.id/statictable/

Batista AP, Niccolai A, Bursic I et al (2019) Microalgae as functional ingredients in savory food products: application to wheat crackers. Foods 8:611. https://doi.org/10.3390/foods8120611

Bhattarai RR, Dhital S, Gidley MJ (2016) Interactions among macronutrients in wheat flour determine their enzymic susceptibility. Food Hydrocoll 61:415–425. https://doi.org/10.1016/j.foodhyd.2016.05.026

Borowitzka MA (2009) Gershwin ME, Belay A (eds) Spirulina in human nutrition and health. J Appl Phycol 21:747–748. https://doi.org/10.1007/s10811-009-9467-0

Cappelli A, Oliva N, Cini E (2020) A systematic review of gluten-free dough and bread: dough rheology, bread characteristics, and improvement strategies. Appl Sci 10:6559. https://doi.org/10.3390/app10186559

Cauvain SP (2012) Breadmaking: an overview, 2nd edn. Woodhead Publishing Limited

Christwardana M, Nur MMA, Hadiyanto H (2013) Spirulina platensis: Potensinya sebagai bahan pangan fungsional. Jurnal Aplikasi Teknologi Pangan 1:1–4

Danesi EDG, Rangel-Yagui CO, Carvalho JCM, Sato S (2004) Effect of reducing the light intensity on the growth and production of chlorophyll by Spirulina platensis. Biomass Bioenergy 26:329–335. https://doi.org/10.1016/S0961-9534(03)00127-2

da Rosa GM, Moraes L, Cardias BB et al (2015) Chemical absorption and CO2 biofixation via the cultivation of Spirulina in semicontinuous mode with nutrient recycle. Bioresour Technol 192:321–327. https://doi.org/10.1016/j.biortech.2015.05.020

de Nogueira AC, Steel CJ (2018) Protein enrichment of biscuits: a review. Food Rev Int 34:796–809. https://doi.org/10.1080/87559129.2018.1441299

Diprat AB, Silveira Thys RC, Rodrigues E, Rech R (2020) Chlorella sorokiniana: a new alternative source of carotenoids and proteins for gluten-free bread. LWT 134:109974. https://doi.org/10.1016/j.lwt.2020.109974

Ekafitri R, Isworo R (2014) Pemanfaatan Kacang-Kacangan sebagai Bahan Baku Sumber Protein Untuk Pangan Darurat. Pangan 23(2):134–145

Fradinho P, Niccolai A, Soares R et al (2020) Effect of Arthrospira platensis (spirulina) incorporation on the rheological and bioactive properties of gluten-free fresh pasta. Algal Res 45:101743. https://doi.org/10.1016/j.algal.2019.101743

Khemiri S, Khelifi N, Nunes MC et al (2020) Microalgae biomass as an additional ingredient of gluten-free bread: dough rheology, texture quality and nutritional properties. Algal Res 50:101998. https://doi.org/10.1016/j.algal.2020.101998

Kowalczewski P, Walkowiak K, Masewicz Ł et al (2019) Gluten-free bread with cricket powder—Mechanical properties and molecular water dynamics in dough and ready product. Foods 8:240. https://doi.org/10.3390/foods8070240
Lafarga T (2018) Potential applications of plant-derived proteins in the food industry. Novel Proteins for Food Pharmaceuticals and …
Lafarga T, Fernández-Sevilla JM et al (2020) Spirulina for the food and functional food industries. Food Research …
Manano J, Ogwok P, Byarugaba-Bazirake GW, Mugampoza E (2021) Rheological, baking and sensory characteristics of bread from wheat-cassava composite dough. J Food Res 10:18. https://doi.org/10.5539/jfr.v10n5p18
Matos J, Cardoso C, Bandarra NM, Afonso C (2017) Microalgae as healthy ingredients for functional food: a review. Food Funct 8:2672–2685. https://doi.org/10.1039/c7fo00409e
Morais Junior WG, Gorgich M, Corrêa PS et al (2020) Microalgae for biotechnological applications: cultivation, harvesting and biomass processing. Aquaculture 528:735562. https://doi.org/10.1016/j.aquaculture.2020.735562
Nuhu AA (2013) Spirulina (Arthrospira): an important source of nutritional and medicinal compounds. J Mar Biol 2013:1–8. https://doi.org/10.1155/2013/325636
Nikoozade H, Taslimi A, Azizi MH (2011) Effect of adding oat bran on rheological properties of bread and quality of Sangak bread. Iranian J Food Sci Technol 8(1):1–10
Pan MH (2017) Processing methods and bio-applications in functional food production. Curr Res Nutr Food Sci 5:180–182. https://doi.org/10.12944/CRNFSJ.5.3.01
Pătraşcu L, Banu I, Vasilean I, Aprodu I (2017) Effect of gluten egg and soy proteins on the rheological and thermo-mechanical properties of wholegrain rice flour. Food Sci Technol Int 23(2):142–155. https://doi.org/10.1177/1082013216665722
Phongthai S, D'Amico S, Schoenlechner R, Rawdkuen S (2016) Comparative study of rice bran protein concentrate and egg albumin on gluten-free bread properties. J Cereal Sci 72:38–45. https://doi.org/10.1016/j.jcs.2016.09.015
Pico J, Reguil6n MP, Bernal J, G6mez M (2019) Effect of rice, pea, egg white and whey proteins on crust quality of rice flour-corn starch based gluten-free breads. J Cereal Sci 86:92–101. https://doi.org/10.1016/j.jcs.2019.01.014
Pinto D, Castro I, Vicente A et al (2014) Functional bakery products: an overview and future perspectives. In: Bakery products science and technology, 2nd edn, pp 431–452. https://doi.org/10.1002/9781118792001.ch25
Quiñones RS, Macachor CP, Quiñones HG (2015) Development of gluten-free composite flour blends. Trop Technol J 19. https://doi.org/10.7603/s40934-015-0003-3
Rabelo SF, Lemes AC, Takeuchi KP et al (2013) Development of cassava doughnuts enriched with Spirulina platensis biomass. Braz J Food Technol 16:42–51. https://doi.org/10.1590/s1981-67232013005000001
Recchia L, Cappelli A, Cini E et al (2019) Environmental sustainability of pasta production chains: an integrated approach for comparing local and global chains. Resources 8:56. https://doi.org/10.3390/resources8010056
Sadeghi A, Ebrahimi M, Raeisi M, Ghods Mofidi SM (2019) Improving the antioxidant capacity of bread rolls by controlled fermentation of rice bran and addition of pumpkin (Cucurbita pepo) puree. J Food Meas Character 13:2837–2845. https://doi.org/10.1007/s11694-019-00204-6
Saharan V, Jood S (2021) Effect of storage on Spirulina platensis powder supplemented breads. J Food Sci Technol 58(3):978–984. https://doi.org/10.1007/s13197-020-04612-1
Salah K, Olkhovatov EA, Aïder M (2019) Effect of canola proteins on rice flour bread and mathematical modelling of the baking process. J Food Sci Technol 56:3744–3753. https://doi.org/10.1007/s13197-019-03842-2
Sanjari S, Sarhadi H, Shahdadi F (2018) Investigating the effect of Spirulina platensis microalgae on textural and sensory properties of baguette bread. J Nutr Food Secur. https://doi.org/10.18502/jnfs.v3i4.166

Ścieszka S, Klewicka E (2020) Influence of the microalga chlorella vulgaris on the growth and metabolic activity of lactobacillus spp. Bacteria. Foods 9:959. https://doi.org/10.3390/foods9070959

Selmo MS, Salas-Mellado MM (2014) Technological quality of bread from rice flour with Spirulina. Int Food Res J 21(4):1523–1528

Singh P, Singh R, Jha A, Rasane P, Gautam AK (2015) Optimization of a process for high fibre and high protein biscuit. J Food Sci Technol 52(3):1394–1403. https://doi.org/10.1007/s13197-013-1139-z

Skendi A, Papageorgiou M, Varzakas T (2021) High protein substitutes for gluten in gluten-free bread. Foods 10. https://doi.org/10.3390/foods10091997

Srikanlaya C, Therdthai N, Ritthiruangdej P, Zhou W (2018) Effect of hydroxypropyl methylcellulose, whey protein concentrate and soy protein isolate enrichment on characteristics of gluten-free rice dough and bread. Int J Food Sci Technol 53:1760–1770. https://doi.org/10.1111/ijfs.13761

Stübler AS, Heinz V, Aganovic K (2020) Development of food products. Curr Opin Green Sustain Chem 25. https://doi.org/10.1016/j.cogsc.2020.100356

Vo T-S, Ngo D-H, Kim S-K (2015) Nutritional and pharmaceutical properties of microalgal Spirulina. In: Handbook of marine microalgae. Elsevier, pp 299–308

Witczak T, Juszczak L, Ziobro R, Korus J (2017) Rheology of gluten-free dough and physical characteristics of bread with potato protein. J Food Process Eng 40:e12491. https://doi.org/10.1111/jfpe.12491

Yang Y, Guan E, Zhang T, Li M, Bian K (2020) Comparison of rheological behavior microstructure of wheat flour doughs and cooking performance of noodles prepared by different mixers. J Food Sci 85(4):956–963. https://doi.org/10.1111/jfds.v85.4; https://doi.org/10.1111/1750-3841.15057

Zhou W, Hui YH, de Leyn I et al (eds) (2014) Bakery products science and technology. Wiley, Chichester, UK

MIX
Papier aus verantwortungsvollen Quellen
Paper from responsible sources
FSC® C105338

If you have any concerns about our products,
you can contact us on
ProductSafety@springernature.com

In case Publisher is established outside the EU,
the EU authorized representative is:
Springer Nature Customer Service Center GmbH
Europaplatz 3, 69115 Heidelberg, Germany

Printed by Libri Plureos GmbH
in Hamburg, Germany